Living Mammals of the World

Living

Mammals
OF THE WORLD

IVAN T. SANDERSON

Fellow of the Linnean Society of London
Fellow of the Royal Geographical Society
Fellow of Zoological Society of London

Photographs by

JOHN MARKHAM

ROY PINNEY

CY LA TOUR

YLLA

VAN NOSTRAND

ERNEST P. WALKER

and others

A CHANTICLEER PRESS EDITION

DOUBLEDAY & COMPANY Inc.

Garden City, New York

PUBLISHED BY DOUBLEDAY & COMPANY, INC.

Garden City, New York

Eighth Printing 1972

PLANNED AND PRODUCED BY CHANTICLEER PRESS, INC., NEW YORK

THE WORLD OF NATURE SERIES

Living Mammals of the World by Ivan T. Sanderson

Living Reptiles of the World by Karl P. Schmidt
  and Robert F. Inger

Living Birds of the World by E. Thomas Gilliard

Living Insects of the World by Alexander B. Klots
  and Elsie B. Klots

*The Lower Animals: Living Invertebrates of the
  World* by Ralph Buchsbaum and Lorus J. Milne,
  in collaboration with Mildred Buchsbaum
  and Margery Milne

Living Fishes of the World by Earl S. Herald

Living Amphibians of the World by Doris M. Cochran

ISBN: 0–385–01445–7

Library of Congress Catalog Card No. 55–10515

PRINTED IN THE UNITED STATES OF AMERICA

This book is humbly dedicated to

the sincerest friend of all the Mammals,

including those of the genus Homo,

DR. WILLIAM M. MANN

DIRECTOR OF THE NATIONAL ZOOLOGICAL PARK

Contents

Introductory

THE Mammals are the most highly evolved and the most important group of animals, at least from the human point of view. Without them, and particularly the domesticated species, Man would probably not be able to survive. Without some of them—many of which are virtually unknown to all except zoologists—we would certainly have to alter our entire mode of living and we might very well succumb in the face of hordes of insects. Yet curiously, although everybody is conversant with the terms fish, reptile, or bird, there is no really popular name for the form of life we are about to describe. The trend today among English-speaking peoples seems to be to call them simply *animals* in a contrasting sense to the terms fish, birds, reptiles, and so forth. This is very misleading and is not to be recommended, first because all material things that are not, as the old parlor game had it, vegetables or minerals are animals, and secondly because science has adopted this word to define a particular group of living things. The Mammals are but one rather minor subdivision of the Kingdom of Animals. Apart from the name *Beasts,* as used in its old and pure sense, none of the other numerous designations such as Quadrupeds, Furred-Animals, and so forth that has been applied to this group is any longer valid as a comprehensive title for the Mammals.

It would at first appear that everybody should know what is or is not a Mammal and be able to differentiate them from all other animals. However, this is not so, and until very recently even our Courts of Law were far from sure of the status of several whole groups of Mammals—notably, the Whales, which they ruled were fish. It is interesting to note that even today the U.S. Customs requires shippers of live animals to declare whether they are importing "Birds" or "Backboned Animals," and that the duty is different on each. Despite these confusions, Mammals can nonetheless be clearly defined, and in a fairly simple manner, for there are three characteristics that they alone display in combination.

No other animals have true hairs, produce milk in their bodies with which to feed their young, and have a four-chambered heart. What is more, with one exception—the matter of hair, which is lacking in some Whales—all Mammals display these three features, for even the Monotremes that—surprisingly for mammals—lay eggs, produce milk and suckle their young. In addition, all Mammals breathe air, and have warm blood and vertebrated backbones. However, they share these three features with the birds; and the reptiles and amphibians also breathe air, though the latter do so only in their adult stages. A vertebrated backbone is common also to the fish.

As we mentioned above, the mammals in reality constitute only a rather minor subdivision of the Animal Kingdom. There are about one million known kinds of animals living on this planet at the present time. Half of these are Insects, and entomologists believe that we have so far discovered only about a tenth of the existing species of these. Of the other half-million animals, about a fifth are backboned or Vertebrates; and of these some 12,000 are Mammals. Nobody has actually counted all the known kinds of mammals for the simple reason that we still do not know just where to draw the line between a *species* on the one hand and mere subspecies, or even individual variations on the other. Thus, until we have a great deal more information about the range of variation of all kinds, among populations of Mammals in their natural habitat, we will not be able to say just how many specific kinds inhabit our earth.

Next to their number, perhaps the most outstanding feature of the Mammals as a whole is their extraordinary range of size, shape, and habits. In these respects they are unique among all groups of animals, and particularly among the Vertebrates. Fish range in size from creatures not much bigger than a housefly to the monstrous Whale-Shark, and may be short or long and either circular in section or flattened from top to bottom or from side to side, but they are all built on much the same basic plan. Amphibians come in only three forms—the frogs, salamanders, and the legless, wormlike Coecilians. Reptiles also come in only three general shapes—the snakelike, tortoise-like, and lizard-like. Birds are really all the same beneath the skin, although some have almost lost their wings. But Mammals come in dozens of completely different forms. There are fishlike ones—the whales; birdlike kinds—the bats; lizard-like—certain tiny Australian marsupial-mice; froglike—Men; tortoise-like—armadillos; and innumerable forms that almost defy description—such as the Pangolins and the Sand-Puppies. They range in size from a shrew that just counterbalances a dime, to the Blue Whale, which has been measured at 113 feet and can weigh as much as 170 tons.

In habits, also, the Mammals are incredibly varied.

They are found throughout the earth, in the air above it, under its surface, in rivers, lakes, seas, and oceans, and on completely waterless deserts and frozen ice-caps. In fact, they have a wider distribution and are more adaptable than any other group of animals except possibly the spiders and their allies. Because of their warm blood, and their extremely efficient mechanism for maintaining its temperature both in considerable heat or in excessive cold, and also because of their varied methods of insulating their bodies, they are able to survive in a variety of environments that are lethal to almost all other animals. It has now been discovered experimentally that not only mammalian tissue but even whole animals can be literally frozen stiff for days and then be revived, but that the same animals can in some cases also survive temperatures that will bake pottery. One of the most adaptable of all mammals, morever, is Man, and this even without all his artificial aids to survival. The natives of Tierra del Fuego survived without clothes or houses in sub-zero temperatures and naked men have survived temperatures far above that of the hottest desert in full sunlight.

Mammals have also diversified greatly in another respect. This is known as phylogenetic diversification, which means, in other words, that they have a very large and complex family tree. Furthermore, they also display, over and over again within this family tree, what is known as adaptive radiation. Reduced to simplest terms, this means that a particular group of mammals started as generalized creatures but then branched out in various directions into all manner of unlike forms, each adapted to an entirely different way of life. Taking the Marsupials as an example, we find that the group contains a central, primitive core of opossum-like creatures but that from these have sprung animals like antelopes—the kangaroos; like wolves—the Thylacine; and like moles, squirrels, badgers, mice, jerboas, rabbits, and so forth. This process has taken place again and again among the mammals throughout their comparatively long history on this earth.

When we say that the history of Mammals is a long one we are speaking in geological terms, for, although the Age of Mammals only began about sixty million years ago, there were tiny, primitive mammals as long ago as 180 million years. During this immense span of time many more kinds of mammals were evolved and became extinct than have survived into the present, so that we are today left with, as it were, only the outer twigs of their family tree. This makes the classification of the mammals a matter of considerable difficulty, for many of them that look alike today are in no way related, whereas many others that appear altogether different from one another prove, on ana-

tomical investigation, to be close relatives. The purpose of this book is not only to describe all the known kinds of living mammals but also to display their relationships, and thus their origins.

Fortunately, there is a recognized procedure for doing this, which is known as *Systematics*. This, in the scientific sense, means arranging things according to their phylogenetic affinities or, in other words, their true position on a family tree. The Mammals could, of course, be classified in a number of other ways, as for instance by size, by habit, or geographically, by their place of origin, but in no arrangement other than the systematic can so many facts about their structure, origin, and relationships be assembled so readily. Classification, in fact, is not just a method of splitting up an unmanageable mass of things; it is also a way of bringing similar things together so that their interrelations may be comprehended. Science devoted the greater part of its time and energies during the past two centuries to the description, listing, and classification of all manner of things, material and otherwise. The science of Zoology is, in fact, founded on the systematic classification of animals, both living and extinct.

Most groups of animals have now been rather thoroughly described and many of them have been adequately classified. Among the latter are, and most fortunately from our point of view, the Mammals, although even among them there is still considerable uncertainty as to many details, and some major problems as to their origin still remain to be solved. There is as a whole, however, now marked agreement as to their general arrangement and this has been embodied in masterly compacted form by G. Gaylord Simpson, of the American Museum of Natural History, in a publication entitled *The Principles of Classification and a Classification of the Mammals* (Bull. Am. Mus. Nat. Hist., 1945, *Vol.* 85). The particular value of this work, especially to students, is that it brings together both the living and the extinct forms in a single series and thus gives the most complete picture presently possible of this whole *Class* of animals, both in space and in time.

The arrangement of the mammals adopted in this book is substantially that laid down by Dr. Simpson, but, for purposes of simplicity, the actual classification used is not in exact accord therewith. In order to arrange all known mammals in an understandable manner, and especially when the extinct forms have to be taken into account, it is necessary to use a very complex system of classification—at least, if the results are to be properly scientific. It is, in fact, necessary to set up a large number of categories of various statuses or degrees of importance. Thus, the *Class Mammalia* has first to be split into several *Subclasses,* and then these, in turn, into categories named technically *Infraclasses, Cohorts, Superorders, Orders, Suborders, Infraorders, Superfamilies, Families, Subfamilies, Tribes, Subtribes, Genera, Subgenera,* and finally into *Species, Subspecies, Races, Clines,* and mere variations. Such a system is not only overwhelming to the non-specialist, but unnecessarily cumbersome for our purposes and may therefore legitimately be reduced to a very simple structure.

As we are neither concerned in this book with the extinct forms of mammals, nor with the technical details of classification *per se,* we have made use only of the minimum necessary to achieve our objective without, we hope, taking undue liberties with established procedure. In order to do this, it has been necessary to omit, though by no means to ignore, what is actually the greater part of the over-all classification of mammals, for that is concerned primarily with extinct forms.

To this end, we first divide all living mammals into nineteen *Orders,* corresponding to those of the established scientific classification. Next, we subdivide each of these—except a few which consist of but one form of animal, like the Aard-Vark, or the Cobego—into two or more groups, to which purely artificial anglicized names are given. These subdivisions are perfectly valid in every case but they display a variety of status in the scientific sense. Thus, the two major groups of Whales—the Baleen and the Toothed Whales—constitute true *Suborders,* whereas those herein entitled Bovines (cattle), Antelopines (antelopes), and Caprines (goats) of the Hoofed Mammals are really only of *Subfamily* rank. These subdivisions, however, represent the popular grouping of the mammals.

Below this category we use only one further principal breakdown. This, on the whole, corresponds in the majority of cases to the *families* of mammals, technically speaking, but it does not do so in all cases. In many instances the group thus designated is really only of *subfamily, generic* or even lesser status—e.g., an arbitrary classification like "The Not-so-Great Cats." Those wishing to know the true scientific status of each group may determine them from the ending of the Latin name given to each. Thus, if the name ends in *-IDAE* it denotes a family; if in *-INAE* a sub-family. Group endings that are of lesser status are explained in each case.

Finally, within these "family" groups the animals are described more or less by *genera,* the number of species contained in each usually being mentioned. In some cases these headings represent groups of genera—e.g., the Voles, in which no less than two dozen recognized genera are covered by one head. Only in the case of certain large and well-known types such

as the Great Apes and the Cats are individual *species* given separate attention.

By this procedure it transpires that there are fifty-nine major subdivisions of the nineteen *Orders* of Mammals, and that these in turn may be broken down into one hundred and eighty-six "families" or kinds. The really overwhelming fact that comes out of this analysis, however, is not the mere number of different recognized forms of mammals but their incredible variety of form, structure, history, size, and behavior. Unfortunately, in a work of this compass, designed to be comprehensive in only one respect—in this case, the *systematic*—there is simply not the space to discuss the last of these aspects of mammalian life—namely behavior.

Actually, we know comparatively very little about the habits of the vast majority of mammals but even that which is known and recorded would fill not a volume but a large library. The closer, more intimately, and more thoroughly any animal is studied, the more facts about it we discover to be unique. Every species of mammal is absolutely distinct, and distinctive, in form, habits and behavior. There cannot really be any end to the marvels, mysteries, and just plain facts that wait to be discovered about mammals, and as yet we have hardly made a start in this field. The wordage that has been recorded on the rat, the guinea-pig, the horse, and sundry other common mammals is probably beyond compute. And yet at least ninety-nine percent of all mammals have to be passed over with the comment that nothing at all is known of their habits or behavior. What is more, the average person probably does not even suspect the very existence of about ninety-five percent of extant mammals.

This, then, is the major purpose of this book: to bring together a comprehensive index of the living mammals, and to describe them adequately enough for anyone, regardless of whether he has specialized knowledge or not, to be able to identify them on sight. In this connection, more than a passing recognition of the illustrations in this book is called for.

Photography, from the day of its discovery and long before its perfection to the point of general usefulness, became more than just a handmaiden to science. For recording reality there can, of course, hardly be a better medium because the camera, even in its most primitive form, sees and records many things that are quite beyond the scope of the human eye. A notable example in the zoological field was the discovery by photography of the true method by which a horse walks, trots, canters, and gallops, matters that had been debated for centuries and which had quite eluded the best of artists.

Almost everybody who has inspected the work of

the classical painters, and more particularly that of the professedly realistic schools, must have wondered whether the artists really saw the persons, objects, or scenes in the manner in which they depicted them. Did the little dogs in Holland always prance about on one front and one hind paw, and did trees always then grow like olive-brown cotton-wool in the shape of a cottage loaf? Similarly, any naturalist with first-hand knowledge of live mammals, who ever looked at the illustrations in the older zoological works must have entertained the gravest doubts as to the reliability of the artist's vision and especially such examples as the grotesquely distorted interpretations of the Audubons. However critically and conscientiously people looked at live animals, it was, before the invention of photography, only the very rare genius who transferred the reality, rather than what he thought he saw, to paper or canvas. In fact, the impressionists of the Old Stone Age far surpassed the best nineteenth-century animal artists in this respect, while the Chinese, Japanese, and certain African tribesmen have been matched only by a few Europeans of the genius of Durer. It was photography that changed all this and not only by giving us its own descriptive record but also by leading the modern artist to depict animals the way they really are.

The final step has come with the development of color photography. Although there is still a very great deal to be learned about the use of this medium and especially for the recording of animal form, nothing surpasses it as a method of describing living things. The illustrations both in color and black and white in this book actually represent an amount of description equivalent to several additional volumes of text. Here are the mammals as they actually are and each in a distinctive pose, something that is of very real significance even in the technical classification of animals. For this remarkable achievement the author wishes herewith to pay tribute to the photographers responsible for these works of scientific art. Just as the field naturalist may see a thousand animals before capturing a single worthwhile specimen so it is with the photographer of animals; his patience, labor, and fortitude are only surpassed by his percentage of disappointments, for a good animal portrait can never be a snapshot and is hardly ever aided by luck in any form.

There remains but one other aspect of this work to be mentioned. We will throughout be constantly referring to geographical zones or areas where this, that, or the other animal or group of animals is to be found. Such references are often contracted to such phrases as "the northern desert zone" or "the south temperate forest belt." Some explanation of this is necessary.

If it were without oceans, the surface of the earth would theoretically (though, of course, not in fact) be girdled by five great climatic belts and capped at top and bottom (i.e., the poles) by two circular areas of similar constitution. Traveling from north to south, these would be (1) a cold, dry, arctic desert, (2) a northern, moist, cool, forest belt, (3) a northern, hot, dry, desert belt, (4) a great, double, hot, wet, equatorial forest belt, girdling the equator, (5) a southern, dry, desert belt, (6) a southern, cool, moist, forest belt, and (7) an Antarctic, cold, dry, desert. However, because of the presence of the ocean in which both cold and hot currents circulate—clockwise in the northern, and counterclockwise in the southern hemisphere—and the reduction of the land surface to three great triangular masses encircling the Arctic Ocean and depending therefrom over the globe in three directions to taper out in the Southern Hemisphere, this neat theoretical arrangement is greatly modified. The five major zones and the two polar caps still pertain but the former are widely displaced to north or south in great undulations. Further, numerous great mountain ranges also disturb the circulation of the atmosphere and so alter the basic climate over wide areas, thus reducing the whole to a great patchwork quilt of different environments. Then again, innumerable local variations of topography, soil, and other features further reduce these patches to a sort of irregular checkerboard of possible habitats. Each of these has a distinctive fauna. Finally, within these are still further subdivisions, ranging from large deserts to places as small as holes in a single kind of tree. Yet, each again provides a different niche for animal life and each in every country is inhabited by one or more particular kinds of mammals.

This is the fundamental clue to the diversity of the mammals, for they, above all groups of animals, have managed to fill almost every niche.

Living Mammals of the World

Egg-Laying Mammals

(*Monotremata*)

THE most remarkable of all living mammals are also the most primitive, most distinctive, and probably most useful from the purely scientific point of view. On the other hand, they are practically useless to man in all other respects, and they were not even known except to some primitive Australian and New Guinea tribesmen until about 150 years ago. They are known as Monotremes because they have a single ventral opening for the elimination of both liquid and solid wastes from the body as well as for sexual conjugation and, in the female, the birth of the young. Birth, moreover, is accomplished in an altogether surprising and, for a mammal, unorthodox manner, by the deposition of small, rubbery, squeezable eggs.

Monotremes are found only in Australia, Tasmania, New Guinea, and sundry neighboring islands. They come in two quite different models known in common parlance as Duckbills and Spiny Anteaters; there are two very distinct kinds of the latter. Although mammals —having fur, warm blood, and suckling their young after they emerge from the egg—Monotremes display a number of reptile-like anatomical features such as a T-shaped bone connecting the collarbones to the chest-plate, or sternum, while their eggs have large yolks and plastic shells like those of lizards, snakes, and turtles. Their blood is not as warm as that of other mammals.

DUCKBILLS

There is only one species, known to everybody nowadays as the Platypus (Plate 1), a curious little creature shaped like an elongated bun and covered with dense, woolly fur. It has a flattened tail not unlike that of a beaver, short limbs with outsized hands and feet, all having five digits armed with sharp claws. The hind feet are completely webbed up to the base of the claws, but in the forefeet the webbing extends well beyond the ends of the fingers and the claws. The animal can roll this webbing under the pads of its hands when it is walking about on land or digging, which it does most industriously and efficiently. Thus far, the animal seems fairly feasible though distinctly odd; the head, however, assumes a quite unforgivable aspect. The head proper is small and flattened, there are no ear-flaps, and the small eyes and ear-holes can be enclosed at the same time between two movable ridges of skin. The front of the head bears a stiff but pliable rubbery structure exactly resembling the bill of a duck but somewhat wider

in front and rising behind into a sort of transverse crest in front of the eyes. It looks as if it belonged to some other animal (or was even man-made) and the Platypus had got its nose caught in the object's hinder opening. This is a bill or beak, and the inside of both it and of the mouth behind it is lined with hard ridges of a horny substance which the animal uses for crushing its food.

The most unexpected feature of all, and one which the Platypus shares with the Spiny Anteaters, is sharp, recurved spurs carried by the males on the insides of their ankles. These are perforated by a duct that leads to a long tube running up the inside of the lower leg, then curving around to the front and passing upwards to large glands on the outside of the upper legs. These glands secrete a poison that is deadly to some small animals and can give intense pain to a human being if he is well jabbed by the spur; and occasionally even more unpleasant complications have been known to set in. These are the only truly venomous mammals known, though the bites of certain of the shrews are poisonous.

Platypuses are found only in the eastern third of the mainland of Australia and in Tasmania. They are semi-aquatic, spending about two hours a day, morning and evening, paddling about in the mud at the bottoms of rivers, streams, lakes, and ponds, looking for worms, insects, shellfish and other food. They inhabit ice-cold streams at 6,000 feet and tropical swamps at sea level. The rest of the time they either wander about on the banks or dig complicated burrows beneath them. Most burrows are inhabited by a pair but when eggs are on the way the female goes off and digs a special retreat from which her mate is excluded. Into this she retires for a period of three weeks and makes a nest. The little three-quarter-inch eggs are incubated in this for about two weeks, the mother holding them on her curled up body. Nor does she leave the nest (which she barricades with a series of loose earth stoppers at intervals along the main tunnel leading to the outside) until the eggs are hatched and the young have learned to suck from her fur the milk that seeps through certain special, enlarged pores in her skin. The eyes of the young open only after eleven weeks and the babies remain in the nest for at least another six weeks before weaning begins. This is completed in three weeks and the young then develop the characteristic colossal appetites of their parents.

SPINY ANTEATERS

These creatures are often called *Echidnas* because that scientific name had previously been applied to them in error. It rightfully belongs by priority to a kind of fish, and should therefore be abandoned. The appearance of these animals (Plate 2) is not quite so appalling to our eyes as that of the Platypus, but it is also very odd, to say the least, and quite startling in the case of the *Zaglossus* or three-toed species of New Guinea.

The species from Tasmania is distinct from those of the mainland and of New Guinea, being covered in a thick coat of brown hair interspersed with scattered short, sharp spines all over the back but having a sort of tonsure of longer, more closely packed spines along either flank and forming interlocking brushes over its rear end. The mainland and New Guinea forms are less hairy and more evenly covered with longer spines. The heads of both are spineless and haired, and their undersides are clothed in shorter, softer fur of a lighter shade. The spines are yellow or white with sharp black tips, and are really much enlarged hairs. Those on the sides and rump are round in section, but elsewhere they are interspersed with softer, flattened, smaller spines. There are five toes bearing stout claws on the front feet and five on the hind, the inner being very small but the next immense, while the other three decrease in size down to a small outer one. The long one is used for scratching, combing and cleaning between the spines. The head is very small and tapers to a distinct, hard, naked beak with nostrils out at the far end. The mouth is a small hole at the tip and food is collected by a long, thin tongue lubricated by a very sticky saliva. There is no tail.

Although the Spiny Anteaters are most competent diggers, they do not tunnel like the Platypus. They live on the surface both in open wooded and grassy areas, but they seem to prefer rocky ground. In order to obtain a sufficient bulk of ants and other insect food they have to lead unexpectedly active lives compared to their lethargic behavior in captivity, where an abundance of food is provided and they spend most of their time sleeping. In their own habitat they are constantly crawling about, snuffling, turning over stones, and probing under things with their snouts. Their strength is prodigious, and even on the slippery floor of a house they can wedge themselves under an average day-bed and shift it right across a room. Their principal method of defense is to dig furiously with all four feet and thus sink quickly down into the ground until only an oblong dome of spines is left above the surface. This usually balks any enemy completely.

Spiny Anteaters also lay small leathery eggs, usually two but sometimes three or only one. These are laid directly into a primitive pouch which develops every year at the breeding season. The mother accomplishes this by curling her body into a ring like a caterpillar, with her hind end innermost so that the egg on emerging from the cloaca slips straight into the pouch. This pouch is lined with coarse hairs, and an extremely sticky substance exudes from the cloaca along with the egg. This dries rapidly in the air and thus gums the eggs to the matted hair. The young hatch out in the pouch and stay there, sucking the milk that exudes from their mother's skin through her fur, until their spines grow long enough to interfere with their movements or irritate the mother. They are then scratched out by her and deposited in a safe hidden spot while she goes out to forage. The young remain in this primitive nest for several weeks before they are weaned.

These animals never have any teeth, but they have a sort of rasping apparatus formed by horny ridges on the roof of the mouth proper and on the upper side of the back of the tongue. These grind up the insect food but use sand and grit in considerable quantities to do so.

This material then passes on into the stomach, which is a sort of gizzard like that of birds, and there the mass is further reduced. The food, though predominantly insects, appears to be very varied. They have been seen to rip open dead animals and lick up blood and other fluids.

The Three-toed Spiny Anteater of New Guinea is an altogether different animal. It stands up on its comparatively long legs instead of shuffling about on its belly like the preceding species, the head is larger, and the beak is very long but not well defined as such, being more an extension of the head and reaching the ground. Its hind feet turn outwards and backwards. The body when young is thickly clothed with hair and small, short spines but aged individuals may lose almost all of both over the back and flanks and come to look like tiny elephantine creatures. One lived in the London Zoo for over thirty years.

Pouched Mammals

(Marsupialia)

IT is difficult, if not impossible, to picture this planet as it apparently was a hundred million years ago; indeed it is nearly hopeless to try to conceive of any such length of time. Yet several lines of research have now confirmed the opinion that at about that time the earth was in its normal state of over-all warmth, so that forests flourished right up to the poles, the continents were virtually without mountains and the oceans spread over much of their area in the form of shallow seas. The predominant form of life was reptilian and the so-called dinosaurs swarmed on land and in the sea. Hidden away in the mouldy floor covering of the forests, in the dense foliage of its canopy, or among rocks and perhaps even in dry desert areas, were a number of comparatively tiny creatures with furry bodies that were internally warmed, gave birth to live young, and then nourished these on milk formed in their own bodies. We have the fossilized bones of only a very few of these obscure creatures, but we have enough to show us clearly that the type of animals we now call mammals had already come into existence.

The strange thing is that already at that early period these little animals were of several distinct kinds—some looked like shrews, others like opossums, and still others like nothing that we know of left on earth. Among the opossums were animals astonishingly like some that are still living today—notably certain South American opossums. There is no doubt of the fact that the living mammals that have pouches in which to carry their young are not only descendants of these earliest known mammals, but are actual survivals, which is something quite different. Marsupials were once found all over the earth, but today they are confined to North and South America, Australia, and a belt of islands north of that continent. They are now of great variety, ranging from the Great Red Kangaroo that can stand seven feet when raised on tail and hind feet to fight, to a tiny Shrew-Opossum of Brazil that weighs less than a half-dollar. Further, those in Australia have diversified in such a way that they have come to duplicate many other kinds of mammals. Thus, there are pouched wolves, cat- and badger-like animals, others which resemble squirrels, flying-squirrels, baby bears, otters, anteaters, moles, rats, mice, rabbits, and various other types. There was once a pouched animal known as *Diprotodon* that grew as big as a rhinoceros and looked like a nightmare rabbit, and the Australian government only a few years ago issued a statement to the effect that a pouched "tiger" may exist in Queensland. The explanation for this curious situation in Australia is that for some reason the mammals without pouches never reached that continent in bulk at any time during the past hundred million years so that these dawn mammals were left alone to fill the earth and all its varied natural niches.

As explained in the Introductory, the world is divided into a specific number of pigeonholes, like a vast storage warehouse, and nature does not tolerate any of these being empty. Thus, when the reptiles began to die away, the mammals began to take their places, and in the absence of any other mammals in Australia, the marsupials filled all the niches. In the Americas it was different. Here, the non-pouched mammals came along to fill up the spaces, and the small-brained, primitive opossums had to get along as best they could in competition with more active creatures of their own and greater size. They did very well and have not only held their own, but in some respects out-survived their more active rivals. Today the Common Opossum of North America is spreading everywhere despite all the rats, cats, dogs, and the native fauna, and is even invading the crowded cities of modern man. The internal machinery of marsupials works at a lower efficiency than that of the non-pouched mammals, but it is therefore often harder to stop. You can literally blow an opossum to pieces before the animal as a whole, or its separate parts, can be pronounced truly dead. On the other hand the Koala of Australia is so specialized in its habits that a seasonal change in the oils produced by the leaves of certain eucalyptus trees on which it feeds can kill the animals, and they have to trek long distances to find other leaves with oils that will not poison them.

Although designated pouched mammals, not all Marsupials have pouches. Some, like the kangaroos, have large bags, others have shallow cups, others only flaps of skin, and some have none at all. Certain kinds have them pointing backwards like miniature observation cars. In all cases, however, the young are born so early that they are no more than half-developed embryos. These usually find their own way to the pouch, or at least to the teats, to which they attach themselves and then remain attached for weeks. In some cases there are invariably more young than there are teats, so that the weaker perish outright. Most marsupials have a tremendous mouth gape and a formidable battery of teeth, but these also have developed in all kinds of ways to meet the diverse needs of the species that bear them.

DIDELPHIDS

There are five groups of marsupials with representatives still living today. The first of these, called the Didelphids because they have two wombs, are found only in the Americas and form a rather tight little group varying in size from the Common Opossum of the U.S., to a tiny, shrewlike animal with a long, pointed nose and a short, mouselike tail that inhabits the leaf mould of the Amazonian forest floor. In all, there are some forty quite distinct kinds of Didelphids that are divided into eleven genera. In South America the so-called banana opossums—*i.e.*, the Four-eyes and the Woolly Opossums —are better known. They all eat just about everything that can be digested, from live insects and other animals, to leaves, fruits, carrion, and not infrequently each other. They move about at night and have poor daytime vision, and are, from our point of view, incredibly stupid, but they all have amazing powers of survival.

AMERICAN OPOSSUMS (*Didelphidae*)

All the opossums of the New World are grouped in a single family by reason of similarities in their anatomical structure. However, they vary in size from that of a cat to a small mouse, and in habits they simulate skunks, rats, shrews, and several other kinds of animals. The primitive Shrew-Opossums could well be separated into a family of their own.

Common Opossums (*Didelphis*)

These animals range from the northern United States to the Argentine, but display a considerable amount of variation in size, shape and color. Those of the north (Plate 4) and colder upland areas are heavy-set, slow-moving, short-tailed, and covered in a thick coat of white underwool and a long grey overcoat. To the south they get blacker and the undercoat sparser, and in the tropics they are long-legged, have very long tails and short fur. Opossums have hands not unlike ours, and their big-toes are constructed like overgrown thumbs. The tail is naked and scaly like that of a rat, and is prehensile so that the animal can hang by it from a branch for considerable periods. They sometimes give birth to as many as eighteen young at one time, but since there are only thirteen nipples, several usually are abandoned at once, while seldom more than half a dozen survive to reach adolescence.

Four-eyed Opossums (*Metacheirus, etc.*)

There are two distinct kinds of Four-eyes, one group has a reasonable pouch for carrying the young; the other has none at all. The former is found in Central and the latter only in South America. They are about the size of house rats. They have naked, prehensile tails the basal quarter of which is furred; the rest is naked and colored brown towards the base, and white for the terminal half. They do not, of course, have four eyes, but two very white, eye-sized spots immediately above the eyes surrounded by black areas giving this curious impression. These spots are to distract striking enemies from the real eyes. They are tree-dwelling, and eat insects and fruits. When their young are old enough to leave the pouch or to detach themselves from the teats, they climb aboard the mother's back and attach themselves to her tail by their own tails, but not as shown in traditional pictures. They usually trail behind her like a bunch of grapes and are violently buffeted about as the mother scrambles through the foliage. The non-pouched Four-eyes makes nests inside things and is a skillful small-bird catcher. They are yellowish brown above, yellow below.

Woolly Opossums (Philander, etc.)

Slightly larger than the Four-eyes are some beautifully colored, woolly-furred little opossums (Plate 5) known by the delightful scientific name of *Philander*—a complete misnomer that somehow got transferred to them from the Malay word meaning a certain small deer. They are also tree-living fruit-eaters though their stomachs often contain insects, snails and other small animals. Their tails are considerably longer than the head and body combined, and are fully prehensile, furred at the base but the rest being naked skin blotched with dark grey and pinkish hues. Their eyes are quite large and bulbous and bright orange; they look like polished pebbles and the pupil contracts to an invisible pinpoint in bright light. They are a bright brick-red above, and usually orange below, but often have a purplish wash about the shoulders and haunches. Two flaps of skin on either side of the mammary swelling which bears the nipples serve as a crude pouch to hold the young.

Mouse-Opossums (Marmosa)

These beautiful little creatures (Plate 3) are about the size of a large mouse and nearly all of them are some shade of rich russet brown or brick-red above, and yellow or white below. Their naked prehensile tails are pink, as are their hands and feet. Their eyes are very large, jet black, and protrude as if the animal were being squeezed to death. The muzzle is long and sharply pointed like that of a tiny fox, and the needle-sharp teeth can give a really terrible bite for their size. Besides, they are able, like all opossums, to open their jaws to an angle of almost 180 degrees. Mouse-Opossums are extremely quick little devils and are completely fearless, opening their mouths at anything that threatens. However, they sometimes forget to close them again for half an hour or so. They are primarily insect-eaters, but will take fruit, scavenge, and fish for small animals in water-filled holes in trees. Their fur is almost as silky as that of chinchillas, and they keep it meticulously clean by industrious licking and combing with the claws and front teeth. They are essentially tree-living animals, but one species inhabits caves in Trinidad, and another scrambles about in the long grass of damp savannahs in Brazil. From time to time they turn up in fruit stores in North American cities, having travelled up in hands of ripening bananas. The young are about the size of a grain of rice. A favorite nesting place of one species is in old, dried-out cacao beans, which are about the size of large pears.

Shrew-Opossums (Monodelphis, etc.)

There are at least four types or genera of these tiny creatures distributed throughout South America and probably all over the moister parts of Central America. However, they are so small, obscure in their ways, and of so little interest to anybody that relatively few have

Yapok or Water-Opossum

ever been caught, and probably less than one person in a million knows of their very existence. Nonetheless, these wonderful little animals probably come as close to some of the dawn mammals of a hundred million years ago as anything living today. They look for all the world like shrews, with mouse-sized but elongated bodies, short limbs and tiny feet, half-length, mouselike tails with sparse hairs but not prehensile, and with long, pointed snouts and tiny eyes. They live under things, making tunnels in forest leafmould, burrowing in loose soil, or infesting hollow trees and holes among roots and rocks. At least one kind lives along stream banks and dives for live food in the water. The smallest is less than three inches long and comes from southern Brazil. The coloring of these animals is varied but, unlike that of their larger relatives, is strange combinations of dark green stippled with yellow and often flushed with russet, red, or orange. Some have black bands along the spine. They do not have pouches.

Water Opossums (*Chironectes*)

One of the most astonishing animals in America is the Yapok or Water Opossum, which has a short, dense, seal-like, pale grey pelt with large, diamond-shaped dark areas over the back. The points of these "diamonds" form wedge-shaped stripes on the flanks. The underside is white. The feet are webbed like those of the beaver, for the animal is semi-aquatic and dives for fish, shellfish, and other stream and pond life. The markings on the head of the Yapok are not to be found on any other animal except occasionally on certain mongrel puppies, and have to be seen to be appreciated.

This animal is found in the mountainous areas from Guatemala to southern Brazil, but it is seldom seen even by the natives, who regard it as a kind of otter. The body is about a foot long and the tail is naked, ratlike, and scaly, but not prehensile, and is a horrid shade of pink. The animal gives off a ghastly smell at times, but whether this is a glandular secretion is not known. Young Yapoks apparently remain attached to their mother when she goes diving for food, but since she does this most of every night it would seem to remain to be explained how they breathe.

DASYURIDS

The Dasyurids are a widely assorted collection of animals. Nevertheless, they all had a common ancestry and are truly related. Two rather abstruse technical points may be used to pin them down; all have eight upper and eight lower front teeth, and all have the toes of the hind feet separate. There are five distinct groups of Dasyurids, but the first four, although different in external appearance, are more closely related anatomically, and constitute a single family. The fifth is most curious. As a whole, the Dasyurids are the carnivores of the marsupial world.

PHASCOGALES (*Phascogalinae*)

These little animals are often called "Pocket-mice"; it would undoubtedly seem easier if we called them "Pouched Mice and Rats," but there are true mice and rats with pouches, although in their cheeks instead of on their bellies, already well known by this name in America. Those animals are Rodents, whereas Phascogales are tiny, kangaroo-like Marsupials. There is an almost bewildering variety of them but they can be divided into seven distinct groups of genera. Although all are small—varying in size from that of a small House Mouse to a fairly large rat—they display a great range of characteristics. It is unlikely that anybody except a keen Australian naturalist will ever see any one of them alive, but then who would ever have supposed that a colony of Rufous Elephant-Shrews from Tanganyika would be established in Annapolis, Maryland. Furthermore, they are very interesting creatures. They go as follows.

Broad-footed Phascogales (*Antechinus*)

These are mouse-sized, mouse-shaped animals, with tapering tails clothed throughout in short hairs (Plate 7). They have very small big-toes without nails, and the pouches of the females vary from a pair of rudimentary ridges to complete pockets pointing backwards. They seem to inhabit the whole of Australia, Tasmania and some islands to the north. Pads on the soles of the feet and hands have ridges to aid in climbing trees and rocks, and one species has actually been seen running across a cave-roof upside down. They eat mostly insects, make large nests of dry stuff, and kill mice. Some store food in the form of fats in their tails. The muzzle is like that of a shrew, the ears are pointed, and the claws are long and sharp. The fur is rather soft and sometimes silky.

Flat-headed Phascogales (*Planigale*)

These tiny animals are unique among mammals in that their heads are so flattened horizontally that the skull is in one case only a little more than an eighth of an inch in depth, *i.e.*, about as thick as a half-dollar. They can thus slip into the most minute cracks in the baked desert surface. One species is the smallest of all marsupials, weighing less than an ounce and measuring less than two inches, not counting a tail of equal length. They live in tussock grass in open places, and appear to feed on crickets. A mother may drag as many as eight young trailing from her teats.

Brush-tailed Phascogales (*Phascogale*)

These beautifully colored, bluish-grey, squirrel-shaped and -sized little animals with the voracious habits of weasels are the commonest of the Dasyurids. They are distributed throughout Australia but are not found in Tasmania. They have even survived the advent of the white man, and attack and kill his poultry, including large ducks. The hinder half of their tails bears a bushy black brush, their ears are large, and their hands and feet are squirrel-like. They build nests in hollow trees, and like American Pack-Rats they steal paper and other materials to line these.

Crest-tailed Marsupial Mice (*Dasycercus*)

These are desert animals, the size of large mice, but with the habits of lizards. They bask in the sun, stalk their prey belly to ground and with lashing tails, and then rush them. Their tails are also lizard-like in form, having fattened bases, though they carry a crest of stiff hairs on the upper side of the terminal half. They kill mice, rats, and even larger animals, and do a neat job of skinning them from nose to tail as they eat the entire body flesh, bones, and internal organs. They are rare and known only from the central and southern part of Australia.

Crest-tailed Marsupial Rats (Dasyuroides)

This is a rare, rat-sized animal from the central deserts. It has a crest of stiff hairs both above and below the tail. Strangely also, it has no big-toe at all. It appears to lead the same sun-basking existence as its smaller relative above.

Narrow-footed Phascogales (Sminthopsis)

Here we come to a new type of animal, one constructed for jumping. The hind feet are narrow and elongated, and the tails, though sometimes fattened as food stores, are much longer in order to balance the animals when they are going in high gear. They also have spring pads at the base of the toes on the underside of the feet and in some species these are covered with tufts of fur. The ears are very large. They eat prodigious amounts of insects, and any other animals they can kill, and make nests underground. Most have complete, backward-facing pouches.

Pouched Jerboas (Antechinomys)

Finally we come to the extreme of the Phascogales, tiny animals that ape the rodent Jerboas of Asia and Africa. Perfectly adapted to life on deserts, they leap about on stiltlike hind legs supplied with thickly-furred feet to gain a purchase in loose sand. Their tails are extraordinarily long and have steering "paddles" in the form of thick tufts at their ends.

Spotted Dasyure or Australian Native Cat

DASYURES (Dasyurinae)

The animals to which the Latin name of *Dasyurus* was given were the first of this family of marsupials to be described. Hence they have passed their scientific name on to the whole group, but they are themselves an odd offshoot from the main evolutionary tree, and are highly specialized as hunters and meat-eaters. They are divided into two very distinct kinds, one comprising a number of spotted, long-tailed animals, the other a black, stocky, digging animal with an outsized head known as the Tasmanian Devil.

Native Cats (Dasyurus)

There are five distinct species, three commonly referred to in Australia as "Native Cats," and two as "Tiger-Cats." They are the size of small cats, except the Common Tiger-Cat which is much larger. The latter have not only the whole body and limbs covered with irregular and variously-shaped light spots, but also the long, tapering tail. The Native Cats have plain-colored tails, either light at the base and dark at the tip, or vice versa. They are fearless little animals that spend the day sleeping in holes, and hunt at night, killing whatever they can overcome. They have well-formed pouches containing six teats but as many as twenty-four babies can be born at one time, which seems most uneconomical since seventy-five percent must perish. Their color varies greatly even within a single litter, from pale mustardy-yellow to reddish-brown, chocolate to almost black or even grey, and the spots may be pure white or yellowish. They were once found all over Australia and Tasmania wherever there were at least some trees, but they have now become extinct throughout wide areas.

The eastern Native Cat, which is still fairly common around Sydney and in other limited areas, has no big-toe at all and no raised pads on the soles of its feet. The species from the central and western part of the country has a big-toe. Those of the far west are larger, white below, and are still plentiful especially along the coast, where they are regular beachcombers. The northern species is only half the size of the preceeding, and has both a big-toe and well-developed climbing pads. The Tiger-Cat of the east is a fully arboreal animal, about three and a half feet long, and is still fairly common in Tasmania.

The Tasmanian Devil (Sarcophilus)

This gruesome-looking but apparently readily tamed and then docile creature (Plate 8) is now known to exist wild only in Tasmania, where it is usually referred to as "The Badger." It lives in burrows, does not climb trees very well, and has a neat defensive trick in that it slips below the surface of any available water if pursued and can cover remarkable distances before silently emerging, usually under thick cover. These animals are armed with a splendid set of badger-like teeth for seizing and crushing prey, and they have stout claws for digging. In color they are black with a few large irregular white marks, usually around the throat and forequarters. The young are born in the southern spring—which is our fall—and are usually two in number. The pouch is a semicircular pocket pointing backwards. They make various ugly noises. The Devil is unexpectedly strong for its size, which is about that of a large cat, and is enormously persistent so that it kills animals much larger than itself and may actually be a menace to young sheep, poultry, and small dogs.

POUCHED WOLF (Thylacininae)

This is the largest known living carnivorous marsupial that has been scientifically examined. There may be a much larger, short-faced animal living in Cape York that has several times been seen and shot, but of which no skins or bones have yet been preserved. The Thylacine is the ultimate development in hunting animals, being exactly like a sleek, short-eared wolf with a tapering tail like that of a dog. It is of a dull brown-grey color, with lighter jaws and throat, but has a number of black stripes starting on the shoulders and increasing in length to the hind legs, and then retreating upwards again to the tail base. A few Thylacines still hang on in the mountains of Tasmania where they live just like wolves, hunting in pairs or small family parties by night, using a permanent lair in the daytime, and being grossly and maliciously persecuted by the farmers and everybody else. They are not as swift as dogs except if seriously chased, when they go into a kangaroo-like gallop, bounding along on their hind legs only. They can open their jaws to almost 180 degrees, and have about the most vicious set of teeth known among land mammals. There are no known cases of their attacking human beings, but the adults, and particularly females with young, will stand their ground, and they are able to kill dogs of all sizes. They have backward-pointing, shallow pouches, and the young are born in the typical semi-embryonic form—blind and naked—and find their own way into the pouch and then attach themselves to the teats.

NUMBATS (Myrmecobiinae)

These beautiful, delicate, quick little animals show some internal affinities to Phascogales, but zoologists don't quite know what to do with them. They are also called Pouched or Marsupial Anteaters. They are of a rich brown color darkening over the rump, which is vividly striped with white. The limbs and underside are lighter to pure white; the tail is grizzled grey and is bushy like that of a squirrel, and the small, pointed head carries an elongated dark eye-mask, bordered by two horizontal white lines. The Numbat is a true "termite-eater" in that its whole feeding apparatus is designed to extract the insects from the galleries they excavate in rotten wood. Real ants and other insects are ignored as long as termites are available. The wood is broken up with the forepaws, and the food collected by shooting the long, sticky tongue several inches into the galleries. They have many small degenerate teeth for crushing the insects. The whole body and head are flattened and the hind legs are rather long. The females have no pouch at all. Numbats were once widely spread over southwest and south Australia, but are getting very rare.

MARSUPIAL MOLES (Notoryctidae)

These incredible mole-shaped creatures (Plate 9) have horny muzzles and naked, hard-skinned, stumpy tails, reduced limbs with paddle-shaped feet, and hands bearing immense digging claws. They are clothed in iridescent silky fur varying from almost pure white to bright tangerine in color. They have no eyes and only minute holes for ears. They burrow under sand and loose earth in the desert areas of central Australia, eat insects and are very rare. They look exactly like the Golden Moles of South Africa, but are true Marsupials. Although more specialized for life underground than any other known mammal, these animals appear to treat the soil more like a liquid than a solid. They dive into it and burrow along about three inches below the surface for a certain distance, collecting subterranean food as they go, but they do not make true tunnels. Instead, they swim through the soil leaving behind what they dig from in front. Every now and then they come up for air, travel a short distance on the surface, and then plunge below again. They eat mostly earthworms, and are furiously active, but are constantly dropping off to sleep for short periods, and indeed so suddenly that they often appear to have "fainted."

CAENOLESTIDS

About a century ago, bones of some strange little animals were discovered in a fossilized condition in the Argentine. They were described as being those of a Marsupial of the Australian type rather than of a Didelphid, which would be more expected in America. Some time later an animal of similar type was caught alive in Colombia. The relationship between this and the fossil bones was not, however, recognized for thirty years, when a second species turned up in Ecuador. Since the discovery of the second, considerable technical wrangling has taken place over these rat-sized, somewhat bug-eyed little beasts, ending up with their being placed in a separate division of the Marsupials all by themselves. It is very doubtful if anybody reading this will ever see one of these animals dead or alive but they nonetheless hold an intrinsic interest. They look rather like small Philanders with minutely scaled tails covered with a few hairs. Most odd are their front teeth, which look more like those of squirrels. They have small shallow pouches but nothing much is known about the way they breed or, for that matter, any other of their habits. The larger species from Ecuador is dark brown and is well enough known locally to have the local Spanish name of raton-runcho. Three genera of them are now known, bearing the scientific names Caenolestes, Orolestes, and Rhyncholestes. They are so obscure and unimpressive to all but zoologists that they are regarded as intensely rare but it is probable that they are quite numerous in the forests where they live.

PERAMELIDS

Like the previous major division of the Pouched Mammals, this contains only one family of animals (the Peramelidae) but its members are much more varied in appearance and habits. All of these are commonly known in Australia as Bandicoots. Their position in the scheme of life lies somewhere between the predominantly carnivorous Phascogales and the more herbivorous Phalangers and Kangaroos. However, it cannot be overemphasized that the whole concept of flesh- or plant-eaters is old-fashioned and invalid. With surprisingly few exceptions, almost any animal will eat everything and anything that is digestible as occasion arises or shortages demand. Bandicoots have the same number of front teeth as the Phascogales, but unlike them, two of the toes on their hind feet, the second and third, are joined together up to the end joints. This curious arrangement is found in all the other Marsupials from here on. Sundry ideas have been put forward to explain this arrangement, the most often heard but the least acceptable being that it was originally developed for tree climbing. No such structure is found in any other treeclimbers, but it is typical of many animals that live on, and have to run over, hard open ground—such as antelopes, deer, jerboas, and, be it noted, kangaroos. Nevertheless, these paired claws are used as combs, rather like the famous Japanese "back scratchers," by all marsupials that possess them. Bandicoots are active little animals ranging in size from that of a rat to a hare, and while some creep about like shrews, others hop like rabbits, and the very odd Pig-footed Bandicoot stands up more like a small Chevrotain. In Australia, they take the place of the Insectivores and to a certain extent that of some of the Rodents and Rabbits of other continents.

Long-nosed Bandicoots (*Perameles*)

There are half a dozen species of these Bandicoots distributed about the whole of Australia and Tasmania.

They have almost exactly the proportions of a rat except for the long, pointed, shrewlike muzzle and a slightly shorter tail. Both the common eastern type and another form from the central area of the continent are plain dull brown in color but the latter has a bright orange rump. All the other species are notable for their peculiar patterns of cross stripes, single bands from hip to hip, or complex concentric half-whorls of dark lines bordered by lighter stripes on their hinder backs.

Short-nosed Bandicoots (*Thylacis*)

These Bandicoots are much plumper in general build, have shorter ears and snouts, and coarser, harder fur. There are about half a dozen distinct species spread over all but the southwestern area of continental Australia, and on various islands around its periphery. The largest species, to which the name of Giant Brindled Bandicoot has been given, appears to be indigenous to the whole eastern third of the country from Cape York to New South Wales. Like the other species of this genus its fur is of a pronounced *agouti* coloration: a phrase that will be used often to denote pelts composed of hairs that are individually banded white, black, cream, and brown.

New Guinea Bandicoots (*Echymipera*)

There is no way in which the layman may distinguish these animals from the foregoing except that they be caught on the island of that name. More annoying still, a species has now been found on the mainland, and other kinds inhabit other islands in the same area.

AUSTRALIAN INFORMATION BUREAU

Long-nosed Bandicoot

Rabbit-Bandicoots (*Macrotis*)

Called "bilbies" in Australia, these long-eared, bushy-tailed Bandicoots, with kangaroo-like hind legs, pointed muzzles, silvery blue-grey fur, and most engaging ways, are viewed with a certain degree of sympathy if not delight even by those who normally consider the destruction of all wildlife a beholden duty. They are nocturnal and true burrowers, digging spiral tunnels several feet deep. They prefer animal food and are death on rodent pests including adult rats. The tail is curled downwards and ends in a horny spine. They sleep sitting up with the head tucked down on the chest and the ears laid back, but with the terminal half folded forward over their eyes. The pouch opens slightly backwards and contains eight nipples, but the average number of young seems to be only two.

Pig-footed Bandicoots (*Choeropus*)

The most exaggerated in form and now rarest of the family, this antelopine little animal was previously found almost all over the continent of Australia. It is altogether different in that it has only the second and third fingers of the forefoot properly developed, with the fourth forming a useless horny spur. The result is a forefoot like that of a tiny pig. The hind foot lacks the big-toe altogether, while the little-toe is but a useless stump. This leaves the combined second and third toes, and the fourth, which latter is so big it looks like a hoof. The animal stands and looks like a tiny deer and has big ears, a sharply pointed head, and a long, upcurled tail bearing a plumelike crest of long hairs on its upper side. It makes nests of grass above ground and is nocturnal. The pouch points backwards and is provided with eight nipples, but apparently only two young are raised at a time.

PHALANGERIDS

This is the fifth major group of pouched mammals and its members are almost as varied in shape, size, and habits as are the Dasyurids. Apart from their name, they are not nearly so bewildering, because after we take out three very distinct types—the well-known Koala, the now almost as well-known (from crossword puzzles) Wombat, and the tiny Honey-Sucker—the rest are all long-tailed climbing animals with woolly fur. Three quite separate types of these have developed parachutes stretching between fore and hind limbs like flying-squirrels (see Rodents). They are found all over Tasmania, Australia, New Guinea, and the East Indian Islands, north and west to Amboina, the Celebes, and Timor. As to their overall and individual names, the answers are not so simple.

The first kind of Phalangerid to be described scientifically happened to be one of the oddest forms, known as the Cuscus, to which the latin name *Phalanger*—or the Fingery-One—was given because of the agile way in which the animal manipulated things with its hands. The name is not inappropriate to all of the Phalangerids with the exception of the Wombats. Unfortunately, the famous Captain Cook not unreasonably applied the name opossum to one of the naked-tailed species because of its superficial resemblance to the common Didelphid of North America, and Australians have ever since tended to call them all "possums" collectively, and by native names individually. Further complication has now arisen from the fact that the word *opossum* does not appear to be an American word after all, but itself an import from the Celebes, where the Buginese name for the Cuscus is *O-Possuh*, meaning a little bag or pouch. It is most inconvenient that the Cuscuses happen to be tagged *Phalanger* while the Common Phalangers have the Latin name *Trichosurus*.

All Phalangerids have the second and third toes joined together up to the base of the claws, and the big-toe is opposed to the other toes—as our thumb is to our fingers—and bears a nail instead of a claw. Almost all have well-developed pouches that point forward and all have five fingers and toes. Only the Koala and the Wombats lack tails.

HONEY-SUCKERS (*Tarsipedinae*)

By far the most extraordinary and one of the most delightful of all marsupials is the incredible little "eager-beaver" known as the Noolbenger that haunts the shrubbery of the extreme southwest of Australia. It is unlike any other animal on earth, being only about the size of a small mouse but with an exaggerated, almost proboscis-like snout from which sprouts a mass of sensitive whiskers. It has small, almost humanly-shaped hands, the typical phalangerid feet with an opposed big-toe, and the prehensile tail of the true opossum, but covered with very short hairs. In color it is a rich brown on the head, flanks, and outer side of the limbs, pale cream below, and grey on the back, darkening to almost black on the mid-back, but with a light longitudinal stripe on either side. Most odd of all is the long, extensible tongue. This is clothed in short hairs with a tuft or brush at the end. The lips have accessory flaps which are used to convert the whole mouth-opening into a sort of suction-pump. The animal feeds exclusively on nectar (honey), pollen, and small insects gathered from flowers by a sort of violent pumping action.

In the area where the Noolbenger is found, some trees or shrubs are always blooming, and the tiny animals rush hysterically around from one to the other at night gathering their food in prodigious quantities. They are extraordinarily lively and nimble, and their tiny size allows them to perform miniature acrobatics on the tiniest twigs in order to reach the blooms. As a result of this diet their teeth, like those of the Numbat, have become degenerate. Neither bees' honey nor garden flowers, even if of Australian origin, seem to provide proper fare for this animal in captivity, but they will thrive on flies, which they catch with unerring accuracy and aplomb by leaping at them while they are in flight. They are gregarious to the extent that large groups invade flowering plants all together and then move on. They make little compact nests like those of the European Harvest Mouse (see Rodents) in bushes, on treetops, or even on the ground. They produce up to four young at a time. They often take over old birds' nests in which they sleep by day sitting up but with their tails curled over them like Dormice.

PHALANGERS (*Phalangerinae*)

We now come to an enormous assemblage of beautiful little animals varying in size from that of a small mouse to a very large cat. All are either tree- or rock-climbers. Members of this sub-family are found all over the continent of Australia and in Tasmania, New Guinea, and the islands to the north, even to the Celebes, Timor, and Amboina beyond the famous "divide" known as Wallace's Line, which officially divides Asia from Australia. All have long tails, the majority prehensile even if fully furred. Others have feather-shaped, plumed tails for use as rudders.

Dormouse-Phalangers (*Cercäertus*)

The smallest and apparently also the most primitive of the whole family are a group of large mouse-sized to small rat-sized animals known to Australians as Pigmy Possums. They are strikingly like the Mouse-Opossums of Central and South America (see above), even to the thickly furred basal portion of the tail. In these animals, however, the rest of the tail is covered with short fur except for the underside of the tip, which is naked. They are nocturnal and sleep during the day, curling up in nests constructed in hollow limbs of trees or under bark. They are very fussy about nesting material and may bring it from up to a quarter of a mile away. They hibernate just like dormice, sometimes doing so twice a year; prior to this they store up fat in their bodies, and noticeably in their tails, which become almost bulbous. They eat insects but have a slight furry brush on their tongues and suck a certain amount of honey from flowers. A species from the southwest has remarkable finger and toe pads that are divided like those of certain lizards and are used more than the claws in climbing. They all have from four to six nipples and appear to breed twice a year. As usual too many young are born to survive.

Striped Phalangers (*Dactylopsila*)

Several species have been identified from the tropical forest areas of Queensland, New Guinea, the Aru and other islands, but they vary individually in color in a most bewildering manner, no two ever seeming to have quite the same pattern. They are fairly large animals vividly marked with black stripes from the top of the head and before the eyes, on either side along the body to the tail, with side stripes leading off to ring the neck and pass down the outer side of arms and legs. The fur is rather long, soft, and fluffy. They give off an overpowering aromatic odor. Their hands are naked, long-fingered and extremely facile, the fourth or "ring-finger" being almost twice as long as the others, and having a spoon-shaped nail rather than a claw. The front teeth are long and recurved like those of the rodents, and meet in a similar manner.

Although these animals appear to eat a variety of foods, the basis of their normal diet is insect larvae which they extract from bark and rotten wood by tapping the outside with their sensitive hands, sniffing with their keen nostrils, tearing the wood apart with their sharp front teeth, and finally extracting the grubs from the deeper holes with the long fourth finger. They are nocturnal and arboreal, and apparently very close to the smaller Gliders.

Feather-tailed Phalangers (*Distöechurus*)

These tiny creatures from the forests of New Guinea should perhaps be placed next to the Dormouse-Phalangers because they appear to represent the kind of animal from which both they and all the other Phalangers could be descended. With thick, very soft but woolly fur, they look like dormice. They are grey to buff above and white below, and their heads have prominent black eye-stripes and naked muzzles that are bright pink. The eyes are large, jet black, and prominent. The hands and feet are simple except for the usual bound third and fourth toes and the thumblike great toe. The tail, however, is most remarkable, being extremely long, furred at the base, but, for the rest, being naked, with a line of long, stiff hairs arranged horizontally along either side, making it look just like a feather. On top and bottom it is clothed in a short fluff, but the long stiff hairs on either side grow in lines in such a way as to give the whole a paddle-shaped outline. These little animals are nocturnal and tree-dwellers, and they make tremendous leaps with outspread arms and legs like the true Gliders. The tail is used not only as a rudder, but appears actually to give the animals a lift to the extent that, by being turned first downwards and then upwards in graceful curves during a leap, it causes the animals to land upright on tree-boles. In some respects they are halfway to Gliders, in others they are an odd offshoot that has developed their own semi-aerial mechanics. They are insect-eaters.

Gliding Feather-tails (*Acrobates*)

These tiny creatures, known also as Pigmy Flying Possums, less than three inches long with tails of just over three inches, are almost identical in structure to the ordinary Feather-tails, but have a narrow, furred parachute of skin stretched from wrist to ankle on either side of the body. This is not nearly as extensive as the parachutes of the true Gliders or even of the Flying-Phalangers (see below), and seem to represent a three-quarter-way stage to the true gliding habit. The tail is again constructed just like a feather, but has a slight naked portion below the tip and is prehensile. There is one species on the mainland of Australia, and there may be another on New Guinea. In color they are dark grey-browns to very dark brown above, and creamy white below with dark eye-stripes, short-furred ears, and naked muzzles. The outer edge of the parachutes, however, bears long, light-colored hairs originating from the underside, to form a distinct margin, and the tip of the tail is also often light. The fingers are slender but bear enlarged terminal pads and long claws for climbing.

Leadbeater's Phalanger (*Gymnobelideus*)

We now come to a rather nondescript little animal of plain coloration that is of great interest to zoologists for anatomical reasons, but of little interest to anyone else.

Flying-Phalanger

There are only half a dozen actual specimens preserved intact in all the museums of the world, though the first one was spotted as far back as 1867. It is of a rich brown color, with a very fluffy tail, but club-shaped and extra long. The muzzle is rather short and the eyes apparently small. It climbs with remarkable agility, aided by enlarged pads on the tips of its fingers and toes. It is known only from the southeastern part of Victoria in Australia, and is virtually a Flying-Phalanger without parachutes.

Flying-Phalangers (*Petaurus*)

These come in a variety of forms known to Australians respectively as the "sugar-squirrel," the "squirrel-glider," and the "yellow-bellied glider," or most misleadingly as "flying squirrels." They have beautifully soft, silky, almost fluffy fur; short, sharp, squirrel-like faces, rather large ears, pink and naked inside and furred on the back; long, bushy tails; small, long-clawed feet, and bright eyes. The "sugar-squirrel" is merely a pint-sized edition of the "squirrel-glider," but is quite distinct. The ranges of the two overlap, the former being found all over the eastern half of Australia and in Papua and Tasmania, and the latter in isolated areas within this region. Both these animals are of a lovely soft blue-grey color above, white below, with black terminal tail portions, a dark stripe along the mid-back from nose to tail, complex light and dark facial markings, and full parachutes that are edged with fluffy white, and, on the upper side, with a pronounced black line. They all make wild screeching noises. They eat insects, some leaves, and much honey or nectar which they lick from flowers. They also apparently eat some flower petals and, even when it is congealed and hardened, large quantities of the aromatic gum that flows from tree trunks, limbs and twigs.

Strangely, they are violent little creatures that fight with real viciousness, yet they are readily tamed. They sleep by day in hollow trees, and make nests of leaves which they transport in bundles held in their curled tails. Mothers are very solicitous of their pouch-young, and have often been observed licking them while suckling and holding the pouch open with their hands. The yellow-bellied species is very much larger and comes from the eastern coastal forests of Australia. It is brownish above and yellow below. It makes prodigious leaps and thus wanders far at night in search of food. It also has a large and varied vocabulary of shrieks, gurglings, mumblings, and hissings, all somewhat unusual for a marsupial.

Ring-tailed Phalangers (Pseudocheirus, etc.)

We come now to a group of small animals that can be divided into four quite distinct kinds—the common, striped, brush-tailed, and rock-living.

The first, the Common, is found in a variety of species all over New Guinea, the eastern and southwestern parts of Australia, and in Tasmania. They are dull grey, grey-brown, or coppery brown animals with rather short faces, small ears, hands and feet, and long tails with white, naked ends. The first two fingers of the hands are opposed to the other three. They are arboreal and nocturnal and many of them make spherical nests of twigs and leaves on branches. They are very slow-witted and insensitive, though most tenacious of life, and really very competent. They are leaf- and fruit-eaters but they will eat almost anything they can digest. Several young are born at a time, but there are only two operative nipples so that few are likely to survive. When developed the young ride on their mother's back like Didelphids.

The Striped Ring-tailed Phalangers (*Pseudochirops*) are fundamentally Papuan though there is one species in Australia. They are predominantly diurnal animals, mostly of a yellow-washed, greenish tinge, and are light below, either pure white or yellow. They have a black line along the back from the top of their head to the tail. The fur is rather long and fluffy, and some species have well-furred tails while others have a large naked strip below. The Australian species alone has the typical white end-portion to the tail. They are agile inhabitants of the upper canopy of the tall equatorial forests.

The Brush-tailed Ringtail (*Hemibelideus*) is a sort of zoological gem that seems to stand at a three-way fork in the evolutionary tree of life. It is a dark brown animal of comparatively large size with soft, woolly fur and a fully furred tail that is bushy right to its tip. In general form it looks exactly like certain of the Great Gliders in their brown phase, except that it lacks parachute membranes. However, it does have a sort of flange about an inch wide along either side of its body between elbows and knees, and it behaves much like the Galagos of Africa (see Primates), leaping wildly about trees and using its tail as a rudder. The tail, however, although fully furred, is prehensile and has a small, naked strip under its tip. It varies in color like the Great Gliders but despite its very close similarity to those animals, it also shows features of another group of Phalangers, namely the Brush-tails.

The Rock Ringtail (*Petropseudes*) comes from far to the northwest near Darwin, and lives among boulders, though it will climb any available trees. Its first and second fingers are not opposed like those of all other species and its tail is very short and has a naked end.

Great Gliders (Schoinobates)

Perhaps the most remarkable of all Marsupials, this large Phalanger which may measure over three feet in overall length, presents us with some extraordinary conundrums. It is found throughout the hilly and mountainous area all down the eastern side of Australia from Queensland to Victoria and more especially in the open gum (eucalyptus) forests. It is, of course, completely arboreal, and thus confined to forests, but it can travel on the ground albeit in a very clumsy manner. Despite the many dangers to which it is then exposed, it will readily cross ground to reach isolated trees that provide its rather specialized food. This consists almost entirely of tender leaves, shoots, and blossoms of various eucalyptus and other trees of the great Myrtle family. The parachute membranes of this animal spread from the elbow to the lead edge of the ankle so that the body of the animal when in "flight" has an outline remarkably like an exaggerated swept-wing jet-plane. The gliding is accomplished by a jump from a high point, a long steady descent, and then a brief upward swoop before landing. They can cover a hundred yards with ease, and have been seen making longer glides. They literally gallop up tree boles, using first both forefeet and then both hind feet together like the African Anomalures (see Rodents). They nest in holes high up and there is normally only one young at a time, though there are two nipples in the well-developed pouch. Their long, extremely soft, fluffy fur varies in color with the individual from black through all sorts of browns and greys to white—an almost unique variation among mammals.

Brush-tailed Phalangers (Trichosurus)

These lovely little animals are the true, common, "'possums" of Australia. They are normally said to have foxy faces but their whole head and their expressions are actually much more like some of the smaller kangaroos. The body is sturdy and compact and looks very plump. The limbs are short, the hands are well clawed but very human in use, while the hind feet have the usual phalangerid thumblike big-toe, and the bound-together second and third toes. Their tails are long and bushy, but have a naked strip under the tip for holding on to branches. Brush-tails are found all over the continent and Tasmania, and although varying enormously in color from a beautiful chinchilla-grey with black tail, to silver, rich coppery red or almost black, they can be clearly divided into two species-groups—namely the long-eared (Plate 12) from all over the mainland and Tasmania, and the short-eared from the hilly and mountainous areas, stretching from Queensland to New South Wales on the eastern side of the continent. While the tails of the former are club-shaped, those of the latter taper towards the tip and have a much more extensive naked area below.

They are common animals and one of the few marsupials that seem to be able to cope with the advent of the white man and his pests and machines. Millions have been slaughtered for their close, firm fur, but at last some measures of protection are being provided for them. They are inoffensive, predominantly leaf-, fruit-, and nut-eaters, and they are night-walking tree-dwellers, though some inhabit miserable dry scrub. Despite their comparatively small size they can put up a terrific fight even against fair-sized dogs, yet, once tamed, they make the most affectionate and complacent pets. One of mine not only knew its name but would come when called.

The Scaly-tailed Phalanger (Wyulda)

There is a large area of northwestern Australia that

still remains, at least comparatively and certainly zoologically, one of the last large unexplored parts of the world. Several new and some unique animals have come from the fringes of this rocky, mountainous region, and not least of them the Ilangurra, a kind of Phalanger that appears to live among rocks instead of in trees, and has a hairless, prehensile tail covered with scales like that of a rat. It is a small animal, white below, with reddish-brown head and limbs, but silvery grey over the neck, back, sides and rump. The eyes are small, and the face ratlike. Its anatomy is close to that of the Brush-tail but seems to point the way to the Cuscuses.

Cuscuses (Phalanger)

These ghostlike night animals (Plate 6) are found all over the Australoid Indies from Timor and the Celebes, throughout the whole of New Guinea to the Solomons, and in the Cape York Peninsula of Queensland in Australia. They have domed skulls, large eyes, close, woolly fur marked in most strange ways, and tails that are part furred and part naked, but covered with a rasplike arrangement of pointed scales. Their color varies so much individually within any one family or local group, and among geographical groups, that scientists have been hard put to it to classify the beasts. It appears, however, that the more or less predominantly spotted species from Australia can be separated from the rest. Even the eye-color varies, one having brilliant red, others orange, yellow, brown, or even blue-green irises. There are isolated populations in various parts of the animals' over-all range where all the members seem to look alike, but even these will interbreed where their ranges meet those of other populations and new color types then spring up. Their habits are completely Lemurine and nocturnal. They are slow-moving tree-dwellers, with blank stares and retiring habits but vicious tempers. To add final confusion to the matter of identification, males are often a different color from their own females, and young from adults, while both sexes may completely change color with the seasons, when breeding, or if they change location or diet.

THE KOALA (Phascolarctidae)

The pert, tubby, tailless little animal (Plate 11) that provided some unknown genius with his model for the first Teddy Bear is unique among marsupials, and is included in the Phalangerids only because it has more in common with them than with any other group. Nonetheless, having a backward-opening pouch, which is an asinine arrangement for a tree-living animal, it is more like some Dasyurids. This poor, harmless little animal has been grossly persecuted for its fur, harassed by dogs, and starved by the clearing of its food trees. It has been almost exterminated in Victoria, reduced to a few colonies in New South Wales, and decimated in Queensland.

Koala have large, rubbery noses, small eyes, fluffy ears, handlike forepaws with the first and second fingers opposable, and the usual opposed big-toe and combined second and third toes. They have rounded ears fringed with hair, rather startled expressions, thick, woolly grey fur, and terribly strong, sharp, claws. They are arboreal and feed exclusively on certain eucalyptid leaves, but only for a certain period each year when that tree is producing specific oils in its leaves. When the type of oil changes, the poor little animals have to go down to the ground and take off in search of other species of trees producing an oil that suits them. One or two young are born at a time, and are only about three-quarters of an inch long. They remain in the pouch for six months and until about six inches long. After this they ride on their mother's back for a year, though still using the pouch as a retreat for three months. When they are really naughty the mother turns them over her knee and spanks them on their bottoms for minutes on end with the flat of her hand, during which time their screams are soul-rending.

WOMBATS (Wombatidae)

The Wombats have developed both internal and external characters rather like some large rodents. They have only two upper and two lower front teeth, and these grow continually like those of porcupines; the other teeth are rootless and also grow all the time, being worn down on top and continuously replaced from below. They are great diggers, some of the holes they make being over a hundred feet long and usually ending in a capacious nest lined with leaves and bark. Their principal diet is grasses, with some roots, shoots, and fungi. In appearance they are rather like bears, with thick close fur, short sturdy limbs and no tails, and they are the size of a very large heavy-set dog with stout legs. They have nail-shaped claws for digging. There are two principal kinds of wombats, one known as the Naked-nosed (Plate 10) and the other as the Hairy-nosed. Both are becoming very rare and are now confined to limited localities. The former originally dwelt in the hills and mountains, while the latter were found in the coastal plains and all over the inland plateaux and lowlands. The Naked-nosed have very coarse, hard fur; the Hairy-nosed soft and silky fur, and longer, more pointed ears.

MUSK RAT-KANGAROOS (Hypsiprymnodontinae)

The three remaining groups of Phalangerids are customarily joined into one family and called collectively Kangaroos. This is, however, misleading, since they are quite distinct. The first is as close to the true Phalangers as it is to any kangaroo, and in any case, the word *kangaroo* is rightfully applicable only to a few species of the last group. The strange little animal which has one of the longest names in zoological parlance, and which has become known popularly as the Musk Rat-Kangaroo, lives on the ground among damp, dense vegetation in certain limited areas of Queensland. They are only about eighteen inches long including a six-inch tail, and are of a bright reddish-brown color turning almost orange below. They look like large rats, but have a very strong, musky odor. These animals feed on insects, and occasionally some fungi, and scratch about for both under leaves, stones, and other litter. To zoologists, their most important features are their hind feet and their tail. The former are very like those of phalangers, having a distinct thumb without claw or nail; the latter is naked and covered with scaly skin that is not found in any wallaby or even rat-kangaroo. These little animals are a true missing link between the two major branches of the Phalangerids.

RAT-KANGAROOS (*Potoroinae*)

There are about a dozen different kinds of these small kangaroo-shaped animals found throughout Australia and Tasmania and associated islands. They are divided into four quite distinct genera. Although looking just like tiny kangaroos externally, they are really a separate offshoot of the same main family, since they eat insects and some other animal as well as vegetable food; they also have well-developed canine or eye-teeth, whereas the wallabies and kangaroos have either minute ones or none at all. Like the latter, however, they have no big or first toes, and their tails are well furred.

Long-nosed Rat-Kangaroos (*Potorous*)

Unfortunately these animals (Plate 17) lost their original native name in the early days of their discovery; for what could have been nicer than "potoroo," to go with wallaroo and kangaroo? The potoroos come very close to the Musk Rat-Kangaroo in that their hind legs and feet are not exaggeratedly long, and they gallop about on all fours more like rats, though they sit up on their haunches to eat their food. They were once widely distributed over Australia, but have now been almost exterminated, two of the species which used to inhabit the southwestern part of the continent being believed to be extinct. Only on the island of Tasmania have they so far managed to hold their own, but even there the swarms of dogs, cats, rats, foxes and other vermin introduced by Europeans are decimating them as well as all the other regional fauna. They are delicate little animals with small pointed ears, small but very bright eyes, slender feet, and tiny, handlike forepaws. The muzzle is almost shrewlike except in one species which may be extinct but which had a short, broad face more like a cat. Potoroos appear always to have lived in damper places, building small nests of grass and scampering about hysterically like rabbits.

Short-nosed Rat-Kangaroos (*Bettongia*)

The Bettongs, Boodies, or Squeakers are on the whole more herbivorous than the Potaroos, but they appear always to have been beachcombers where lake or seashores were to hand, and they have become pronounced scavengers around human habitations. One of the four species is a regular carrion-eater, but another is almost wholly vegetarian. Bettongs are reported to dig extensive burrows in one district, live in rabbit warrens in another, do a little scratching in a third, and never dig at all but make runways under grass and herbage in another. Although persecuted relentlessly since the white man colonized Australia, the little bettongs have so far and somehow managed to survive, and that they have done so may in no small part be owing to their ability to adapt their habits to changing circumstances. They have taken to residing with the imported rabbits in their burrows; they have learned to live on human garbage; they have even taken to marauding in the face of that deadly Australian pest, the domestic cat. Bettongs are small, dusty grey, hopping animals, with sharp snouts, little rounded ears, kangaroo-like hind limbs, and long-furred tails, the hairs of which get longer towards the end and may become bushy or even crested. The tail brush is usually darker to black, but may be white-tipped.

Desert Rat-Kangaroos (*Caloprymnus*)

This is a richly colored little desert animal of kangaroo appearance, but having, according to anatomists, much in common with the Phalangers. It lives in the central desert areas of Australia, but although it does not appear to drink, it keeps out of the true, waterless sand desert. Although it digs for roots with its hands, it does not use burrows, but constructs grass nests on the surface under scrub-bushes. Records of prodigious runs by these little animals are recorded, with one said to have outdistanced two horses before giving up the flight.

Rufous Rat-Kangaroos (*Aepyprymnus*)

This is the most kangaroo-like of all, being much longer and stouter in build, with long legs and feet and a sturdy tail. The fur is harsher and brindled. The face has that remarkably inquisitive look of the wallabies. The muzzle is covered with fur, the backs of the ears are jet black, and there are dark facial stripes from nose to ear base, enclosing the eyes. It is a true grass-, root-, and herbage-eater, and was once very common all over the eastern seaboard, but is now found only in some areas of Queensland.

KANGAROOS (*Macropodinae*)

The name kangaroo presents a number of problems that always cause a great deal of confusion. The origin of the name is definitely Australian, having been taken from one of the languages of the aborigines, but there is a delightful story, doubtless untrue though far from improbable, that one of the first white men to see these animals noted that whenever one popped up and made off, the local natives would yell "kang guru," but it was years before anybody unraveled the local lingo and discovered these words meant simply "There he goes." Be this as it may, the name still has no precise meaning, since it covers well over half a hundred different animals ranging from two to nine feet in length and including even a few species which live in trees. Among this general assemblage, moreover, various animals are also known, almost indiscriminately, as Wallabies, Wallaroos, or Pademelons, while the scientific names given to the sundry groups are often at variance with these.

Tree-Kangaroos (*Dendrolagus*)

This name may at first sound a little mad, since one would no more expect to see a kangaroo than an antelope in a tree. Nonetheless, these anachronisms (Plate 15) are quite common in the forested mountain fastnesses of the great island of New Guinea, on certain other islands, and in parts of northern Queensland.

There is some doubt as to whether they should be included among the kangaroos at all, or be given a separate grouping of their own, since, quite apart from climbing trees, they have so many features that are unique. However, anatomists are satisfied that all such characteristics as their large, heavy-clawed hands, short legs, broad feet, small, fixed ears, and long, tufted tails, are only special developments that have come about to aid the animals in climbing trees. There are over half a dozen different kinds, of colors varying from almost black to pale fawn or greyish and sometimes of complex and beautiful patterns, dark above and light below, and with contrasting face, hands, and feet. They spend much time on the ground but sleep in trees, sitting up with their heads bent between their legs, and they feed on all manner of fruits, leaves, ferns, and even grubs. They get about trees very well but are by no means perfectly adapted to do so; they descend trunks backwards and are rather awkward among small limbs. However, they make prodigious leaps especially from tree to ground with the tail stretched stiffly behind as a rudder. Leaps downward of as much as sixty feet have been measured. They are undoubtedly ground wallabies that have returned to the trees.

Hare-Wallabies (Lagorchestes)

These are the smallest of the "kangaroos" and are harelike both in coloration and habits. They have slender pointed ears, immense hind limbs with long narrow feet and slender tails. Everything about them is adapted for speed and high-jumping and some of the jumping feats ascribed to them are, to say the least, sensational. They used to be found widely throughout the south, central and western parts of the continent of Australia as well as on several islands off its coast, and some species ranged to Queensland and New South Wales. They have now been driven back to a few small isolated areas. All but one are brownish above and lighter below, with a reddish wash on the flanks; the exception is known as the Banded Hare-Wallaby and comes from the southwest. It is grey, with a series of dark crossbands down the midback. They are nocturnal and although one kind digs a little, they all make "forms" or nests like hares.

Rock-Wallabies (Petrogale and Peradorcus)

The coloration of these remarkable animals is impossible to describe since it is immensely varied, one having a banded black and yellow tail, one being generally grey, another reddish above and cream below, and three out of the dozen distinct kinds having dark eye-stripes and most complex bands of light and dark on other parts of their bodies. They are the chamois of Australia, living among rocks and performing among them like acrobats. They sleep in caves where these are available and make polished paths across rock-faces by their endless coming and going. They are slender animals with very long legs and feet and long thin well-furred tails becoming bushy towards the tip, which they use as rudders when leaping. They give alarm signals by thumping on the ground like rabbits.

Nail-tailed Wallabies (Onychogale)

Out on the grassy plains, the place of the last group is taken by these silky-furred little kangaroos, not much larger than hares, which also have very long legs and feet and whiplike but furred tails. One from the northern deserts is a pale, sandy color but the others are grey with reddish areas about the shoulders and curious white stripes curving round the arms and enclosing dark brown areas in the armpits. They are grass-eaters but dig with their hands for roots. Most remarkable of all is the tip of the tail, which ends in a horny spur, unique among kangaroos but somewhat like the spur on a lion's tail.

Pademelons (Setonyx and Thylogale)

Pademelons are found all over Australia, Tasmania, New Guinea and on many associated islands. In contrast to the foregoing species and those to come, they have compact bodies, proportionately much shorter hind limbs, feet, and tails. Their noses are pointed and their ears are rather small. They are rat-shaped and on an average about two feet nine inches long with a tail half that length. They are all colored a rather dull brown. They make tunnels in dense undergrowth and swamp herbage and graze on open grassy places nearby at sunrise and dusk. Their soft fur was once used commercially. They usually have only one young at a time and like all kangaroos carry this "joey"—as all young animals and even human children are sometimes called in Australia—in a forwardly directed pouch.

True Wallabies (Wallabia)

Also called Brush-Wallabies, these are the medium-sized animals that stand between all those that have been described above and the large animals more popularly known as kangaroos. To the average person they are just small kangaroos. They are, however, lighter in general build than the two remaining groups, with very long, round tails that taper gradually and are furred throughout. The forepaws are handlike, but are armed with strong claws, and in most species they are black or at least darker than the forearms. Wallabies have large, deerlike ears which can be turned about and their feet are very long and narrow, usually white or at least lighter than the legs and bear stout black hooflike claws. They come in all manner of colors and color combinations (Plate 16)—pale grey; grey and white; grey, brown and white; sandy; brindled with reddish shoulders; dark chocolate brown; and so forth. Their habits are not as varied, all being grass- and herbage-eaters, and almost all of them staying in the taller brush and especially in those areas where the trees stand somewhat apart from each other. The species depicted in Color Plate 14 (W. elegans) has a delightful and descriptive popular name—Pretty-faced Wallaby.

Wallaroos (Osphranter)

This group is also known as the Rock-Kangaroos, a properly descriptive title since they are confined exclusively to rocky territory and are in several ways adapted to life in such places. They are stocky, compact kangaroos with rather short, broad feet and specially roughened pads to give them a sure footing. There are about half a dozen species and with one exception are hardly distinguishable by any except experts. The one exception is a very large species known as the Euro, which is smooth dark grey with white chest, belly and lower hind limbs; it once occupied the whole central and southern area of Australia. The others are all reddish to dark brown in color and are distributed around this area in a great arc from the southwest to the north and thence down to the southeast. Today all of them have been greatly reduced in numbers and their ranges have shrunk to isolated tracts of country away from man and his settlements. You may distinguish Wallaroos from other kangaroos by their hairless muzzles—though getting close enough to do this is difficult even in a zoo.

True Kangaroos (*Macropus*)

There are two principal groups of species of true kangaroos, plus one or two island races and a variable number of other types that may be regarded as distinct, depending on the way you define a species. First, there is the Forester or Great Grey Kangaroo, and secondly the Great Red Kangaroo (Plate 13), which is the largest marsupial we know of, standing seven feet on the tripod formed by its toes and its mighty muscled tail. The former may be clearly divided into eastern and western forms and is, as its name indicates, predominantly grey, although washed with rufous in the western form. The male of the Great Red is a most curious rich reddish color with a mauve sheen on the back, a grey face and white throat, chest, underparts and feet, and with the insides of the legs and arms and the underside of the tail also white. The female, known colloquially as the "Blue Flyer," is normally a beautiful smoky grey above and white below, but reddish-tinged adults have been recorded. These animals are the Australian equivalents of the deer and antelopes of other continents, being grass-grazers and brush-browsers. They travel about in "mobs" under the watchful eye of an old male or "boomer" who maintains strict discipline and defends his position against younger male rivals in no uncertain way. He does so by biting and boxing while reserving the terrible slashing down-cut of his tremendous back claws for the coup de grace. The Great Red is found all over the interior of the continent; the Great Grey on the east and west sides and in Tasmania. The appearance of these kangaroos is well known to the people of all countries where there are zoos, books on animals, movies or television, but their true habits are known to few even in Australia where they are still fairly common. Space does not permit me to elaborate on this, but any who are interested should read a little book entitled *Kangaroo* by Henry G. Lamond.

Although large active animals of rather complicated habits, they are not really very intelligent—at least, by our standards. They are easily panicked and then, if not able to escape by headlong flight, usually make all the wrong moves, rushing into fences, running in circles, or stopping to look at their pursuers. They are placid grazers and will seldom stand and fight.

Insect-Eating Mammals

(*Insectivora*)

THE strange assortment of curious little animals that make up this order of mammals are all basically related but less closely so than the members of any of the other nineteen orders of living mammals. They form a sort of zoological catch-all into which a number of very ancient forms—some very primitive in structure, others exceedingly specialized—have been tossed. As was noted in the introduction to the Pouched Mammals, the earliest mammals of which we have records (in the form of fossilized bones found in very ancient strata) are separable into three kinds. One of these very closely resembled certain living Insectivores. These tiny creatures have apparently just gone on and on since time immemorial without any substantial alteration in bodily structure.

Living Insectivores fall into four major groups but these are very unequal both in number of species and in individual members. They are each, moreover, of different homogeneity; that is to say, the members of two of the groups are very closely related and those of another are composed of far-distant cousins. Thus, the curious tenrecs of Madagascar are isolated both in space and ancestry from the Giant Water-Shrew of West Africa and the absurd-looking Solenodons of the West Indies. On the other hand, both the Hedgehog group (which includes the Gymnures) and the little Elephant-Shrews, or Macroscelids, form very compact little units. The so-called Soricids are an enormous conglomeration of small animals that can be laid out in a more or less continuous series starting with a shrew at one end and finishing up with a mole at the other.

Perhaps the most unexpected thing about the Insectivores is the enormous number of individual animals in this order. Not only are they found all over the world (except in the greater part of South America, the whole of Australia, the polar regions, and some of the driest deserts) but many species occur in untold millions over truly vast areas. Yet it is probably an understatement to say that not one person in a thousand has ever heard of a shrew, not one in five thousand has ever seen one, and not one in ten thousand actually knows what the animal is. Nonetheless, these fragile little creatures are everywhere, even in city parks, suburban yards, farm fields, and woodlands. In the tropics they not only live almost everywhere, but occasionally swarm in untold billions. At such a time, moreover, some of them display the extraordinary ability to grow individually in bulk to many times their normal size and weight. Why this occurs is not known.

It should be mentioned here—though this will be

elaborated later—that, despite their fundamentally primitive make-up, the Insectivores may well contain among their number our direct ancestors. So direct is this line that the animals which used to be known as Tree-Shrews were, until a few years ago, classed as Insectivores. These are now considered to be Primates and are placed in that order, which includes the lemurs, monkeys, apes and ourselves. They are now called by their native Malayan name, Tupaias.

Among the living Insectivores are found some of the oddest anatomical structures in mammals and some of the strangest habits. For this and other reasons this order of mammals is of particular interest to zoologists.

TENRECIDS

The first major group of Insectivores has its headquarters in the great island of Madagascar off the east coast of Africa in the Indian Ocean. Only three related animals have been found elsewhere and two of these live in the most unexpected places—namely, the Solenodons, which are found only in the islands of Hispaniola and Cuba in the West Indies, and the Giant Water Shrew or *Potamogale,* which inhabits a large area of northwest Central Africa. Despite their widely scattered distribution these animals are more closely related to each other than to any other animals. Among them are the most primitive living mammals—apart from the marsupials which some of them resemble in certain respects. Tacked on to this group is a fourth type of exaggeratedly specialized form, known as the Cape Golden Moles, which inhabits the southern part of Africa.

TENRECS (*Tenrecidae*)

These curious animals vary in size from that of a large mouse to a bulbous, prickly beast that may measure eighteen inches from the tip of its long snout to the place where its nonexistent tail should be, and are thus by far the largest of all Insectivores. There are eight quite distinct kinds of Tenrecs varying widely in appearance and habits but all living only on the island of Madagascar. They are leftovers from very ancient geologic times and are so close to the basic mammalian stock that they even show anatomical characteristics otherwise found only among the marsupials.

The Common Tenrec (*Tenrec*)

Also known as Tanrecs or Tendracs, these are very queer-looking creatures. They have long, pointed snouts, small, beady eyes, and small, triangular, perked ears. The legs are short and both hands and feet are armed with sharp claws. There is no tail at all and the body is almost globular in shape so that the whole animal looks as if it is the front half of a much larger animal. When young these animals have three narrow rows of yellow spines running down the back, are quite slender, and in no way resemble their parents. When the animal becomes adult these flexible spines are concentrated in a sort of collar around the back of the neck, and the rest of the body is thickly clothed in a mixture of fine spines, bristles, and coarse hair of a nondescript grey-brownish-yellow hue. The full complement of teeth for a mammal

is forty-four but very few, especially when full grown, carry the whole number at one time. Young Tenrecs have forty, but when the animals reach maturity four "wisdom-teeth" appear, an almost unique development among mammals and one that greatly interests zoologists. The teeth are very sharp, and the males develop large tusks. The fact that the cheek teeth are triangular in shape stimulates anatomists because such teeth are otherwise found only in one rare kind of primitive dog and among marsupials. Nor is this the only characteristic of these bizarre creatures that shows them to be close to the pouched mammals. Their skulls are astonishingly like those of some opossums, and like those animals also they give birth to enormous litters of young, the record until last year being twenty-one at a time, but with twenty-four reported more recently. This exceeds even the best any marsupial can do though twenty-five nipples have been counted on one opossum.

These animals are entirely nocturnal and live in the dense fern forests of the Mascarene mountains. They eat mostly worms and insects but, again like some opossums, they seem to be able to digest almost anything. In May they dig deep burrows into which they retire to hibernate during the southern winter and they often do not appear again till December. Before retiring they store up fat in their bodies and become almost spherical. For some reason these animals have a habit of yawning at regular intervals all the time; and when they do so it looks as if their throats were going to fall out because the hinder parts of their palates for some unknown reason blow up into large pink bulbs.

Striped Tenrecs (*Hemicentetes*)

These seem to be the ancestors of the common tenrec that have got, as it were, stuck in the immature condition. They are the size of small rats without tails and the lines of yellow spines found on the young of the above species are developed all over the body and retained throughout life, the intermediate areas being covered with coarse black hair, giving the whole a streaked effect. The next group seem to be a further extension of the spiny defensive form.

Hedgehog-Tenrecs (*Setifer*)

These look for all the world like small hedgehogs (Plate 19), being covered with small sharp spines, and being able, to a certain extent, to roll themselves into a ball for defensive purposes. They have little stumpy tails and very sharp snouts with which they grub about for insects in drier places during the night. They also dig burrows and go into partial hibernation. Still another genus of spiny tenrecs (known technically as *Echinops*) appears to stand halfway between the two latter, but forms the end of a series in that its members have only thirty-two teeth.

Rice-Tenrecs (*Oryzorictes*)

In a limited way, the Insectivores of Madagascar have done just what the Marsupials of Australia did long ago; they have branched out into all sorts of forms. Thus the Rice-Tenrecs have become molelike, with paddle-shaped front feet for digging, stout claws, rather spindle-shaped bodies and flabby tails. They have a mouthful of teeth for crushing insects and they do a lot of damage to rice crops by their tunnelling. They are covered in soft fur.

Long-tailed Tenrecs (Microgale, etc.)

Most surprising of all tenrecs are these mouse-sized and mouse-shaped, furry creatures which, for some unknown reason, have the longest tails of any mammal—that of one kind being more than twice the length of the combined head and body, and having no less than forty-seven vertebrae to support its length.

Water-Tenrecs (Limnogale and Geogale)

Almost equally surprising are the two rat-sized, close-furred animals, with webbed feet and long tails flattened from side to side, that inhabit the streams and rivers in some of the more out-of-the-way parts of Madagascar. Internally they are built like tenrecs but they have the habits and much the external appearance of the next member of the Tenrecids, found in west and central Africa.

THE GIANT WATER-SHREW (Potamogalidae)

Throughout the equatorial forested area of Africa from Nigeria to the Congo in certain river systems and notably those that arise in high mountains but by no means exclusively confined to these, may be found a sleek otter-shaped animal about two feet in overall length. The head is shovel-shaped like that of a shark, with the mouth underneath; the eyes are tiny, the ears mere flaps, and the fur very short, dense, and glossy. The tail is compressed sidewise into a blade-shaped structure, and is covered in rubbery black skin with a plush of fine fur. The feet are not webbed, and the soles of the hind feet have strange fins rising from their outside edges. They eat frogs, freshwater snails and clams, and occasionally fish, and live the lives of otters, making holes in river banks with an entrance under the water. They are nocturnal.

SOLENODONS (Solenodontidae)

Solenodons are like gigantic rats, but have immensely long, tubular snouts bristling with sensitive whiskers. Their tails are scaly, but short, stiff hairs sprout from between the scales. The front feet are armed with extremely long, somewhat curved, slender claws; the hind feet have just long claws, and the animals walk only on the tips of their toes. However, when eating or scratching and combing their long, coarse, shaggy fur, they sit up on a tripod formed by the whole soles of the feet and the base of the tail, as do kangaroos. The Cuban species is reddish-brown, turning black on back and throat, and is found only in the Bayama Mountains at the east end of the island. That from Hispaniola is sandy brown with black thighs, pale cream underside, and a yellowish face the long hairs of which extend back in the form of a ruff over the shoulders. It is now found only on the south-western peninsula of Haiti and in the northeast of Santo Domingo. Solenodons are vicious, irascible creatures that fly into sudden rages, when they bite indiscriminately and let out screams of rage. They eat almost anything, but prefer insects, carrion, and lizards.

GOLDEN MOLES (Chrysochloridae)

Throughout that part of Africa south of the equator there are areas where the earth is seen to be criss-crossed by little meandering ridges. These are made by mole-shaped animals with brilliantly metallic golden fur, bulb-shaped bodies, and pointed snouts armed with a horny prow. The limbs are inside the body, the hind feet have four toes with long claws and the front feet two immense talons placed side by side, and a tiny rudimentary

[continued on page 49

Haitian Solenodon

1. *Platypus*
AUSTRALIAN INFORMATION
BUREAU

2.
*Five-toed
Spiny
Anteater*
AUSTRALIAN
INFORMATION
BUREAU

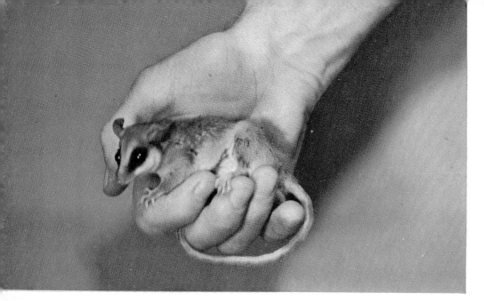

5. *Woolly Opossum*

MARKHAM

→

3. *Mouse-Opossum*

E. P. WALKER

4. *Common
North American
Opossum*

LA TOUR

6. *Cuscus*

AUSTRALIAN INFORMATION BUREAU

7. *Broad-footed Phascogale*
AUSTRALIAN INFORMATION BUREAU

8. *Tasmanian Devil*
AUSTRALIAN INFORMATION BUREAU

9. *Marsupial Mole*
AUSTRALIAN INFORMATION BUREAU

13. *Great Red Kangaroo*
AUSTRALIAN INFORMATION BUREAU

→

10. *Naked-nosed Wombat*
AUSTRALIAN INFORMATION BUREAU

11. *Koala*
AUSTRALIAN INFORMATION BUREAU

12. *Brush-tailed Phalanger*
AUSTRALIAN INFORMATION BUREAU

14. *Pretty-faced Wallaby*

15. *Tree-Kangaroo*

16. *Red-necked Wallaby*

AUSTRALIAN INFORMATION BUREAU

17. *Long-nosed Rat-Kangaroo*

AUSTRALIAN INFORMATION BUREAU

18. *European Hedgehog*

MARKHAM

19. *Hedgehog-Tenrec*

LA TOUR

20. *Elephant-Shrews*

E. P. WALKER

21. *Common North American Mole*

MOHR FROM NATIONAL AUDUBON

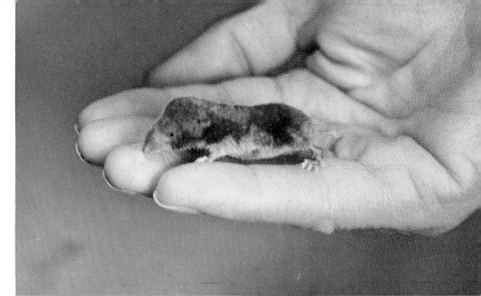

22. *Lesser Short-tailed Shrew*
E. P. WALKER

23. *Long-tailed Shrew*
MASLOWSKI & GOODPASTER FROM NATIONAL AUDUBON

24. *Star-nosed Mole*
E. P. WALKER

26. *Vampire Bat*

LARSON

27. *American Fruit Bat*

MOHR FROM NATIONAL AUDUBON

25. *Flying Fox*

MARKHAM

28. *Serotine Bat*

E. P. WALKER

29. *Free-tailed Bat* (*Tadarida*)

MOHR FROM NATIONAL AUDUBON

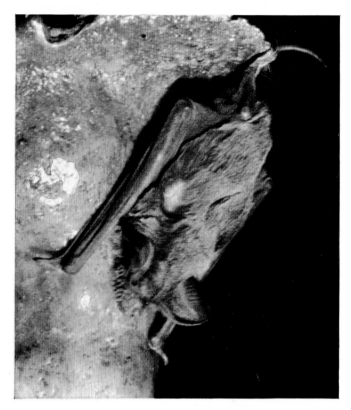

31. *Cinereous Bat* (*Lasiurus*)

E. P. WALKER

30. *Red Bat* (*Nycteris*)

PINNEY

33. *Tree-Shrew*

PINNEY

32. *Tarsier*

LA TOUR

34. *Moholi Bushbaby*

MARKHAM

35. *Black Lemur (female)*

LA TOUR

36. *Slow Loris*

MARKHAM

37. *Potto*

MARKHAM

claw on either side and underneath these. The eyes are under the skin and there is no external ear. They feed on earthworms and are wholly subterranean. Four quite distinct kinds have now been found.

ERINACIDS

This, the second division of the Insectivores is composed of two groups of animals known respectively as the Hedgehogs and the Gymnures, which look totally unlike each other but which are closely related. There is even an intermediate form tucked away in the remote parts of China. These are also extremely ancient creatures in that their anatomy still retains features of the dawn mammals that lived alongside the Dinosaurs. Fossilized bones of types that differ very little, if at all, from some of those living today have been found in rocks that were formed on lake bottoms as much as thirty million years ago. The miracle is that they have survived at all, yet they are among the commonest animals throughout a very large part of the world.

HEDGEHOGS (Erinaceinae)

There are five distinct genera of hedgehogs, or as they were once called, "hedgidogs," individual species of which are spread over an enormous area comprising all of Europe south of the boreal pine forests, almost the whole of Africa except for the really wet areas of equatorial forest, and the whole of Asia with the exception of southern China, Burma, the Indo-Chinese peninsula, Malaya and the Indonesian islands. Their range northward in Asia is limited by the extent of the deciduous woodlands. As a whole, hedgehogs are very much of a oneness, but the best known and one of the most widely spread of those found in Europe (Plate 18) happens to be one of the most extreme forms. They are small animals with tiny naked stumps for tails, heavy, bloated bodies, small, pointed heads, short limbs with small feet armed with short fingers and toes but slender sharp claws. The head, throat, and undersides, including the limbs, are clothed in fur which may be sparse and coarse or abundant and fluffy; the rest of the body is covered in an interlocking mass of short, hard spines with exceedingly sharp points. Each of these is longitudinally grooved and in some kinds the ridges intervening between the grooves are serrated like the teeth of a blunt saw. Although some dig, they do not make burrows but live in the cover of tangled undergrowth where they rummage about for their most varied diet of insects, eggs, small animals of all kinds, fruit, fungi, roots, and other vegetable items.

The habits of hedgehogs vary according to their kinds, the climates they live in and the physical conditions of their habitat so that some in India are wholly nocturnal while others, notably in the moister open forests of Africa, move about by day. Their most outstanding trick is a defensive action that has probably been wholly responsible for their survival. If alarmed, even by a modest squeak or a thump on the ground, the creature gives a little leap into the air with all four limbs so that its domed back of sharp spines will strike anything in the immediate vicinity: it then immediately snaps its body into a tight ball, using certain muscles under the skin and packing its four legs and head against its belly so that it presents an almost spherical and quite impossible conundrum to any assailant.

They all either hibernate or aestivate to avoid cold, hot, or excessively dry periods when food is scarce, and almost all of them hole up during dry weather. They have four young in a litter but as many as six may appear at a time. The young are blind and almost naked except for tiny soft white spines that harden in two or three days. The European species has tiny rounded ears but in Cyprus and North Africa there are species that have long doglike ears and stand on comparatively tall legs. The color is usually a sort of stippled olive brown and cream resulting from the banding of the spines but some desert forms are white with red spines and one from southern India is almost jet black all over with white-tipped spines. For centuries there has been a belief in Europe that hedgehogs carry off ripe apples by impaling them on their spines. The idea was scoffed at as an old wives' tale but has recently been observed regularly and even filmed.

GYMNURES (Echinosoricinae)

To our eyes these appear to be among the most revolting of all mammals. They are rat-shaped with long, scaled, and bristled tails, large naked long-fingered and -toed feet, small ears and large heads drawn out to excessive snouts rather devoid of hair but bristling with whiskers. The mouth is very large but sharp, and filled with pointed teeth. They are clothed in a thick underwool and an overcoat of coarse bristles and they smell atrociously somewhere between a skunk cabbage after rain and a rotten garlic bulb. They take the place of their relatives the hedgehogs in southeast Asia and the greater Malayan islands. There are four distinct kinds, the most interesting being a small rat-sized animal that is clothed in reddish fur, has a long stout tail and lives in Sze-Chwan in west central China. This looks like the rare gymnure of the Philippine mountains but internally is constructed more like a hedgehog. The Long-tailed species (Gymnura) from Burma, Malay, and Sumatra is black and white in color, and another species from Borneo is pure white. The short-tailed types (Hylomys) are only half as big, being about five inches long, and are found all over Malaya, Sumatra, Borneo, and Java.

MACROSCELIDS

Throughout the continent of Africa from Algiers to the Cape, but predominantly in dry areas, and excluding large areas in the west and in the central wet equatorial zone, there exist countless numbers of remarkable little animals having the appearance of miniature kangaroos but with sleek fur and small trunks. These have become known as Elephant-Shrews. They range in size from that of a large mouse to a medium rat and come in a large variety of colors and color combinations. Zoologists have divided up the group into five genera, depending on such details as the number of fingers and toes, and the number and form of the teeth, but, with the exception of a large, rather short-legged type from East Africa known as the Rock-jumper, there are no popular English names to use in differentiating them, but they vary greatly in external appearance.

They are delicate little animals (Plate 20) with comparatively long limbs, the hind ones being almost kangaroo-like in proportions. Their heads are rather outsized, the eyes very large and the ears prominent, in one case very large indeed. They are very active, nervous little creatures and when really alarmed take off in a whirlwind of dust or dead leaves by a series of prodigious leaps, using their hind legs alone, and their long tails as balancing organs. The most distinctive feature of these animals is the snout, which is so long and flexible that it may truly be called a trunk. With this they probe into nooks and crannies for insects and particularly into the tunnels made by ants and termites in rotten wood. The food is whipped into the mouth with a long worm-shaped tongue. The Rock-jumper (*Petrodromus*) shows something unique among mammals; the terminal half of its tail is covered with quarter-inch bristles, each ending in a club-shaped bulb, the purpose of which is not known. Probably because of their active habits combined with their highly developed powers of sight, hearing, and smell, they have disproportionately large brains for their size. The best known species (*Macroscelides*) is found in North Africa and was known to the Romans, who first gave them their popular European name of "trumpet-rats." The group that inhabits equatorial Africa (*Rhynchocyon*) is often very beautifully colored, being various combinations of rich russet red, warm browns, and orange below, darkening over the back, and adorned over the rump with half a dozen lines of large yellow or white spots extending from the shoulders to the base of the tail. Altogether, the Elephant-Shrews form one of the most highly evolved mammals but are withal an ancient sprig on the tree of mammalian life.

SORICIDS

We come now for the first time to a group of mammals that, although still virtually unknown to the public at large, is not only of considerable interest to the specialist but of the utmost importance to Man. If by any unlikely chance something should suddenly happen to all of either one or both of the groups of animals included in this division of the Insectivores, namely the shrews or the moles, the whole economy of nature throughout the woodlands and farmlands of the world, outside of Australia and South America, would be thrown into a turmoil at the least. Many of our most important crops, various types of plantation, and many valuable stocks of trees, would suddenly be deprived of one of their greatest natural allies in their everlasting struggle against insect pests and the ravages of slugs, snails, and a variety of other marauders. The loss of moles would, on the surface appear to be less of a menace than the elimination of shrews, for moles eat mostly earthworms and these lowly creatures tend to be beneficial. They also eat a lot of beetle larvae and these are harmful. The importance of shrews and moles in the balance of nature

cannot be overemphasized and is probably second only to that of rats and mice among mammals.

The Soricids, like the rest of the Insectivores, are an ancient stock but unlike the others they have become enormously widespread and numerous. They are found in all moist areas south of the subarctic frost line throughout Eurasia and North America, and in Africa; one species of shrew spreads through Central America to northern South America. The moles are not found in Africa or Southeast Asia. They are all primarily insect-eaters but shrews will tackle almost anything alive and can subsist on carrion and some vegetable matter. Shrews live under things on the surface of the ground, moles burrow below the surface. Both exist in countless millions and although they are seldom seen and little understood they are almost everywhere.

SHREWS (*Soricidae*)

There are literally dozens of different species of shrews arranged by specialists in over twenty types or genera. With a few exceptions they are all very much alike but they can be divided into three major lots, two of considerable compass, the third reserved for a single incredible little animal found only in parts of Central Africa. The only simple distinction between the first two groups is the color of their teeth, one being reddish-orange-tipped, those of the other the normal white. The first three lots to be described are red-toothed (*Soricinae*), the next four white-toothed (*Crocidurinae*)

Common Shrews (*Sorex, etc.*)

What is said of these animals applies to all other shrews unless otherwise indicated, for they are all very much alike. Of three kinds of shrews included under this heading one is found in Asia, Europe, and North America, one only in the first, and one only in the last. They are all small animals, the Pigmy shrew of America being positively minute and much smaller than any mouse, the largest having a head and body length of just six inches. In addition, shrews are light-bodied and very fragile, with short limbs and delicate little five-fingered hands and feet. Common shrews have fairly long tails covered with fine fur and some longer fine bristles. Their heads are in no way like those of rats or mice, being drawn out into a long pointed snout bristling with whiskers. The eyes are minute, and the ears little, much-crumpled affairs close to the side of the head. The fur is very short, fine and silky; and almost all Common Shrews are some shade of dark grey to black with some individuals showing a wash of brown or silver. If there remains any doubt as to whether an animal is a shrew or a mouse one has only to get a look at the teeth. Mice have two large recurved teeth in the front of their mouths, top and bottom; shrews have a continuous row of needle-sharp pointed teeth, in this case colored bright orange or red-tipped.

All shrews are excessively nervous and irascible little furies and their tiny bodies tick at such a high rate that their output of energy and thus their fuel requirements far surpass those of any other mammal and almost any other animal. Some of the smaller kinds are so high strung that a sudden sound and even a paltry squeak will cause them to leap into the air, knock themselves out, fall in a dead faint, or even drop dead. Their ribs are not much thicker than pigs' bristles and all their bones are very light so that a tap, squeeze, or bump that a mouse would hardly notice, may kill them instantly or wound them mortally.

They have to eat all the time and must do so at very short intervals; otherwise their whole high-speed mechanism runs down and stops. The amount they eat is prodigious, being more than twice their own weight in a day, and may when food is plentiful be almost unlimited except by the rate at which they can chew and swallow. And they will eat almost anything digestible, animal or vegetable, alive, fresh, tainted or putrid, including their own young and each other if other food runs out. They do not appear to breed in their first year of life but in the second they have up to four litters in a season between early spring and late summer, and there may be as many as eight young in a litter. Then they seem to die in their second year. The color illustration (Plate 23) shows a common American species known as the Long-tailed Shrew (*S. dispar*).

Short-tailed Shrews (Blarina, etc.)

The second lot of red-toothed shrews are little different in appearance or habits from the last genus but they are predominantly North American, with one genus in eastern Asia. On the whole they have much shorter tails, are a little more chubby and tend to stand up higher on their limbs (Plate 22). Their muzzles are not quite so long and pointed. Apart from a very rare type from the southwestern part of the United States, known as *Notiosorex,* which has large ears and a white stripe down the hind back, this group tends to have very small ears concealed in the thick fur. Like the common shrews they live under trash and matted vegetation or dead leaves, and in the tunnels and burrows of mice and other small creatures. They make small nests in which to hibernate, rest or give birth to their young, and they eat insects and any other small animal food they come across. They also give off powerful musky odors, and make shrill partially supersonic squeaky or keening noises, and shriek when enraged, which is very often.

Common Water-Shrews (Neomys)

A species of the Common Shrews in America has taken to living in and out of water in mountain streams and lakes. It is silky grey above and white below and has developed a fringe of stiff bristles along the outside edge of its rather broad hind feet. In Europe and parts of western Asia an entirely different animal (*Neomys*) of almost identical form occupies this natural niche. The differences are only of concern to specialists but it is of interest to see two animals so closely related yet not basically the same coming to look alike through adopting the same habits. Water-Shrews spend a great deal of their time in the water hunting for water insects, small snails and so forth. They make tunnels into banks and although they hibernate they may be active when rivers are frozen and there is deep snow on the ground. To aid in swimming, there are fringes of bristles on the hind feet and also along the underside of the tail. However, progress through the water is actually effected by a kind of wriggling motion combined with an alternate paddling of the back feet while the forefeet are kept pressed back against the chest, and the tail is held straight out behind to counteract the wobbling. These animals have three little flaps on the ears provided with special muscles so that they may be closed like an efficient swimming-cap. Curiously, these animals appear to have some form of hydrostatic apparatus, meaning that they can regulate their density to conform with that of the water and so either float at the surface, suspend in mid-water, or sink to the bottom where they can walk about as if on land.

Musk-Shrews (Crocidura)

These and the shrews still to be described have white teeth and belong to a larger group, none of which are found in America. Their headquarters is in Africa and includes the giants among shrews. Musk-Shrews are almost the commonest mammals in parts of West Africa, especially on cleared land and around native villages. They live amongst grass and tangled vegetation and are entirely nocturnal. They are altogether sturdier than other shrews and they display one extraordinary quality that, although not entirely unknown among other mammals, is in a degree unique. From time to time the number of these shrews begins to mount in some area and then, as among certain lemmings (see Rodents), larger litters begin to come more frequently and in a short space of time swarm conditions pertain. But with the Musk-Shrews this is not all: the individual animals increase in size and not only by generations but by succeeding litters. Individuals may end up by being over twice the length and five to ten times the bulk of the average of their parents.

Forest-Shrews (Sylvisorex, etc.)

Just to keep the record straight, and in case anyone counts the number of genera of shrews we mention, or goes collecting shrews in Africa or south Asia, we must point out that there are half a dozen genera of small shrews to be found in the deep loam of the equatorial forest floor. They do not look very different from small Common Shrews, and lead much the same lives except that they do not as far as we know hibernate.

Fat-tailed Shrews (Suncus)

These are given a special note not because they look in any way different from the smaller common shrews nor even because of their very short bulbous tails but because the European species is the smallest known mammal. Adult males are smaller even than the minute Pigmy Shrews of either America or central Europe. Other species occur in Africa and throughout west and central Asia in limited areas. These minute beasts are just over two inches long.

Mole-Shrews (Anourosorex)

Throughout a wide area in eastern Asia, stretching from Assam and northern Burma through the mountains to inner Tibet and Sze-chuan in China, may be found colonies of curious little fat, almost mole-shaped shrews with bluntly pointed heads, minute eyes covered by fur and no visible ears. They have tiny stumps for tails and very short limbs. They live in rather special locations, where the forest floor is of a particular consistency or composition but are nowhere common. Like moles and gophers they can run as fast backwards as forwards. Not only do they look like moles, they seem to be very close to them in structure.

Asiatic Water-Shrews (*Chimarrogale*)

We remarked on the fact that a Common Shrew in North America, and another kind altogether (*Neomys*) in Europe and central Asia, had come to look alike through taking to the water. That is not all, for still another quite separate animal has done the same in the Oriental region of Asia. Species of this group or genus are known from the Himalayas, the mountains of Borneo, and Japan. They are large shrews, sleek black above and white below, with tiny eyes and small ears provided with flaps hidden in the fur. Neither fingers nor toes are webbed but the latter bear fringes of stiff white bristles, and there is a line of the same along the outside of the hind foot and along either side of the tail. They also can walk on the bottom of streams.

Web-footed Shrews (*Nectogale*)

To complete the picture and perhaps thoroughly confuse the inquiring layman we then come to a fourth distinct kind of shrew that has taken to the water. This one is found in the icy mountain streams that descend on to the central Plateau of Tibet and it is fully adapted to life in such waters, having rather long, immensely thick silky fur of a dark chocolate-brown color, with an overcoat of shining white longer hairs. They are about four inches long, with a tail somewhat longer. The feet are fully webbed and the soles bear large pads. The feet and tail are fringed with white bristles and the pointed muzzle carries long, white, feeler-like whiskers. There is no external ear-conch and the eyes are completely hidden in the fur. The young are brown all over but the adults are usually pure white below.

The Girder-backed Shrew (*Scutisorex*)

This animal usually but misleadingly called the Armored Shrew is unique and constitutes the third group of the family. Externally the animal is not especially odd in appearance but internally it, or at least part of it, is altogether fantastic. This is the animal's backbone. For some unknown purpose this has developed into a massive sort of complex girder filling half the back and looking, when cleaned of flesh and ligaments, like a piece of steel bridge-work conceived by an engineer obsessed with the idea that his structure will be unequal to the strain it will be subjected to. The thing is quite impossible to describe even by anatomists and it is best to resort to a photograph. Why this tiny obscure animal should have this amazing structure defies imagination but it would appear to be devised to support loads far in excess of anything the paltry little limbs or the viscera underneath could take without being completely squashed. Nature is full of surprises. It is found only in the forested areas of the Congo basin of central Africa.

MOLES (*Talpidae*)

Moles in the technical sense, which is to say the animals included in the family named *Talpidae*, are much more varied in appearance than the shrews. They may be clearly divided into five groups, two of which are properly molelike in form and habits, two of which are not at all so, and the fifth of which is altogether bizarre and in its way just as unique as is the Girder-backed Shrew. Moles generally speaking are found all over Europe and Asia north of the Himalayas. In other words they are not found in what is called the Oriental Region. One is reputed to be resident in the Atlas Mountains of Morocco but no valid specimens have been forthcoming from any part of Africa. They also inhabit both sides of North America but there is a moleless belt running down the Rockies.

Asiatic Shrew-Moles (*Uropsilus and Nasillus*)

This is one of the most confusing details of mammalogy and really need not concern anyone unless he proposes to indulge the unlikely pursuit of collecting moles in Tibet, Japan, and on the west coast of North America; yet, it may prove the point that attention to detail is essential if we are to get any real idea of the mammals that inhabit our world. There are three quite distinct animals that are in appearance exactly halfway between shrews and moles. One lives in America, one in Japan, and one, which we are introducing here, in Tibet. This almost entirely shrewlike animal, with normal feet and a long scaly tail appears to live on the surface of the ground. However, internally it is technically a mole.

Desmans (*Desmana and Galemys*)

Much odder and just as unmole-like are two rather bulky, plush-furred animals, found in Russia and Spain respectively, which have revolting-looking flexible trunks and spend most of their time in ponds, lakes and small rivers. The first can measure as much as eighteen inches, including a rather long tail which is flattened from side to side, is covered with short fur, and is narrow at the base. The Pyrenean form from northern Spain is less than half this size. Their bodies are indeed molelike in shape and have the same extraordinarily soft fur; the front feet are spadelike, with short widely spaced claws and very short fingers. The eyes are minute and there are no visible ears, but the front of the face is almost hidden by a mass of whiskers and bears a naked trunk with a trumpet-shaped end. The mouth is a small round hole. The hind feet are relatively huge, fully webbed and bear a brush of plumed bristles along their outer edge. Internally, Desmans are fairly primitive, having the full complement of forty-four mammalian teeth and a skeleton that shows certain ancient features. They spend most of their time in the water probing for food but dig tunnels into banks and maintain a dry nest above water level for resting and breeding. The fur is a rich reddish brown above and silvery below and was once an article of commerce. They have large glands opening under the tail that let out an overpowering musky stink. Besides being much smaller, the Pyrenean species has a round rather than a flattened tail.

Eurasian Moles (*Talpa*, etc.)

The true moles are divided into two great clans, one composed of the original European mole and three allied genera in Asia, one of which spreads to the Pacific. The other, which we shall meet next, comprises eight genera spread around the Northern Pacific, with an outlier of two types in eastern North America. Moles have since time immemorial intrigued human beings. There is something quite unexpected about a mammal living underground; and the appearance of the animal and its extraordinary strength have given rise to all sorts of weird beliefs about its habits and potentialities. Moles naturally prefer areas where the soil is loose and by long association with man they tend to infest his gardens and farmlands. There they burrow along about two inches below the surface in search of food. This consists mostly of earthworms but they will eat any small animals, including mice, shrews or even their own kind. The latter happens when they bumble into another mole's run and one is defeated in the battle that inevitably takes place. They also maintain large dwelling-burrows placed at a lower level and usually under the roots of a tree. This is lined with leaves, moss and other soft material and has an escape exit.

Moles go out to eat morning, noon, and night, literally, for when they are full they fall asleep but invariably wake up in about six hours and unless they start eating at once soon collapse of starvation. They put away about twice their own weight per day. Their bodies are bun-shaped, the head pointed, the eyes minute and either covered by skin or buried in the fur, and there are no visible ears though they have very acute hearing, especially for earth-borne sounds. The tail is short and naked but carries a few sensitive bristles. The hind legs are very short but the feet fairly normal and made for shoving the animal forward; the front feet stick out sidewise, are very short but sturdy, and end in huge handlike paddles with their palms pointing backwards. The nails are tremendous and there is a "sixth finger" made of an extension of one of the wrist bones. When digging, the spade-hands are shoved forward in front of the nose alternately and the earth is then scooped backwards past the head and under the body. The hind feet then take over and shoot it on backwards into the tunnel. Every now and then the animal makes a vent and erupts the excess out on to the surface, thus making a molehill. Some species, however, manage to get along without these periodical eruptions and all moles hard-pack the walls of their tunnels.

The fur of moles is very soft and silky and it grows straight up so that its lay will not hinder the animals in going either backwards or forwards. Moles, like many Shrews and other Insectivores, swim very well and, in fact, delight in entering water either in search of food or when moving from one locality to another. At this time, those moles whose eyes are not covered by skin make good use of what sight they have. The speed with which moles can get about even on the surface of the earth is almost unbelievable, but underground they can move faster than a man normally walks above! The love-life of the common mole is remarkable. Normally, he lives alone but at the appropriate season he either burrows into the tunnel system of a female or one breaks into his. Whereupon he promptly takes her captive and if any other male appears he seals her up in a side-hole, then quickly hollows out an arena and goes to work on the interloper. He fights until death and the winner takes the female. Meantime, she is busily engaged trying to dig her way to freedom. In time the pair settle down together and work alongside each other in the endless pursuit of food. When the young are due, the female hollows out a nest at a crossroads in the tunnel system where there are plenty of escape routes, and lines the hole with soft material. Three to seven young are born at a time; their eyes are closed and their ears are covered with skin and they are naked. Moles make rather solicitous parents and take a long time to wean the babies. There is always a great dearth of females! The true moles are distributed all over Europe and Asia and one species is found on the southern slopes of the Himalayas and in Assam, but none occurs down on the plains of India or in the Indonesian region.

Pacific Moles (*Scapanus*, etc.)

These animals do not look very different from the preceding group though specifically there are differences that may be noted by the non-specialist as well as the expert in insectivorous odontology. The host of species are divided into eight genera. Of these, two are typical moles in appearances and behavior. One is large and comes from the west coast of North America and is known to science as *Scapanus;* the other is considerably smaller and comes from the central plains and the east of North America and is known as *Scalops* (Plate 21). Then there is another much smaller type from the central and eastern part of the United States that has a bulbous tail covered with scales and a thick coat of stiff short hairs, and a naked muzzle. This animal is about six inches long, has very close fur, and seems to spend much time on the surface. Its toes are not webbed like the other Pacific Moles and it prefers hilly or even mountainous districts. It is known as *Parascalops*. After this, however, we come to a number of little-known animals that have been unearthed in the most surprising places.

One, a typical mole found in the Province of Kan-su in China, is almost identical to the North American Western Mole, though smaller, and has been called *Scapanulus*. Then, two more forms—named *Scaptonyx* and *Dymecodon*—have been found in Sze-chuan Province of southwestern China and in Japan respectively. Both are typical moles in habit but they are internally very closely related to the moles of America and have nothing to do with those of the rest of Asia and Europe. Finally, there are a pair of most difficult beasts, one found in Japan and the other on the west coast of North America, that are just about halfway in appearance and habits between the moles and the shrews but neither of which has anything to do with the Shrew-Mole of central Asia. They are known as *Urotrichus* and *Neurotrichus,* and both have very large hands armed with great claws, but shrewlike bodies with rather fat tails that are of modest length and are clothed in both scales and a fuzz of short bristles. They are diggers and true burrowers but their front feet are not modified to serve as digging paddles as in the true moles and they are not set sidewise to the body. Their noses are long and naked.

The Star-nosed Mole (Condylura)

Every now and then one comes across an animal that is truly unique. It is as if Nature has an inexhaustible fund of wonders and almost nothing is impossible to her. The backbone of the *Scutisorex* is a genuine shocker to an anatomist, but the front end of a Star-nosed Mole (see below and Plate 24) surprises even naturalists and amazes the uninitiated. Here is an animal looking like a rough-furred mole, with a comparatively long tail and rather long slender back feet, that has enormous paddle-shaped hands with short fingers and stout nails, and a typical pointed head with tiny eyes, but then carries a structure on its snout that is quite out of this world. From this snout grow twenty-two bright pink fleshy fingers arranged radially like the petals of a small flower. They are mobile and very sensitive and can be partially collapsed or retracted and appear to be some form of tentacles to help the animal find its way about. The animal lives in damp soils near water and is semi-aquatic, diving for food, and using its hands like flippers. Star-nosed Moles are found only in the eastern half of temperate North America.

Star-nosed Mole

Flying Mammals
(Chiroptera)

THE second largest order of living mammals, both in point of actual numbers and number of kinds, is the Bats. Apart, of course, from those orders that contain only one or two types, they are by far the most homogeneous group both in general appearance and general anatomy. They are an old group in point of time, fossil bones of recognizable bats having been found in Middle Eocene rocks which were formed on lake bottoms at least fifty-five million years ago, and other remains that are almost certainly those of bats, in the Palaeocene rocks which were the first formed after the age of the dinosaurs. Bats' ancestors must, then, have been flying around during that age but since they live over land for the most part and are very fragile little things, they seldom get themselves fossilized even if they drop dead in small lakes or pools. They give every appearance of being an early offshoot from the Insectivore stock developed from some tree-climbing forms that took to volplaning after airborne insects. The necessity for going upwards in pursuit of these probably induced a clawing action with the front limbs just as a drowning man will claw at the water to get up for air, and it is interesting to note that bats, unlike birds, do actually *swim* through the air, reaching forward with their widespread paddle-like hands, curving their fingers around a piece of air, and then pulling their bodies past it—as can be seen in the remarkable series of photographs shown in these pages. That the smaller and more primitive bats are still predominantly insect-eaters is also of significance.

The changes that have come about in the anatomy of bats, although exceedingly radical in order to make true flight possible, are not, however, really quite so farfetched as their external appearance would at first lead us to believe. If you take the body of a freshly killed Mastiff Bat of one of the species that has comparatively short wings and long legs, cut away the flight membranes of the wings, and snip off the immense fingers a few millimeters from the wrist joint, you will find that you can stand the little corpse up in such a pose that it will look very like a small, fat shrew. Internally, however, all sorts of changes in the musculature and the bones to support it have taken place, and the skull has in many cases been altered in the most exaggerated manner for special purposes.

Bats can be very clearly divided into two great groups normally called, both in popular and scientific parlance, the Greater, and the Lesser Bats, but which we refer to as the Fruit-eating and the Insect-eating. Both pairs of names are somewhat misleading since the Great contains many species that are much smaller than many of the Lesser, while there are Lesser bats that eat fish, fruit, or, in two cases, nothing but fresh blood. There is even evidence that some Great Bats eat a certain amount of insect food. Nonetheless, the distinction is real, for there are structural differences between the two, notably in the construction of the shoulders which, in the case of the fruit-bats, is more like ours and aids steady long-distance flight. Fruit-Bats are found only in the tropical and subtropical parts of the Old World, including Africa, India, the Oriental Region, Australia, and the Pacific archipelagos. Insect-eaters are distributed throughout the world between the Arctic and Antarctic circles, even on New Zealand and many isolated oceanic islands.

Ecologically, bats may be divided into three classes—commuters, metropolitans, and isolationists. The first live either in caves or in vast tree-roosts whence they go roaring off on a very precise time schedule, all together, every evening, to pre-selected feeding grounds, and from which they return all in a rush to their sleeping quarters every morning. The second live in vast communities in caves, large hollow trees, or buildings, from which they emerge about dark individually and to which they may return throughout the night since their labors outside are within easy flight distance. The third kind are rugged individualists shunning their own and other kinds, moving about extensively by night and resting by day either alone or in small family parties in isolated hideaways in nooks, crannies, or under leaves, in small holes in trees, or sometimes in isolated parts of the very caves inhabited by a vast army of metropolitans. The number of bats of one kind that may live together in caves or in roosts is almost inconceivable. It has been estimated that the insect-eaters that once lived together in Ney Cavern in Texas numbered upward of twenty millions. Their evening flight led to the discovery of the cave because the emerging hordes looked like an enormous smoke-cloud and blackened the sky for miles. There were roosts in tropical Australia that covered hundreds of acres wherein every bit of space on every branch of every tree was festooned with a seething mass of chattering bats to the point where their sheer weight often brought large limbs crashing to the ground.

Almost every cave in the world—and in the tropics almost every hollow tree—houses some bats, but there are many caves that contain millions. Recently, a serious decrease in the number of such metropolitan communities and of the numbers of bats in them has been noted in several parts of the world and particularly in North America. This may be partly a natural cycle or it may be connected with certain overall changes in climate or with disturbance caused in these resting quarters by cave explorers and vandals, but it may also have much more to do with the widespread use of insecticides which are not only cutting down the bats' food but poisoning the little mammals themselves. This is a serious matter, since bats fill almost as vital a role as shrews and moles in both the natural and our own economies. They are actually much better insecticides than any chemical spray, and they are selective. The incidence of malaria in certain areas has been shown to vary in exactly inverse proportion to the number of bats in local mass-resting places—the more bats the less malarial-carrying

mosquitoes. The case of fruit-eating bats is, of course, different, since commuting hordes may pick on a fruit orchard or some other specialized man-made crop and destroy it in one night: and not even smokescreens or floodlighting will deter them.

The most interesting thing about bats is their development of *sonar*. When this capacity was first discovered and then identified as normal and age-old behavior among certain bats, it was believed that it was a unique habit among animals as a whole. We now know that quite a large variety of animals employ not only sonar but radar, and underwater radar at that—vide: certain fish (*Gymnarchus*) that store up electromagnetic energy in true batteries, release it in controlled bursts into the water, and pick up reflections from surrounding solid objects with special sense-organs arranged about the outside of their bodies. There are other mammals that are believed to employ sonar but none in which it is so well developed as in certain insectivorous bats.

It was long ago noted that covering the eyes of bats did not impair their powers of flight nor cause them to collide with things while in the air. They maintained their uncanny and unerring ability to avoid even hundreds of piano wires stretched at all angles across a small room, but if one ear was taped down, they flew round in circles and bumped into obstacles, and if both ears were taped down, they could hardly fly at all and crashed into the first obstruction. Nothing being known of supersonic sound in those days, it was believed that the bats sensed things ahead by increased air-pressure as they approached, and that their ears were the devices they used to do so. Later researches, however, demonstrated that bats keep uttering tremendous bursts of supersonic sound (comparatively much greater in volume than that made by the largest propeller-driven plane), ranging as high as 30,000 frequencies per second and of as little as a 200th of a second each, and maintained continuously while the animal is in flight. These sounds are broadcast through the nostrils, which are often surrounded by most complex flaps and other structures that beam the sounds. The sound waves on striking an obstacle—including, be it noted, an insect in flight—are reflected backwards to the bat, which picks them up with its extremely complex ears. These are often not only enormous, folded, creased in all manner of ways, and supplied with one or even two false ears standing up in front of the main conch, but sprout all manner of sensitive hairs, bristles, and true vibrissae.

So incredibly sensitive are bats and so instantaneous are their reactions that they have time to draw in a wing or completely alter course after receiving these tiny sound reflections from only a few inches ahead. As a consequence, half a million bats with a wingspan of a foot may mill around for hours in total darkness in a cave without a single collision. A film of this performance, slowed down for human vision, is one of the most incredible sights provided by Nature; if viewed against a light background, the entire screen may appear solid for brief periods, so closely interlocked are the animals. The method of flight is, as we mentioned above, actually a kind of crawling or clawing motion, using the hands

Long-eared Bat in flight

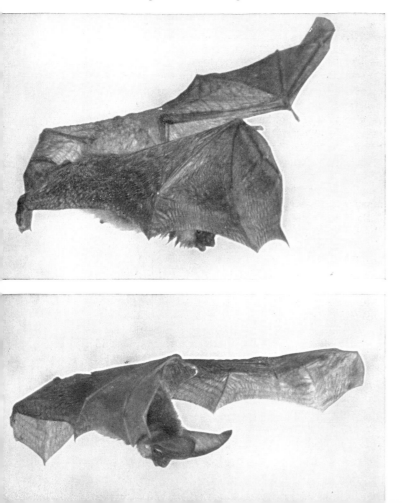

with the interdigital membranes to scoop back pieces of air. The result is that the wings of a bat may adopt almost any attitude, meeting above the head, in front or below it, over the back, or below the body. Further, most bats have tails, and the flight membranes stretch from these to the hind legs and thence to the little finger. It has now been discovered that these animals catch flying insects in the cup made by curving this tail-membrane forward under the body, and that they then bend the head down and eat the food while on the wing.

The aerodynamics of bats drive engineers mad and are quite beyond us ordinary mortals, but it may be said in partial explanation that there is one physical law that explains much of the problem. This is, that the surface area of a solid object increases in relation to its volume as the size of the object decreases, so that the smaller a thing is, the slower it falls. The net results of this is that if a donkey falls down a hundred foot well, it literally bursts; if a man falls, he breaks every bone in his body; if a cat falls, it breaks its legs; but if a mouse goes down, he will get up and run away almost unhurt. Bats are so small they almost float in the air.

FRUIT-EATING BATS

These are known technically as the *Megachiroptera,* an outsize word meaning only "big bats." Quite a number of them are small and several of them eat flowers, pollen, and some insects, so that even the above designation is only approximate. There are well over two hundred described kinds and they are now divided up among two score genera. We divide these into three principal and ten subsidiary groups for purposes of simplicity and to help those interested in finding the kind they want. This is not quite so unlikely as it may sound, for these animals are a great pest in certain countries where some of our most economically valuable crops are grown. They range from West Africa, south of the Sahara, to Egypt and thence north through Palestine to Cyprus and Syria. They inhabit all the islands of the Indian Ocean and its coast, south of the desert belt, to India, the Oriental Region, Indonesia, Papua and forested Australia, and the Pacific islands east to Samoa and Tonga. One species almost reaches Japan. Their thumbs are well developed and bear a big claw, and the index-finger has three joints and is enclosed in the wing mem-

Long-eared Bat resting

brane, but usually also bears a claw. A few have short tails underneath the membrane between the legs, but most have none at all.

SMALL-TONGUED FRUIT-BATS (*Pteropinae*)

The classification of the numerous Fruit-Bats is a matter for experts because they are very much of a oneness. However, they can and must be divided if only for the purposes of cataloguing them, and they can be separated into a number of valid families, the members of which are truly related. When we attempt to do this non-technically we are forced to make use of characteristics that are often quite picayune and, in this instance, actually of no use in identifying any of these animals, either in the field or even if held in the hand. A combination of features plus a knowledge of where the animal came from must be taken into account. The length of the tongue is actually only comparative, half of it being attached to the floor of the mouth in this family, and its tip not being capable of extrusion to any great length.

Fox-Bats (*Pteropus, etc.*)

This is by far the largest group of Fruit-Bats and contains the largest known individuals, creatures with a wingspan that has been known to exceed five feet and bodies as big as ravens. A large number of these are also known as Flying Foxes (Plate 25). They are distributed all over the Eastern Pacific, Australia, Indonesia, the Oriental Region and north almost to Japan, throughout India, and west around the Indian Ocean to the little island of Pemba off the coast of East Africa just north of Zanzibar, the Comoro Islands and Madagascar. For some strange reason they have not established themselves on the mainland of Africa only a hundred miles hard by. They come in all sorts of colors and color combinations, usually with a ruff of a complementary shade around the neck that forms a sort of cape over the shoulders. Apart from size, they all look much alike, having (as can be seen in the color photograph) a heart-shaped body, large clammy-looking wings, and a face not unlike a fox, with bright brown eyes, small pricked ears, and a simple muzzle.

Most of them live in colonies and are commuters, flying forth at sundown to feeding grounds that may be twenty miles away and even on islands across the sea. They travel in long lines or streamers and fly slowly and purposefully with measured strokes and are one of the most impressive sights in the clear evening skies of an Oriental sunset, passing by overhead in their countless millions regardless of man's works below. Both at the feeding grounds and back at their roosts they must fight every day for a place to hang upside down, for they crowd together and are bad-tempered creatures lashing out at all comers with their taloned thumbs, and biting savagely with their sharp teeth. They clamber about the branches upside down and bear their young in that position, subsequently carrying them clamped to their bellies. Some live in caves and some have taken up quarters in buildings, notably old temples and abandoned plane hangars.

E. P. WALKER

Big Brown Bat climbing

Tailed Fruit-Bats (*Rousettus, etc.*)

These bats are closely related to the last but as a whole have a more western range and although they lack the cape of the Fox-Bats, have a sort of collar of longer, hairy fur. They are found from India and Burma west via the Persian Gulf to Palestine in the north and the whole of tropical Africa and Madagascar in the south. They are by no means the only Fruit-Bats that have tails and what they have are poor efforts, being attached below to the inter-leg membranes. The presence of a tail serves to separate them from the Fox-Bats. They rest more frequently in caves and buildings and on the whole are smaller in size. Many live in the great matted heads of large palm trees and, where the natives bore holes in these to draw off the sap into calabashes or pots suspended from the bottom end of a hollow bamboo, these bats often get so drunk on the semi-fermented liquor that they drop to the ground in a stupor. They eat soft fruit and are very destructive to guava, sweet plantain, and other plantations.

Epauletted Fruit-Bats (*Epomophorus, etc.*)

Some of the smallest of the Great Bats are included in this large group, all of which have strange sacs on the sides of their necks out of each of which sprouts a tuft of long, pale hairs, giving the animals the appearance of wearing epaulettes. They also have a light patch of fluffy fur at the bottom of the ears and their faces are like those of long-nosed dogs with the tip of the nose cut off. They have rather repulsive looking lips that can be extended and then somewhat everted to form a sort of suction cup like the device used in administering anesthetics. They are also soft-fruit-eaters for the most part and since they eat while hanging upside down they have a special organ to prevent them from being drowned in the juices—a strange inflatable sac or pouch provided with circular muscles to close it off from the throat. They appear to use this as an air storage tank to supply the lungs while sucking on the fruits. They are entirely African.

Short-nosed Fruit-Bats (*Cynopterus, etc.*)

The small Fruit-Bats of the Oriental Region, that is from the Philippines through Indonesia to India, have short, rounded faces with a curious groove or gutter running from the middle of the upper lip up between the nostrils to the top of the muzzle. This is another device for carrying off fruit juices when hanging upside down and preventing the liquid from streaming into the nostrils and suffocating the animals. Many of this group are isolationists in that they sleep alone or in small parties in crannies among rocks or in thick foliage in palm and other trees. Most of them are rather dull brown in color and about four and a half inches in body length. They have short tails attached under the leg membranes and have the more fluttering flight of the Lesser or Insect-eating bats. There are some odd isolated species in Burma.

Bare-backed Fruit-Bats (*Dobsonia, etc.*)

These most remarkable bats from the New Guinea and west Pacific area have the wing membranes joined right across the back and then attached to the backbone, or rather the mid-back skin, by an upright flange. Thus, the body of the animal is like that of the original monoplanes, having a continuous wing right across the top and a sort of nacelle, their body, suspended below. The tiny tail is also slung underneath the hind inter-leg membrane, its tip being free. The membranes are covered with very short, fine, satiny fur, the body itself having a thick coat. In color they are dull brownish greys but they often appear to be washed with yellow and have yellow undersides. This coloration usually denotes a diurnal habit and these bats are often abroad in full sunlight. They are cave dwellers and vary in size from that of a large sparrow to a pigeon. An extreme form from the Philippines is called the Harpy Bat and is often put in a special family.

The Hammer-headed Bat (*Hypsignathus*)

The most incredible of all Fruit-Bats is found in West Africa and is closely related to the Epauletted Bats, though it lacks the neck pouches. Its wings are very short but broad and the body is large and heavy but the head is almost beyond belief. That of an adult male looks just like a cartoon of a horse when viewed from the side. The eyes are very large and have greenish irises, the ears are horselike, and the muzzle is much deeper in front of the eyes than behind them. Looked at from in front there are all manner of folds around the mouth and lips. The head is about a third the size of the body and hangs down when the animal flies. They live in caves, eat fruit but for reasons unknown do a lot of dipping into water.

LARGE-TONGUED FRUIT-BATS (*Macroglossinae*)

The only distinction between these and the former group is that they have narrower tongues, attached only for about one third of their length to the floor of the mouth but extrusible to a really startling length. They are confined to the Oriental, Indonesian and Australian Regions, with extensions to the Fiji Islands, and one, *Megaloglossus*, is isolated far away in West Africa.

Long-tongued Fruit-Bats (*Macroglossus, etc.*)

There are seven genera of these small Fruit-Bats, all with long pointed snouts, simple muzzles, and a pad of long papillae, or little, pointed, recurved pimples, at the end of their tongues. They are pollen-eaters rather like the marsupial Honey-sucker. The smallest is not much bigger than the average insect-eating Lesser Bat. Most of them are profound isolationists and sleep in caves, hollow trees, or under palm fronds, but there is a magnificently colored species in the Solomon Islands which had iridescent orange fur and almost black wings, rests in large colonies on certain trees in the second tier of the rain forests and looks, when so sleeping, like a cluster of brilliant fungi.

TUBE-NOSED FRUIT-BATS (*Nyctimeninae*)

Apart from the Hammer-headed Bat these are the oddest and most grotesque of the Fruit-Bats. They are found in the southeastern islands of Indonesia and in north Australia. They are really very extreme forms of the Short-nosed Fruit-Bats but so outstanding as to warrant separate notice. Their bodies and wings are fairly normal but their faces are exceedingly short and abruptly rounded in front like a sort of fish. They have a deep median groove between the nasal apertures, which jut forwards, outwards, and slightly downwards in the form of half-inch, naked tubes, just as if two bits of black rubber piping had been stuffed up their nostrils. This again is a device for keeping either liquid or dust out of the nose while feeding, for the animals are eaters of fruit, pollen, and probably also of insects. Their eyes are very large and their lips are lined with small round bosses so that when closed they look as if they were zippered.

INSECT-EATING BATS

The number of different kinds of these bats is positively staggering. Although authorities differ in their definition of the various groupings, there are at least sixteen different families, broken down into about one hundred genera which together represent several hundred valid species. Unless born with an overpowering passion for these animals, no one can be expected to be much interested in the appearance of each or the differences between them, but they *are* all different and they display a remarkable variation in behavior and habits. We refer new recruits to the science of chiropterology to the numerous published works of this order; we can mention only the really distinctive animals and the greater oddities in each family.

Almost all of these animals look singularly alike. They have small, bun-shaped bodies clothed in soft thick silky fur usually of some shade of grey or brown and often lighter below. There are a few parti-colored ones, black and white, and some are pure white, or vivid golden orange. The flight membranes are usually naked but sometimes partly furred and, although extremely thin, are composed of two skins containing blood vessels, and are very strong. Most of these bats have a membrane stretched between the back legs and this may contain a tail; but some have no tail, others practically no membrane, and in still others the tail either is free or projects through a sheath out of the top of the membrane. Most of the differences between these animals are based on details of their internal anatomy or their teeth, but these concern experts only. Otherwise, it is the faces of the bats that are the best guides to their identity but without very clear enlarged drawings of all of them, it is useless to attempt to describe them. Even photographs of some are just unbelievable. The ears vary enormously in shape and size, and all manner of incredible flaps, leaflike structures, and, in some kinds, even little balls raised on slender stems like those seen outside pawnbrokers' shops, sprout from their lips, nose or forehead.

Many bats hibernate, and sometimes even for more than half their lives; others migrate south in winter and north in summer; and it has now been discovered that there are species in which one sex migrates while the other hibernates. They almost all live in caves, hollow trees, houses, or under leaves but some sleep in the open on the sides of trees or rocks, and nearly all hang upside down when doing so. There are day-flying species and many that run about on all fours with their wings folded up.

MOUSE-TAILED BATS (Rhinopomatidae)

These most peculiar-looking little bats are distributed from Egypt through southern Asia to Burma and Sumatra and are considered to be the most primitive of the Insect-eaters. They have large ears joined across the top of their heads by a sort of flange, and a most peculiarly shaped muzzle bearing a fleshy pad on the nose. Behind this the top of the snout is hollowed out into a sort of gutter that ends above in a deep pit in front of the ears. The shoulder joint is very like that of the fruit-eaters and another primitive feature is the index finger which still has two joints. Most remarkable are the tails of these bats, which are as long as the combined head and body, very thin and mouselike, and free except for a tiny portion at the base which passes through the inter-leg membrane. Before the winter in certain areas these bats store up quantities of fat under the skin around the lower body and tail base. Upon this they apparently live during a period of semi-hibernation. They are isolationists.

SHEATH-TAILED BATS (Emballonuridae)

These comprise a large number of kinds divided into eight different genera. There is little that these have in common externally except that those with tails invariably have a curious arrangement whereby the tail lies in a sheath in the membrane between the hind legs, its tip emerging on the upper side to form a free feeler. The tail itself can be extruded or withdrawn and, in the former case, its vertebrae act as a hoop, curving upwards, downwards, or to one side or the other, and thus shape the membrane in sundry ways to form a rudder. This naturally provides the animals with an exceedingly efficient steering device. Collectively, they are found over wide areas of the Old and New World tropics and in some places are one of the commonest bats—notably, the so-called Tomb-Bats (Taphozous), which seem to have a predilection for ancient memorials like the pyramids of Egypt. Some small species

(Rhynchiscus) common throughout tropical America are remarkable in that they are completely diurnal, sleeping in the open on the sides of trees and chasing insects over rivers and ponds in the midday sun.

A number of others (Saccopteryx, etc.) have pouches on the lead-edge of their wings on the underside near the elbow joint. These may be lined with wrinkled white skin that can be everted or may secrete a foul-smelling, blood-colored fluid; the purpose of this fluid is not known unless it be to attract the females, since they are better developed in the males. The Tomb-Bats have pouches under their chins, and another remarkable group from tropical America have a similar sac under the inter-leg membrane. Still another member of this family (Diclidurus) is among the most spectacular of bats. having pure white wings and membranes, and fur that, although dark at the base, is pure creamy white or silver on the surface. They sleep by day among the leaves of certain trees like the wild papaya which are covered with silvery hairs below.

HARE-LIPPED BATS (Noctilionidae)

These are really very odd tropical American bats with faces that look for all the world like those of tiny Jack-Rabbits. The upper lip is cleft in the form of a "V" with a nostril at the top of each tine. Further, the middle front upper teeth are comparatively enormous and cover the next smaller pair on either side, so that they look just like those of a rodent. These bats have large feet with exaggerated claws. Strangest of all, they seem to be truly carnivorous in that they catch not only insects, including certain flying "water-scorpions" that are twice their bulk, but also tree-frogs, small bats, and mice and, oddest of all, shrimps and fish which they pluck from rivers and lake-surfaces. One species often lives in caves on the seashore and fishes in salt water.

HOLLOW-FACED BATS (Nycteridae)

These are rather nondescript little bats with their ears joined together across the tops of their heads by a slight wall of skin, and peculiar guttered muzzles not unlike the Mouse-tailed Bats but with little petals of naked skin on either side. They have long tails. Strangely, although mostly African, one kind is found in Malaya and Java on the other side of the Indian Ocean. They are tropical except for one that occurs in Egypt.

FALSE-VAMPIRE BATS (Megadermatidae)

If certain bats are sensational in appearance, these reach just about the limit of credibility. The ears are positively immense, each being many times the size of the face, and are joined together over the crown half way up to their tips. Inside these ears grow a kind of double secondary ear. Then, on the nose is erected a fabulous leaf-shaped structure lopped-off on top, and in front of this stands another leaflike spear of naked skin. They have no visible tails and no upper front teeth. They are known as False-Vampires because, although they have nothing to do with the true Vampires (Desmodus) and are not parasites or exclusively blood-eaters, they appear to be completely carnivorous. Report after report by eye-witnesses tells of these bats carrying tree-frogs, small bats, mice, and other back-boned animals as well as insects back to their sleeping quarters or to special feeding lairs. They even attack small birds and have been known to break into canaries' cages.

HORSESHOE-NOSED BATS (*Rhinolophidae and Hipposideridae*)

Over a hundred kinds of small bats divided between these two families and comprising eight genera are spread over a wide area, including Europe, the whole of Africa, India and the Oriental Region to northeast Australia and the Solomon Islands. They are all normal little bats except for their faces which with one exception bear a horseshoe-shaped dish of naked skin below the nostrils, and in many cases another structure above the nostrils which is similar in shape but inverted. The ears are usually tall and pointed and their bottom outer edge is folded back into the conch. There is no false ear. They are insect-eaters and are mostly isolationists, sleeping alone or in small communities in hollow trees, houses, and under palm fronds. The oddest (*Anthops*), the Flower-faced Bat, comes from the Solomon Islands and has to be seen to be believed. The whole face is covered with rosettes of petal-like skin flaps from which rise three little balls on thin stems, doubtless some supersonar device.

LEAF-NOSED BATS (*Phyllostomatidae*)

This family comprises no less than thirty-five genera split into some two hundred species. It is wholly tropical and confined to the New World. One genus (*Macrotus*) is found in the southeastern part of the United States. They are all distinguished by a pointed leaf- or spear-shaped structure that arises from the nose. There are two odd types that do not have this structure: one, the Leaf-lipped Bat (*Chilonycteris*), has sort of lappets of loose, folded skin hanging down from the lower lip, and the other, the Centurion Bat (*Centurio*), has the whole face thrown into a series of grotesque folds linking mouth, nose, eyes, and ears to a sort of bowl in the center of the forehead in which is cupped a spherical shiny gland.

The main body of the Leaf-nosed Bats can be divided into seven sub-families of which one is the Leaf-lipped Bat. The next, covering eleven genera, contains (besides a number of ordinary looking types) the largest bat both of the New World and of all Insect-Eaters, though it happens to be a fruit-eater itself. This is known as the Javelin Bat (*Vampyrum*) which has a wingspan of up to two and a half feet, huge ears and a tremendous spear-shaped noseleaf. It is found in the Amazonian forest area. A smaller edition of this bat (*Phyllostomus*) sometimes becomes infected with rabies as a result of being bitten by rabid Vampire Bats in its sleeping quarters and may then, in turn, attack men or other animals. As a result, it was long believed that these were true bloodsuckers; they are not, and most of them are predominantly fruit-eaters.

The third group of species is typified by the small Long-tongued Leaf-nosed Bats (*Glossophaga*). These are fruit-, pollen-, and insect-eaters having long pointed snouts covered with bristling whiskers and tongues that can be extruded to twice the length of the head and bear an expanded pad at the tip covered above with hardened, backwardly directed, spinelike papillae for rasping. Like many Leaf-nosed Bats these species often hold fruit in their "hands," twiddling it around dexterously with their thumbs. The fourth group is known as the Short-tailed Leaf-nosed Bats (*Carollia*). They are both metropolitan and in some places commuters in the true sense, living in vast numbers together in caves. The fifth group (*Sturnira*) consists of tailless fruit-eaters with teeth like the true Fruit-Bats and with little epaulettes of stiff hairs on their shoulders.

The next group of eight genera is well exemplified by the large *Artibeus* known as the American Fruit-Bat (Plate 27) which live in small isolated communities and sally forth before dark to their feeding grounds. They have comparatively large eyes and often sleep in places where direct sunlight may penetrate. One type (*Vampyrops*) is a beautiful little coffee-colored animal with vivid white stripes down the back. The last group (*Phyllonycteris* and *Erophylla*) are West Indian and have very long snouts and tongues, again with terminal rasping pads.

VAMPIRE BATS (*Desmodontidae*)

The true Vampire Bats (Plate 26) are unique in habits, being the only mammalian true parasites in that they can feed only on the blood of other animals, ranging from man to toads, which they take from live victims. They are gruesome-looking little bats with almost spherical bodies, pointed ears and a naked nose rather like that of a bulldog. They fly well, but also scuttle about on all fours like great spiders, holding their wings tightly folded back along the forearm and having large, padded thumbs.

They do not suck but lap blood. The method of attack is to flutter in front of the animal victim (when they may exude a partially soporific or calming scent), then land on the ground or a branch nearby, run towards the animal on all fours, and then jump to its body. There, they slash a shallow gutter in the skin with a downward stroke of the head using the two large triangular chisel-shaped upper front teeth, then skip around so that their head points upwards and their lips can be pressed to the bottom of the wound. The tongue then pumps the blood into the stomach. They then retire to special digesting retreats that are marked by their inky, tarlike excrement. They sleep in caves, hollow trees, houses and culverts, and they exist in countless millions throughout an enormous area from Argentina to the border of the United States.

These bats have always fed on human beings and horses as well as other animals and as they can carry rabies, Murrina, Chagas Disease, and other fatal inflictions for long periods in their bodies, they have had a very serious effect in South and Central America. There is a belief among the Amerindians that they were responsible for the extinction of the original American horses. In the last quarter of a century they have decimated the cattle herds of South America and caused serious outbreaks of human rabies in Trinidad and Mexico. The danger is that they are slowly moving northward with the progressive warming up of our climate.

ODD FAMILIES

For purposes of simplification and in order to fulfill our expressed objective of covering all living mammals even if only by a mention, we here group together five small *families* of bats containing six genera only. Among them are some of the oddest of all bats. The first, known as the Long-legged Bats (*Natalidae*) are tropical American, with one small kind found in Cuba known as *Nyctiellus*. They have enormously domed heads, tiny, puglike faces with wide, fleshy-lipped mouths, and huge ears arising just above the tiny eyes. Inside the ears are little spirally twisted false-ears. They are insect-eaters like the next family, which have been called Furies (*Furipteridae*), and are also tropical American. Of them, only one thing of non-technical interest need be mentioned, and this is that they are the only bats in which the thumb is reduced to a mere knob and is not used at all.

The next two families are each represented only by a single form, both most remarkable and in much the same manner, though one, the *Thyropteridae*, lives in Brazil and north to Honduras, while the other, the *Myzopodidae*, is found only in Madagascar. The so-called Tricolored or Disc-winged Bats (*Thyropterus*) are most odd in that they have large clawed thumbs, the index finger is reduced to a mere stub, and there is a large disc under the thumb that acts as a sucker with which the bats can attach themselves even to the underside of a glass surface. Their feet are tiny, with only two joints to each toe and the fourth and third toes joined by the bones. Underneath the ankles are other large sucker-like discs. The Madagascan animals (*Myzopoda*) also have suction cups on thumbs and ankles, and their toes are bound together. These bats, although isolated in the Indian Ocean, are definitely related to their South American cousins and the Natalids.

The last of these odd little families consists only of the Short-tailed Bat of New Zealand (*Mystacinidae*), one of the only two mammals indigenous to that country, and is close to the great group of Mastiff Bats, or *Molossidae*. It has a somewhat doglike face with a small nose, and ears like the Evening Bats, but has a small, naked tail projecting from the middle of the interfemoral membrane. Oddest of all are the claws of its thumbs and toes, which have double talons, one above the other.

EVENING BATS (*Vespertilionidae*)

We now come to what is by far the largest group of bats, one that has no less than twenty-five genera but an even larger number of species than has the Leaf-nosed Bats. Moreover, most of them are distressingly alike so that their identification is a pastime for experts only. They are insect-eaters and are found throughout the tropical and temperate regions of the world. They do not have nose-leaves, and the nostrils are simple. The ears always contain a false-ear or tragus. Their tails are contained wholly within the membrane between the hind legs and always extend to the hind edge of that structure. They may be broken down into six groups, five of which contain only one or two genera, the sixth no less than seventeen.

Starting with the last we find ourselves in the midst of a bewildering number of small brownish bats most of which are known indiscriminately in various countries as common bats, brown bats (Plate 31), little brown bats, or by slightly more specific names such as Barbastelles, Serotines (Plate 28), Pipistrelles, silver, hoary, red (Plate 30), long-eared, lump-nosed, and so forth. None of these names are of much value except to emphasize the general homogeneity of the group. It is interesting to note that the name of the whole family is derived from the Latin name for bats generally, *Vespertilio*, meaning "the night-flying ones," but is now restricted to a particular genus found only in Europe. The outstanding genus is that of the Long-eared Bats (*Plecotus*) of which there are species in both the Old and New Worlds, those in the latter commonly being referred to as the Lump-nosed Bats, and being listed under the Latin name of *Corynorhinus*. They do in point of fact have enormous ears that are joined together over the top of the head and contain very large tragi. Further, there are lumpy swellings on their noses that point the way to the development of nose-leaves such as those of some Horseshoe-nosed types.

Of the other five groups or genera, the Long-winged Bats (*Miniopterus*) are exceedingly light fliers. They are found in southern Europe, northern Africa and southern Asia. The crowns of their heads are curiously swollen. Most distinctive of all Vespertilionids are perhaps the Tube-nosed Bats (*Harpiocephalus*) of eastern Asia ranging from Tibet to Japan and Indonesia. They are ordinary in all respects other than the nostrils which stick out from the snout in the form of two small rubbery tubes. The next group is African and south Asiatic, and its members are called the Painted Bats (*Kerivoula* and *Anamygdon*) and some of them certainly do appear to be just that. The body fur comes in a variety of colors, from smoky black to brown, coffee, and even golden, and the wings and other membranes show a parti-colored and distinct pattern. The so-called Big-eared Bats (*Antrozous* and *Nyctophilus*) form another group found in parts of Australia and the Oriental Region and in North America. The last group consists of a bat from Peru named *Tomopeas* which is most interesting because, although it has the general external appearance of a typical little Vespertilionid, it displays internal anatomical features that link it with the next and last great family of the bats, the *Molossidae*.

All this host of little Evening Bats have surprisingly similar habits. Most of them sleep in hollow trees, houses, sometimes in caves, and even under banks or amongst matted vegetation; few of them are truly metropolitan though many are communal. Only at times of hibernation do they normally mass together in vast assemblages, usually in caves, and although thousands may live in a single building they tend to nest together in little separate parties. They constitute the commonest bats in almost all countries.

[61]

MASTIFF-BATS (*Molossidae*)

The least batlike of bats, though in point of structure perhaps the most advanced, specialized, and thus profoundly of the bat class, are the Molossids. They are dog-faced, naked-tailed, short-furred, folded-eared, bright-eyed little creatures (see Plate 29, the Free-tailed Bat, *Tadarida*). Of these there are six genera spread over all the warmer areas of the world. They have large tails that protrude from the back edge of the inter-leg membrane but that are loose within the skin and can be pulled in or pushed out. Their ears are indescribable, being folded inwards all around and joined across the top of the head; in some species the two ears run continuously from side to side in one piece. They have two large upper front teeth and very large, wide feet. The thumbs are large and bear prominent pads below. Their mouths are large and their lips wrinkled and floppy. Most outstanding is the shape of their wings, which are comparatively small, very pointed, and narrow from front to back. This shape denotes to aeronautical engineers both speed and great powers of maneuverability. However, these bats can also pack their wings and membranes away more completely than even the Vampires, and are thus free to spend much time running about on all fours, which they customarily do.

Most of them live in large communities and in one house which they thus occupied (in which the author lived) they would crawl all over the beds, the dinner table, and even people at all times of the day and night. They have a brush of curved bristles over the claws of the feet and strange sensitive hairs on their muzzles that have club- or spoon-shaped globular tips.

Oddest of all the Molossids and certainly one of the strangest bats, with habits unique among mammals, is the Naked Bat (*Chiromeles*) of the Malayan region. They are repulsive-looking creatures with bloated bodies, the skin around the neck thrown into fatty folds, and the head just like that of a tiny pig, with large, lobate, forwardly directed ears and huge open nostrils. Under the chin is a roll of skin extending on to the chest; in this the mother was once believed to carry her young. However, the male has a similar pouch and both sexes have pouches at the junction of the wing membranes with the body.

Gliding Mammals

(*Dermaptera*)

EVERYTHING about these extraordinary and abstruse mammals is "wrong" either by our concepts of a mammal, or by simple definition. Nonetheless, they are one of the most interesting and objectively important of all living types. Although they have been assigned to the Bats, the Insectivores and the Primates via the Lemurs, nobody has yet decided quite how to classify them. Dr. Simpson, who brings the advantageous evidence of fossil types to bear upon the problem, points out that even in Palaeocene times—some 55 to 60 millions of years ago—there were distinctly Dermapterine forms (named *Planetetherium*) and in lower Eocene times others which are named *Plagiomene*. These mammals stand all alone but somewhere at a crucial crossroads linking the Insectivores and the Primates with, possibly, the Bats.

Before we come to the nature of the beast we must, however, make some attempt to disentangle its nomenclature. First of all, it forms an Order all by itself, and this is called *Dermaptera,* which is from the Greek *derma* or *dermatos* meaning skin, and *pteron,* a wing. (The name is *Dermaptera* and not, as sometimes given, *Dermoptera*.) However, the generic name presents equal problems. It was first given as *Cynocephalus,* meaning "Dog-headed," by Boddaert in 1768, but was changed to *Galeopithecus,* meaning "The Cloaked-Monkey," by Pallas in 1780. Both names are wholly inappropriate, since the animal is not at all like a dog in facial conformity or expression and although "cloaked" in the manner of a Roman legionnaire crouching below his shield, it is in no way a monkey. Nonetheless, by the laws of precedence it must remain *Cynocephalus* and thus for all time deprive a group of truly dog-faced monkeys from the use of that name. All of this may serve to demonstrate the complexities of naming animals, the slight absurdities sometimes encountered therein, and the relative failure of zoology on this occasion as compared with the natives of various countries, who call it the Kobego, Cobego, Colugo, Kobugo, or Kaguan.

They are about eighteen inches in length with a ten-inch tail and are truly singular beasts. The face is distinctly foxy and extremely like that of certain Malagash lemurs, notably the Ringtail. The body is somewhat elongated and streamlined, and the limbs are rather long and all of equal length. The tail is comparatively long and gradually tapering. However, there is a double-skin, furred patagium or parachute extending from the neck to the front paws, which are completely webbed and thus continue the chute, and thence to the hind feet which are also fully webbed and spread when in flight, and behind the feet from the little toe to the tip of the tail. Thus, when the animal leaps from a tree and spreads its limbs it opens a furry kite much more extensive than that of any other mammal apart from the bats, which do the whole thing on quite other principles. The whole animal is covered with short, thick, excessively fine and silky fur almost of the consistency of the Chinchilla, and this is colored grey to various browns above with irregular tashes of silver, simulating the occurrence of lichens on the tree trunks on which it spends most of its time. Underneath it is yellow to reddish brown. On forefeet and hind feet there are large, sharp, even-length claws, all five on each foot being held parallel and close together.

There are two very distinct types: one found in upper

Malaya, Siam, Sumatra, Borneo, and Java; the other in certain of the Philippine Islands. The latter is much smaller. They are of great interest to anatomists and thus to anyone truly interested in animals, for their internal structure is quite extraordinary in the scheme of mammals. They are basically Insectivores but their brains have two curious folds not so far observed in any other mammal. On the other hand, they have a primitive kind of "undertongue" like the lemurs, but their stomachs and intestines are unique in construction, comparative size, and proportions. This may be owing to their leaf, flower, and fruit diet. Most odd of all are their teeth, the incisors or "front" ones being very wide and slotted like combs though growing on a narrow base, and the two outer, plus the upper dog-teeth, having two fangs inserted in the bone of the jaw; something known in no other mammal, though certain Insectivores have double-fanged incisors.

There is apparently but one young at a time and this is naked and very foetal, almost marsupialian in this respect, and it clings to the mother's chest for a long time. Meantime, she makes things difficult for it by leading her normal life, gliding from tree to tree in the forest and sometimes covering as much as a hundred feet between them, losing only about one part in five in altitude in the flight. Kobegos sleep hanging upside down, with all four feet together on the branch above.

These mammals are not in any way like bats and show no affinities with them or disposition to use their fore limbs in a similar manner. They are probably nearer the Tupaias, which we shall meet next.

Top Mammals

(*Primates*)

THIS great order of mammals to which we ourselves belong was in past times placed at the top of the whole scheme of life, and still retains a name appropriate to that exalted estate. However, all researches conducted along many different lines in anatomy, palaeontology, and even histology and psychology, have now combined to alter this status. The Primates, in fact, are the end-product of an ancient offshoot of the mammalian tree that branched off way down the trunk almost at ground level, and all of them retain what is called a considerably primitive or basic makeup.

Primates do, however, appear to stand on top in point of what we call intelligence and ingenuity, but this is because certain of our ancestors somewhere about the Tarsioid stage started the special development of their hands and their brains. Many other kinds of mammals could have done likewise, even some lowly marsupials like the opossums, and many different kinds have started well with their hands, but did not happen to find a large brain necessary at the same time. There are others like the Elephants which evolved enormous brains, but continued to use their hands simply for walking. The bats produced the brains, but wasted their hands on flight.

The factor that brought about the lucky combination in the early primates was a tree-dwelling habit combined with the necessity for great speed and agility in pursuit of food or to avoid enemies. This led to the development of tactile pads on the ends of the fingers and toes, increased the size of the eyes, ears and other sense organs, and produced an increase in certain parts of the brain. The next stage came automatically, and was of the utmost importance. The nimble little hands devised for retaining a firm grip on branches became dexterous enough to catch and hold food and to bring it up to the mouth.

The importance of this seemingly trivial action cannot be overemphasized, because bringing food to the mouth instead of having to put the mouth down to the food leaves an animal's eyes and other sense organs free to remain on guard against enemies while he is eating. Thus the hands, sense organs, and brain working together, developed together, making possible among other important steps—such as an upright gait—the marvelously maneuverable hand, wrist, and whole arm, which in man can reach in almost every direction without the body being moved, and at the same time employ a rotary action amounting to almost a complete circle. Further, by standing upright, the head in order to balance properly moved to a new position, which altered the pressure of gravity upon its various parts, the blood pressure, and various other matters which happen to be of an advantageous nature.

Finally, the eyes came together on the front of the head so that parallaxial, and thus stereoscopic vision were possible, an enormous advantage to any living thing in judging not only distance but the form of objects. It is notable that most lower animals seem to be unable to detect an enemy or other animal when it is stationary. None of this is theory; it is the result of correlating the observed facts with basic physical, mechanical, and optical laws, the findings of biochemical and physiological research, and the results of many practical experiments.

The Primates are an enormous group, and exceedingly varied. Their living representatives display all the states from the wholly bestial, clawed, tree-climber using its forefeet solely for locomotion, and its mouth for catching food, to busy and ingenious twentieth-century man who can live in a wider variety of environments than any other living thing because of the versatility of his body and the activity of his hands.

So varied are the Primates that it is necessary to divide them first into no less than eight groups. The normal such division is into three, namely the Lemur-like, the Monkey-like, and the Manlike, but modern researches and particularly those published by W. C. Osman Hill have clearly demonstrated that this is altogether too simple and altogether misleading. Other experts agree with his breakdown, except that he excludes the Tree-Shrews from the Primates proper. In this only we do not follow Hill, and we raise the status of another group, the Marmosets, to a position equivalent to that of the other major divisions, which is justifiable for our present purposes but is not, it must be understood, common practice.

In the subdivision of these eight major groups we have introduced considerably more by way of innovation. Hill's work is not yet completely published, and until it is, any of a number of classifications are possible and permissible. Since our primary objective is simplification without distortion, we feel that the breakdown is justified, provided each step be explained, keyed to established nomenclature, and identified as to its original authorization.

Except for man and a few monkeys that reach Europe and northeastern Asia, the Primates are today a tropical group.

TUPAIOIDS

The first division of the Primates consists of small squirrel-shaped animals found in the Oriental Region; these in no way resemble the rest of the order externally. Even specialists can hardly bring themselves to believe that they are Primates, and many still prefer to place them with the Insectivores. Their position in the scheme of life is somewhere between these two orders, which most clearly demonstrates the fact that the Primates sprang directly from insectivore-type animals. There are two kinds of Tupaioids.

Tree-Shrews (Tupaia, etc.)

There are a number of species now divided among five genera which are distributed as follows: (*Tupaia*) (Plate 33) in India, Burma, Siam, Indo-China, South China, the greater East Indies, and the Philippines; a second (*Anthana*) from India only; a third (*Dendrogale*) from Indo-China and Borneo only; the fourth (*Tana*) from Borneo and Sumatra; and *Urogale* from the Philippines alone. The word "Tupai" means in Malayan any quick little squirrel-shaped tree animal and is pronounced "tup-pie" to rhyme with "two pies." Tana, on the other hand, is the specific Malayan name for the Tree-Shrews. Most of them are so like squirrels in behavior and color that it is impossible to tell them apart in the trees where they constantly associate. They have long, shrewlike snouts with lots of whiskers, four upper and six lower front teeth, five fingers and toes all with sharp claws, and long bushy tails, of which the fur on the underside is short and harsh. Like shrews they are irascible and fight and scream among each other, but in all other respects they are much closer to the Lemurs. For instance, the eye is surrounded by a complete bony ring, some have a tiny appendix, and all have a rudimentary second, cartilaginous tongue underneath the ordinary fleshy one, looking very like that of the Aye-aye (see below). They are very fond of water and bathe regularly in water-filled hollows between tree branches. They hold food in their hands, and sit up when eating; they are completely omnivorous, and their behavior points towards the great Primate evolution mentioned above.

Pen-tailed Tree-Shrews (Ptilocercus)

Much odder and more surprising are mouse-sized little arboreal animals found in Borneo and on the Malayan Peninsula; these are in most respects typical Tupaias, but have shorter, softer fur, dark grey above and yellow below, with naked, black, scaly tails, the terminal third of which have feather-like fringes of long, stiff, white hairs on either side. These are rudders to aid the animals in balancing on tiny twigs. They are definitely of basal Primate stock, though highly specialized.

LORISOIDS

This strangely assorted group of lemur-like mammals has been separated by Osman Hill from the True Lemurs on a great deal of demonstrable evidence, and especially that of fossil types. Although they come in two very different models—the slow-moving Lorises and the agile Bush-Babies or Galagos—they form a small, closely-knit group which split off from the main lemurine and primate stem at an early date and have remained distinct ever since. Today, two of the Lorises are Oriental and two are African, while all the Galagos are African. All other lemurs are found only in Madagascar.

LORISES (Lorisidae)

The name "loris" comes from the Hollandsche "loeris," meaning a clown, and was originally given to the Slow Loris (*Nycticebus coucang*) of Indonesia because of its tragicomic countenance. The name, as will be seen from color plate 36, is fully justified. The two African forms, though not simply other forms of lorises, are closely enough related to permit their inclusion in a single family.

The Slow Loris (Nycticebus)

This, the largest species, measuring up to sixteen inches, is almost tailless. It is found in ten distinguishable forms from the Bramaputra River in Assam east to Tongking and thence south to Singapore and beyond to the islands of Sumatra, Java, and Borneo. It is very corpulent, and has short, sturdy limbs and a rather small head with soulful-looking eyes placed close together. In color it is brownish cream with a silver wash, and it has russet-colored facial markings and a median dorsal stripe. Its thumbs and great toes are so widely opposed that they point almost directly away from the four fingers and other toes. The index finger is small, the thumb enormous. All the digits bear small nails except that next to the big-toe, the second, which, as in all Lorisidae, bears a large, recurved claw. It is nocturnal and eats a wide variety of animal and vegetable food, but notably certain large insects which it catches with its hands while holding on with its feet, sometimes suspended from a branch.

The Slender Loris (Loris)

This is a much smaller animal—about eight inches long—with comparatively very long and extremely thin legs and enormous eyes placed very close together. The muzzle is sharply pointed. It is found only in the forested part of southern India and in Ceylon. It is also extremely slow-moving and cautious in all its actions and moves about the trees only at night; it is confined to the lowland forests. The color is much the same as that of the Slow Loris but there is no dark dorsal stripe and the fur is much closer and somewhat softer. Its hands and feet display a further stage in specialization; all the fingers and toes are shorter than in the species discussed above, the second one on the foot bearing the claw being reduced to a tiny, one-jointed affair. There are large dark areas around the eyes. They eat much less vegetable food than the Slow Loris.

Pottos (*Periodicticus*)

Found throughout the forested areas of West, Central, and East Africa. They are a little longer but slimmer than the Slow Loris, have furred tails about a quarter of the length of the head and body, and are gingery or yellowish-brown in color, sometimes with a grey or silvery overwash (Plate 37). Their eyes are fairly large and the hands and feet go one stage further in that both the second fingers and toes are reduced to mere stumps. They are known in West Africa as "Softly-Softlies," which aptly describes their deliberate movements, but they really hustle along when alarmed, and can make very swift passes with their hands in taking food. The spines of their neck vertebrae protrude into and sometimes through the skin and are capped with sharp horny processes. The animal uses these in defense by suddenly flipping the head down and then butting with the back of the neck. They eat fruits and insects and also catch sleeping birds by creeping up on them.

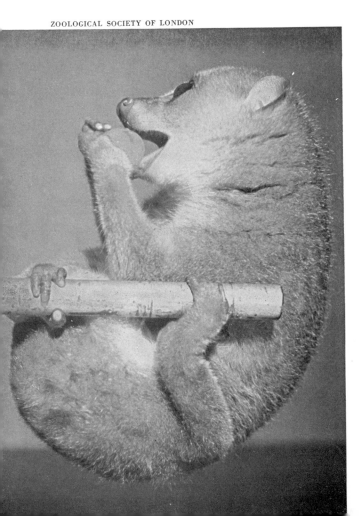

Angwantibo

The Angwantibo (*Arctocebus*)

This is less than half the bulk of the Potto and has a more foxy face and smaller eyes placed together. This animal has rather prominent, furry ears, and the second finger has gone altogether, giving the hand the appearance of a parrot's foot. The second toe also is a mere stub. They are found around the "corner" of West Africa from the Niger to the Congo Rivers. The tail is about half an inch long and concealed in the fur.

BUSH-BABIES (*Galagidae*)

These are altogether different animals, having long, bushy tails, very large ears that can be voluntarily wrinkled individually and turned about in the most extraordinary manner. They have very sharp faces and large, staring eyes; the pupils are contractile to the point of invisibility. Their thumbs and big-toes are large and opposed, but not so much so as in the lorises, and the fingers and other toes are long, fairly slender, and provided with huge, fleshy, terminal clinging pads. The fur is woolly, soft and very thick. There are three distinct kinds, the first and principal type (*Galago*) occurring throughout the forested areas of Africa.

Common Bush-Babies (*Galago*)

There are three great groups of this species, one greyish washed with brown in general tint, with reddish limbs and a not very fluffy tail that is almost black, and with very slender fingers and toes. It is found only in the Cameroon and Gabun in west central Africa. The second, which is composed of the largest species, occurs all down the forested parts of the East African coast from Kenya to Natal and over another large but isolated area from Lake Tanganyika through the southern Congo to the Angola coast. The third group, typified by the little Moholi Galago (Plate 34), which has become quite well known as a pet, is spread in a variety of forms all over the savannah country between the forests inhabited by the other two, from Senegal in the west to Kenya in the east and south down to the Limpopo River. The last two species vary much in size and color from place to place, but all are for the most part nocturnal and eat much more vegetable matter than the lorises.

Pigmy Bush-Babies (*Galagoides*)

These tiny lemurs, only about a foot in over-all length, are found all over the forest area from West to inner East Africa. Though looking externally like the preceding species, they are internally quite distinct and their habits are different. The differences are subtle and can be appreciated only when one studies them in forests containing one of the larger forms. Some of them seem to move about by day in the high canopy and these eat more insects. They are extremely agile and so light that they make even faster time through the treetops than do squirrels. The third, fourth, and fifth toes are very long and slender, and the big-toe looks just like a very big human thumb.

Needle-clawed Bush-Babies (*Euoticus*)

This is a different-looking animal from all the preceding. It inhabits the tall forests from the Niger River in the west to the Congo and perhaps farther east to the Rift Valley. It is best distinguished by its nails, all of which, except those of the thumb, first, and third toes, have a ridge running down the center that projects beyond the rest of the nail in the form of a very fine, sharp, needle-like claw. These animals are heavy-set, with woolly, orange-brown fur, rather short, wide faces, and enormous eyes with bright orange irises. The hands and feet have flesh-pink soles with large pads on all the fingers and toes. From my own observation in West Africa, they would appear to be wholly eaters of fruit and green nuts. All the Lorisoids and Lemuroids have a structure under the tongue known as the *sublingua,* which means simply an "under-tongue." In the Needle-Claws this structure is very well developed and reaches to the base of the bottom front teeth. Like that of other lemurs, it is made of rigid cartilage and the front end bears a lot of small, sharp points like a tiny comb. The purpose of this strange organ has been much debated, and although it may be used for a variety of purposes it is certainly used by *Euoticus* not only to comb the fur, but also to free the fine hairs from between the bottom front teeth which do most of the preliminary unravelling of tangles in the fur, and the removing of burrs and other foreign matter. These animals can make prodigious leaps from tree to tree with little loss of altitude; during such leaps they hold their limbs spread-eagled.

LEMUROIDS

We now come to what may be described as the True Lemurs, some score of different animals, all related and all found only in the great island of Madagascar and its natural dependencies, the Comoro Islands. These differ enormously in shape, size, and habits. The smallest is slightly larger than a mouse, the largest as bulky as a big dog and about four feet in overall length. The smaller and more primitive are very like the Bush-Babies of Africa; the larger are monkey-like and not unintelligent by our standards. All except one (the Indri) have long bushy tails, and all have pointed faces with the nostrils at the tip of the snout. Their lower front teeth are directed forwards and form a comb, the outer pair being the eye-teeth, and the lower tusk really a cheek-tooth with two roots. With the exception of the first very odd animal, they all have nails on all fingers and toes except for the second toe, which carries a stout curved claw. They may be divided into four groups.

THE AYE-AYE (*Daubentoniidae*)

The most peculiar and one of the oddest of all lemurs, if not all mammals, is found in two widely separated areas in the northwest and central eastern forested parts of Madagascar. Previously another giant species dwelt in the dry southwest, but it is now extinct. These animals could be separated as a major primate group all their own, so different are they from all other Lemuroids, did they not show some internal affinities with some of the Silky Lemurs. They are about the size of cats, but have huge bushy tails that they carry in a stiff downward curve. They are clothed in a thick underwool covered by a long coat of coarse black hairs sprinkled with white like a Silver Fox. The throat is often yellowish and the underside reddish. The face is very catlike, but the ears are extremely large and naked. They inhabit the dense forests and giant bamboo brakes, and eat insects, eggs, some small animals, and the pulp and juices of fruits, canes and bamboos. Only one young is born at a time and the mothers make large, spherical nests in which to raise them. Most odd are the teeth, hands, and feet of the Aye-aye.

For a long time this animal was thought to be a Rodent because it has only two upper and two lower front teeth that grow continuously and have chisel tips formed by flinty enamel in front and softer ivory behind. Further, they have no canines, and the cheek-teeth are only four in number above and three below, and flat-topped like those of rats. Their feet have opposed big-toes bearing a nail, but the other toes are short, are held together and bear claws. All the fingers bear claws including the opposed thumb; the middle finger is only about a fourth as thick as the others and is used for tapping on logs to disturb insects in order to extract them from their holes. All the fingers and toes bear pronounced terminal pads.

SMALL WOOLLY LEMURS (*Cheirogalaginae*)

There are three genera of small woolly-furred lemurs that have elongated ankle joints—giving the hind legs a four-jointed appearance—pronounced finger and toe pads, and large eyes. These are all nocturnal, and of very small size.

Aye-aye

Mouse-Lemurs (*Microcebus*)

The smallest Primate, the Lesser Mouse-Lemur, is less than four inches long and has a tail of the same length. It and other species known as Coquerell's Lemur live in the tall damp forests, on the open scrublands of the south, and even in open reed beds. They build very neat, round nests of twigs and leaves, and leap about just like Bush-babies. Very curiously for a Primate, they spend part of the year in a comatose condition after storing up masses of fat under the skin about the hind legs and base of the tail. They retire for the hot dry season. They are principally insect-eaters but also take honey and some plant saps. The fur is brownish-grey.

Dwarf Lemurs (*Cheirogaleus* and *Phaner*)

The remaining small species are all much alike in form and color, have shorter, more catlike faces, but vary in size from that of a large rat to some three feet in over-all length. One (*Phaner*) has been separated as being intermediate between these and the Mouse-Lemurs in that it jumps rather than runs, and builds nests. It also has a vivid dark stripe along the mid-back that divides on top of the head to form a fork descending to the eyes. The three remaining species, known as the Hairy-eared, the Fat-tailed, and the Greater, appear to live in holes in trees by day, to have—like all Woolly Lemurs—two young at a time, and to be more squirrel-like in movements. They are predominantly insect-eaters and are found all over the island. Only the Fat-tailed species, which stores up such huge masses of semi-liquid fat under the skin that it may swell the tail to almost two inches in diameter, appears to aestivate for short periods in some regions. The others are active at all times of the year. The largest, the Greater, is reddish-brown in color.

LARGE WOOLLY LEMURS (*Lemurinae*)

Here again there are three genera, two of which are fairly consistent, having two distinct species in each genus, but the third of which presents zoologists with the most bewildering problems. They are bulky, agile animals that move about in large family parties or tribes, both by day and night and often descend to the ground to forage even in forested areas.

Weasel-Lemurs (*Lepilemur*)

The two species have been given the rather delightful English names of the Weasel and the Sportive Lemurs, and are found on the west and south, and on the east coasts of Madagascar respectively. They are both nocturnal animals that move about the forests in large parties and build complicated nests of leaves and twigs. They are exclusively leaf-eaters and possibly this is the reason they have no upper front teeth at all. The color of the fur varies considerably, but is usually some shade of grey with a yellowish wash, and silvery grey below, but those from open forest and dry areas are often reddish above and yellow below. In length they are about ten to twelve inches, and the tail is slightly shorter.

Gentle Lemurs (*Hapalemur*)

The endless apparent duplication and reduplication of

Lesser Mouse-Lemur

little, obscure animals, may appear irksome or even ridiculous to the average person, but nonetheless every one of these creatures is so different from all others that it requires some special mention. These two species of Lemurs are nothing whatsoever like the last two, and all four are quite different in shape, color, movements, and habits from the following. The differences are much greater in life than among, for instance, poodles, dachshunds, and police dogs. The Gentle Lemurs have short, catlike faces and small ears of remarkably human appearance, are comparatively large animals—some reaching four feet in length—and are of a metallic grey color with a strange sulphur wash. On their wrists one kind displays a naked area covered with short, stiff bristles; the other, the so-called Broad-faced Lemur, does not. Both of them live in dense thickets of Giant Bamboos and feed on the growing shoots of these plants by stripping off the hard outer layers. The Broad-nosed species is also found among the great marshes, and has adopted the unique and extraordinary habit of living on masses of floating reeds, felling and eating the pithy stems of the plants which form its floating home. These animals have curious heart-shaped nails with a raised central ridge ending in a sharp point.

Common Lemurs (*Lemur*)

There are six distinct species of Common Lemurs, but apart from the grey, ring-tailed, black-spectacled, rock-dwelling species of south-central Madagascar, and the startlingly colored black-and-white Ruffed Lemur, they are very hard to describe and still harder to identify. The first species, having the delightful Latin name of *Lemur catta* is so originally marked and so much better known than the others that it may be recognized by anybody. The tail is bushy and composed of vivid black and white rings. They are ground-living and run about on all fours with the tail held straight up in the air, but they sit on their haunches to eat, hold food in their hands and bite it with their back teeth. In captivity they have an inordinate love of water faucets, and will sit with their tummies under a dripping tap for hours.

The Ruffed Lemur comes in a variety of color patterns, but is normally black and white, though no two are exactly alike. There are races with either more or less black or white, and others in which the white is replaced by orange, red, or russet. They are the only Common Lemurs that make nests; they are also more nocturnal, but have the strange habit of sunning themselves in the morning. They eat mostly leaves. The face is black with a white muzzle, and a white ruff round the ears and around the throat. The fore-body and upper arms are black; the forearms white, and the hands black.

The Black Lemur (*L. macaco*) is entirely arboreal though it will drop to the ground when pursued. The female (Plate 35) instead of being all black is reddish brown with paler ruff and rump.

The Brown Lemur is even more bewildering, no two being exactly alike, but there are usually light areas on the sides of the muzzle, and it has neither ear tufts nor cheek fringes. They are forest animals and move about in large parties by day. In color they vary from almost grey to deep brown, and sometimes bright reddish, but there seems to be no set rule about their appearance, many different colors being found in the same parties and no one color-type appearing to be typical of any particular area.

Still more confounding are the smaller Mongoose-Lemurs, which have a more woolly fur and more bushy tails but also vary in a hopeless manner. The muzzles of this species are, however, much shorter and are almost always white, and there are marked differences between the sexes, the males having white and the females red cheeks. They do not have collars or ruffs, and the ears are small. They move about by day and are mostly vegetarians, feeding off leaves and fruits at the tops of forest trees. There is much confusion between this and the preceding species, and in nature they may live side by side in the same forests but keep quite apart. Mongoose-Lemurs are found only in the northern part of Madagascar and on some of the Comoro Islands.

The last species is known as the Red-bellied Lemur, and this adds the final confusion to the whole matter since it appears to be distributed all over the island in appropriate locations and, although keeping to itself, may look very like some of the other types. It is the smallest species of all and the fur is more silky, the tail more bushy. It has very pointed nails with a distinct central ridge and appears to be more wholly diurnal. Altogether, the Common Lemurs are one of the most difficult groups of mammals to disentangle, but since they are fairly common animals in zoos—not to mention crossword puzzles—more questions are asked about them than about many commonly known groups.

SILKY LEMURS (*Indriidae*)

The remaining four living lemurs are divided into three quite distinct genera with the pleasant-sounding names of Sifakas, Avahis, and Indris. They have short, broad muzzles and widely spaced eyes that give them a more monkey-like and less foxlike appearance. The face is naked. They are slower-moving and more precise than the Common Lemurs, descending trees backwards with great care and walking about on the ground on their exceptionally long hind legs or hopping along like kangaroos. They are pure vegetarians and they bear only a single young at a time. Their hands and feet are proportionately very large; the thumb is small and only slightly opposed, but the big-toe is enormous and diverges at more than a right angle from the other toes, which are held together by webbing and act as a single unit. They are large animals having silky coats; the Sifakas and the Indri are diurnal and go about in small parties. The Avahi is nocturnal, solitary, and has a partly woolly pelage. The Indri is tailless.

Sifakas (*Propithecus*)

There are two well-marked groups of species known as the Diademed and Verreaux's Sifakas found on the east, and on the northwest central and southwest sides of Madagascar respectively, but it is almost impossible to describe either of them or the differences between them. They have naked black faces and large eyes with rather long lashes; when resting they carry their tails coiled like springs. The coat is silky, rather long, and fluffy. They come in a bewildering assortment of colors, patterns, and color combinations varying in both species from almost pure white with a yellow or pink tinge to pure black. Basically the skin is black and the fur white-tipped but beyond this any worthwhile description would take a disproportionate amount of space.

The Avahi (*Lichanotus*)

This is a smaller animal with, rather annoyingly, woolly as opposed to silky fur. They are completely nocturnal and sleep curled up by day. They are found all down the east coast and in a limited forest area of the northwest. The face is covered with short hairs and the ears are concealed in the head fur. The eyes are huge and stare straight ahead. They are grey-brown with a light rump and white legs, forearms, hands and feet, and an orange-reddish tail; but some individuals are almost all white and others red.

The Indri (*Indri*)

This is the oddest of all Lemurs, having no tail, long, slender hind limbs and enormous hands and feet that look as if they were inserted into woolly gloves. The face is foxy and the large ears bear a long fringe. Though slender, they are very large, sometimes measuring over two feet in length. In color they vary greatly but the face is usually dark with a light topknot and throat collar. The body is dark, but the elbows, flanks and hind limbs are light and there is much variation in pattern and in color from black and white to brown and cream.

TARSIOIDS

In the great equatorial forests of Sumatra, Java, Borneo, Celebes, and the Philippines, and in a number of associated smaller islands, painstaking search in holes at the tops of trees or among matted conglomerations of vines, dead leaves, and parasitic plants may sometimes bring to light pairs of incredible, somewhat hysterical little animals about the size of rats (Plate 32). These combine as odd an assortment of strange features as can be found in the world of mammals. These obscure little beasts have in recent years become comparatively well known through the medium of picture magazines, popular films, and even toy representations. They are known as Tarsiers (*Tarsius spectrum,* etc.) and their intrinsic interest is twofold.

First, their general appearance is not only spectral but quite out of this world: secondly, they are of exceptional scientific interest since they stand alone as the sole survivors of a most ancient lineage. Away back, some sixty million years ago, their ancestors must have been something like the Tupaioids, but they seem to have become distinct even at that early date, and although they diversified somewhat later, all of their kin died away and left them alone to survive until the present. In basic anatomical structure they are unlike any other Primate, but in most superficial features they most resemble some Lorisoids and Lemuroids. In some respects they may be considered nearer the basal stock from which all the Primates arose.

They are insect-eaters and wholly nocturnal, but catch their food with their hands and bring it up to their mouths to eat. Their faces are dominated by a pair of colossal eyes, completely circular in outline, placed close together and directed straight forward. Proportionately large eyes in man would probably be a foot across! The muzzle is a ridiculous little puglike affair squeezed between the eyes and the small mouth, and there is practically no chin. The ears are small, naked and erected on either side of the round, flat-topped head.

The body, forelimbs, and the upper part of the legs are proportioned like those of a squirrel, but the ankle bones are elongated to such an extent that they form an additional joint that is almost as long again as the lower leg. The fingers and toes are all very long and thin and each ends in a grotesque, bulbous, fleshy pad, shaped like the discs on the feet of tree-frogs. The thumbs and the big-toes are somewhat opposed and bear small nails; so also do the other fingers and the fourth and fifth toes; but the second and third toes carry small claws that stick almost straight up.

The fur is very short, thick, woolly and soft, and the tail is very long, furry at the base, then almost naked; it bears a terminal club of long hairs.

HAPALOIDS

There is still another group of Primates somewhat more closely related to the monkeys, but still not sufficiently close to be classed with them. This is a South American group composed of several scores of very small animals with squirrel-shaped bodies, but faces more resembling miniature Sifakas. They come in a great range of colors and include some of the most exotically clothed of all mammals. Anatomically they are more like the Ceboids, or South American Monkeys, but they are highly specialized, have lost the four hindermost pairs of teeth, and have retained a number of primitive features, notably claws on all fingers and toes except the big-toe, which bears a nail in all but the two most monkey-like types. Neither the small thumb nor the tiny big-toe are opposed. They may be divided into three very unequal sub-groups.

MARMOSETS (*Callithricidae*)

The largest marmoset, if Goeldi's be given a separate standing, is not as bulky as a large Black Rat, and the smallest is as small as, if not lighter than, the smallest Mouse-Lemur, an adult male weighing only three ounces, and may thus be the smallest Primate. Marmosets, which have been known since the discovery of South America as the most desirable of all exotic pets, have puzzled zoologists and laymen alike since they were first carried by wealthy European courtesans in their bosoms. They have the bodies of squirrels but voices, vocabularies, and facial expressions that far surpass in flexibility and versatility all other mammals except the apes. They are all diurnal, tree-dwelling, gregarious animals with exceedingly bright and intelligent little eyes, naked faces, mouths filled with needle-sharp teeth and pronounced tusks. It is customary to divide them simply into two genera called Marmosets and Tamarins, according to the size of these tusks, but current researches have shown this to be no valid criterion, while other features, and notably geographical distribution, necessitate a much more extensive breakdown. Whether the nine recognizable groups constitute full generic status is debatable, but all are clearly defined.

Pigmy Marmosets (*Cebuella*)

These tiny sprites with furred but not bushy tails a little shorter than their bodies, live in the upper canopy of the giant, primaeval, equatorial forests of a small area in Amazonia around the Napo and Uayacali river valleys on the border of Ecuador, but on the eastern slope of the Andes. There may be another light-bellied form in Southern Brazil. They have a brindled, greenish-yellow coat of fur so fine and soft it cannot be felt if touched when one's eyes are closed. The head is rounded, the face small, the eyes wide apart and set obliquely. The very human little ears are almost covered by the long plumelike fur which is brushed smoothly back over the whole head and forms a close mane over the neck. They breed in holes in treetops, and as with all marmosets, twins are the rule; at birth they are about the size of an uncooked navy bean, though their eyes are open and they are fully furred. A lens is needed to see their fingers. They are carried by the father, one in each groin, pointing outwards and backwards, and are only given to their mother for nursing at regular intervals.

Pigmy Marmoset

feet, and the limbs are very long. The tusks are huge, but in voice, face, and movements they are closer to the Pigmy than any other type.

Plumed Marmosets (*Callithrix*)

In the same area, but spreading over a much more extensive territory—namely, the tall, moist tropical and sub-tropical forests, known to Brazilians as the *Tupi*, which stretch along the southeast coast of their country and which are isolated from the Amazon forests by the dry uplands known as the *Caatinga*—are found a whole aggregation of small marmosets that are most distinctive. They have small, compact bodies and comparatively short back legs so that they run and hop rather than leap about the trees. Their faces are small and usually bear a triangular light blaze on the forehead. Their ears are large but partly concealed by the head fur and partly by a drooping tuft or whisk of long plumelike hairs that sprout from the back of their cheeks. There are about half a dozen species, all clothed in very soft fur that is black at the base, then orange, and then tipped with black and white. Their tails are ringed, the feet small, the teeth moderate. The two best known (*C. aurita* and *penicillata*) (Plates 39, 40) have pure white and pure black plumes respectively.

Ruffed Marmosets (*Hapale*)

The first known Marmosets that gave their name to this whole group of Primates, came from the forests north of the *Caatinga,* from Bahia to the Amazon and thence westward along its southern bank. They are slender, very long-legged leapers that come in a bewildering variety of colors from the brown-and-cream brindled effect of the common *Hapale jacchus* of Bahia, to pure creamy white, silver, parti-colored white and buff or russet, and sundry other combinations. Further, several appear to interbreed, while others show great individual variation, and some have been known to change color completely after normal or pathological moults. They all, however, have the ears individually ringed by a corona or ruff of long, white or light, rather stiff hairs.

Bald Marmosets (*Marikina*)

In a comparatively small and narrow belt north of the Amazon and extending about a third of the way up its course in central Brazil, live two or three small Marmosets with completely bald face, chin, and throat, and with the hair only starting behind the crown of the head. They are parti-colored and have small ears like the following types, but the moderate teeth of the former.

White Tamarins (*Mico*)

To the south of the Amazon, extending from the coast about its mouth to the borders of Peru in the west and south to the Matto Grosso, the above are replaced by a closely allied group of Tamarins with large tusks, small wrinkled ears that are carried laterally and are naked and often bright pink like their faces. One is pure, glistening white with a jet-black tail and a vividly pink face which has nasty-looking, irregular red spots.

Black Tamarins (*Tamarin*)

Tucked away up in the Guiana massif north of the

Maned Marmosets (*Leontocebus*)

Far away to the southeast in two river systems, the Ribeiro and the Parahyba, that debouch into the Atlantic on the Brazilian coast, live two of the most colorful of all mammals and the most exotic of Marmosets. They are known as the Golden Lion Marmosets. They also have manes, but these are much more exaggerated, and the whole pelage is longer and more silky. In one species (Plate 38) it is brilliant, metallic gold all over; in the other the face, body, and arms are black, and the tail, except for a yellow crest, is dark grey. They have exceptionally long, slender hands and

Amazon and the range of the Bald Marmosets, are a group of naked-eared, beetle-browed, low-crowned Tamarins with jet-black skins and faces. In all of them the coat is soft and jet-black, though usually tipped with reddish cream about the hinder back and on the tail, which is bushy and normally carried curled into a tight, springlike spiral. Some have bright orange-red hands and feet, looking just as if they were wearing gloves and socks. They are very irascible creatures, biting savagely with their long tusks but they appear to live in company with bands of Ceboid Monkeys. They eat green fruits and nuts and spurn insects, even locusts.

Moustached Tamarins (*Tamarinus*)

Far to the west and beyond the range of both the Bald Marmosets and the White Tamarins, but covering the range of the Pigmy Marmosets, lives a huge assemblage of large-tusked, naked-eared, flat-headed, small Tamarins with comparatively short back legs like the Black Tamarins. Their skins are pink, but their faces usually black and clothed in short hairs, and all have some kind of white facial markings, the most exaggerated being a huge pair of Bismarckian moustaches found in a species named *T. imperator*. In some respects they resemble the little Plumed Marmosets, but their tusks are longer. About a dozen different species have been described.

Pinchés (*Oedipomidas*)

The last group of Marmosets are known as the Pinchés, and they are the most exaggerated of all. There are three species, all found on the west or Pacific side of the Andes in the high equatorial forests of Colombia, Panama, and in one case even north of that isthmus west to Costa Rica. They are large, slender, and excessively long-legged, with naked faces clothed in a fuzz of very short, light hairs extending over the forehead to the base of the ears and under the chin to the throat. They have immensely long, well-furred tails and are white, grey, or at least lighter below and over the arms and legs and dark above, often brown with a bright brick-red area on the flanks. One species, the Cottonhead, has an immense, plumed, topknot of pure white fur that looks like the ostrich-feather headdresses affected by some African chieftains. The Panamanian species (Geoffrey's Marmoset, or *O. spixii*) has a triangular white skullcap.

GOELDI'S MARMOSET (*Callimiconinae*)

This exceedingly rare and little-known animal is customarily placed in a sub-family all its own, but as a sub-division of the Ceboids because it is said to have an extra back tooth on either side of both jaws, unlike the Marmosets and tamarins, but like the South American monkeys. However, although twice the size of the largest Tamarin, it agrees in every other respect with the Hapaloids. While this was being written a live specimen became available for observation and confirmed the fact that this animal is in every way a Hapaloid, not only in general shape but in movements and voice—the latter a most distinctive and, among all mammals, a unique characteristic of the group. Like the Marmosets, also, all the fingers and toes except the big toe bear long,

Cottonhead Pinché

sharp claws. In color, the young are jet-black all over. The fur is silky and long and forms a domed cap on the head, a separate erectile mane over the neck, and a slight cape over the hind body. The adults have the cape tashed with gold, and the black tail develops two remarkable white or cream rings near the base, one twice the width of the other. Nothing is known of their habits.

TITIS (*Callicebinae*)

This group, usually constituted only as a genus of the Ceboids, comprises a host of small monkey- or large marmoset-sized animals that have been divided into

[71]

some forty species. They appear to range all over the vast Amazon basin and could probably be broken down into a number of distinct sub-groups. However, very little is really known about them, and there is much confusion as to the exact identity of most of the species. They agree with the other Hapaloids in having claws instead of nails, but even the big-toe is so armed. They are more monkey-like in posture than the Tamarins, but they have the movements and much of the vocabulary of the true Marmosets, while their small, naked faces and somewhat silky pelage is typically Hapaloid. They are forest-dwellers and go about in small tribes, eating a mixed diet of insects, snails, lizards, green fruits and nuts, and some flowers and leaves. Unlike monkeys—except the Douroucouli, which is very like the Titis and is usually grouped with them and which also bears clawlike nails—they sleep and nest in holes in trees. In color almost all of them are brindled on the back and flanks while over other areas they are usually flushed with rich brick-red. In some, the face is surrounded by a white ruff; others are predominantly grey; still others have an overcoat of stiffer hair.

CEBOIDS

The so-called Monkeys of South America are equivalent to, but only very distantly related to the true monkeys

Dusky Titi

or Simioids of the Old World. Either the two groups diverged at a very early date, or they had quite separate origins among the lemuroid-insectivores. It has been suggested that the Ceboids sprang from very ancient and long extinct American Lorisoids. Whatever their origin, this branch of the primates diversified in an amazing way, sending off first an offshoot that gave rise to the Hapaloids, then, later, another that has ended in creatures that are not quite monkeys and that we here call The Half-Monkeys (*Pithecinae*), and finally producing the Hand-tailed varieties (*Cebinae*). Just where to place the Squirrel-Monkeys (*Saimiri*) is anyone's guess since they have as much in common with the Half-Monkeys as with the Hand-tailed group. There are at least ten very distinct genera of Ceboids but an entirely new one was found last year, and there may well be other quite novel types still to be found in the vast forests of the Amazon.

HALF-MONKEYS (*Pithecinae*)

There are four and perhaps five clear-cut kinds of monkey-like animals in South America that have thickened, and in at least two cases, clawlike nails on both fingers and toes. They have very close, thick fur—in one case with a coarse overcoat—and bushy tails that may be clubbed, may taper like those of dogs, or be cylindrical like that of a cat. Their fingers are long, slender, and very human in form and arrangement, the thumbs and big toes being widely opposed. They have small faces, and the eyes are close together and point forwards. Stripped of their fur, their bodies are extremely elongated and slender, and the limbs are very long. Their skulls are small and rounded, and the lower jaw is very deep and powerful at the back. The lower teeth in front tend to shelve outwards and are almost completely horizontal in the Sakiwinkis.

Douroucoulis (*Aotes*)

The first genus is entirely nocturnal, an altogether unmonkey-like habit, and is very closely related to—or has come to resemble very nearly—the Titis in many respects. Half a dozen different species of Night or Owl Monkeys have been described, but it appears that there are only two basic stocks, one found on the north and west side of the Andes in Colombia, and the other on the south or Amazonian side. Both have bodies about a foot long, covered with close, rather woolly fur, neatly rounded heads and very flat, naked faces. The eyes are enormous, circular, very close together, and owl-like, having highly contractile irises of a bright golden color. The muzzle, mouth and chin form a round dome, and the ears are small and fully furred, and almost concealed in the fur. The tails are very long and thickly furred, but neither bushy nor clubbed, and the hairs on the underside at the base are stiff and form a sharp keel. All of them have vivid white "false-eyes" above the real eyes, and these may extend upwards to form white stripes which may meet on top of the head; otherwise the whole front and top of the head is black. The northern group is more monkey-formed, with large heads, and short fur; these (Plate 41) are greenish golden-brown above and bright orange below and have black tails.

The southern group is grey above and dirty white below; they walk with their rumps elevated above their head and shoulders. The former are docile, the latter ill-tempered. They are fruit-eaters but take insects.

Sakiwinkis (*Pithecia*)

Even more astonishing in appearance are the Sakis, which defy adequate description. Their limbs are long and slender; the legs and the hands especially are grotesquely human while the feet are large. They are covered with long, coarse hair that forms a sort of toque on the head and a cape over the body. The tail is bushy and forms a brush like that of a fox. The face is like that of a Pug dog. The narrow area between the rather small eyes, extending to the broad nose, muzzle and chin are naked, but the rest of the face is clothed in all manner of strange ways. In the male of one species, which is otherwise pure black, it is clothed in close, recurved, short, bristle-like hairs forming a kidney-shaped mask like that of a Barn Owl. This is pure white centrally but stained with bright yellow around the edges. The female is, however, brindled yellowish-grey all over and very shaggy, has a face clothed in a short black fluff, but sports a pair of yellow dundreary whiskers. Sakis vary in color in a bewildering manner, and one type often blends imperceptibly into one, two or three other kinds. Those described above come from the Guianese coast; in the upper Amazon are pure black species with a sort of ruff or boa encircling the face, and almost pure silver kinds with black faces, long capes and short beards. These animals are diurnal fruit-eaters that move about in troupes and sleep sitting up on open boughs. Like the Douroucoulis they have an enormous vocabulary with special alarm calls that can be heard for miles. They are prodigious jumpers and habitually run along branches standing upright on their hind legs.

Bearded Sakis (*Chiropotes*)

These are really in no way like the preceding animals either in form, habits, or behavior. They are larger animals, up to two feet in head and body, with limbs of more equal length than the preceding species, and very dense, woolly coats. The hands and feet are shorter, and the digits thicker, and the thumb is more truly opposed. Their tails are incredible, being longer than the head and body, clothed in a dense long wool of uniform length from base to tip, so that the whole is circular in section. Their small faces are naked, the eyes round, and the whole muzzle is protuberant but rounded, giving the animals a perpetually grumpy expression. On the top of the head grows a preposterous mushroom-shaped mop of long hair radiating in all directions, and falling down over the neck, ears, and forehead to the eyes. To complete the picture, a perfectly formed beard, either in one piece or divided down the center, grows from the cheeks and chin just as in men. The commonest species is known as the Red-Backed Saki, which is very dark reddish brown to black, with a yellowish-grey body. Others are jet-black, greyish, or brown and red, and one has a vivid, pure white nose, but they also seem to vary much and to blend spatially one into the other. They are fruit-eaters and slow, deliberate climbers.

[73]

Silvered Sakiwinki

Uacaris (*Cacajao*)

Just as ridiculous in appearance, if not more so, are a group of monkey-like animals clothed in long, scraggly, sparse, red hair, and with enormous, naked, pink feet and absurd little stumps of tails. These are known as Uacaris. The sparse hair on their heads is parted in the middle or grows in a whorl from the top of the

Red-faced Uacari

crown; it hangs down all over the large, naked ears and cadaverous face. This gives the animals a woebegone expression. Under the cape of long hair, the body is sickeningly slim, and the face is generally emaciated, with deep-sunk eyes and prominent features and is clothed in naked skin which is either bright pink or a deathly grey. There are several recognizable types, but again they blend one into the other over a wide area north of the Amazon between the Andes and the Guiana massif and thence north to the Orinoco valley. They live in troupes in the tops of trees, but spend some time on the ground collecting fallen fruits. One type is completely bald and looks revolting.

HAND-TAILED MONKEYS (Cebinae)

The remaining Primates of the New World are categorically monkey-like in appearance. In fact, with one exception, they are more wholly so in the eyes of the public than any Old World form, including even those that have been known to man since before the dawn of history—the reason being that these alone among all Primates can hold on by their tails. No Old World monkey can do this, ten million tales and illustrations to the contrary. There are six known genera, only one of which —the first—does not have a prehensile tail, and its status is therefore uncertain.

Squirrel-Monkeys (Saimiri)

In the gallery-forests that line the great waterways of the Orinoco, Guianese, Amazon, and associated watersheds, the commonest primates are small, long-legged monkeys with very long club-shaped tails (Plate 44). They go about in large troupes, sometimes numbering hundreds. They are called in pidgin-English "Monkey-monkey Monkeys." Nevertheless, they are most unmonkey-like in many respects, having elongated ankle

bones like some lemurs, and strangely formed skulls that bulge out behind the neck almost as far as the face does in front. The face is small and white, with a black muzzle and dark rings round the eyes. The ears are large and furred, and often adorned with tufts or plumes. The fur is close and straight, though woolly to the touch, and is a mixture of brown, gold, and bright green, blending to vivid yellow-orange on the flanks, forearms, legs, hands and feet. In some, the head and shoulders are black, and the tail is slightly clubbed and black towards the end. The underside may be white or yellow, or gold. They have exaggerated tusks. The average member of a troupe is only slightly larger than a large squirrel, but they are very prolific, while their mortality rate is exceptionally high. At the same time, they seem to be exceptionally long-lived so that those which do reach full maturity appear like giants, being four times the size of the average adult, and measuring almost two feet when sitting up. These giants have comparatively much longer tails. Squirrel-Monkeys are insect- and fruit-eaters and they make continuous croaking noises unlike any other mammal, and quite unlike other Ceboids.

Capuchin Monkeys (Cebus)

These are "monkeys" par excellence, including, as they do, the typical companion of the street organ-grinder of old, the commonest monkeys of almost all zoos, and the principal forms offered for sale in all American pet-stores. However, there is the most appalling confusion and an almost universal misconception about their habits, affinities, and nomenclature. They are average-sized monkeys, the head and body measuring about fourteen inches to two feet when adult, but they come in a large number of forms that are distributed from Guatemala in Central America to the southern limits of the Amazonian forests in South America. Their tails are fully furred throughout, and are partially prehensile—that is to say, the animals use them as a steadying organ and can hang by them for a limited period, but they cannot be used as a fifth hand like the tails of the monkeys reviewed below. Dozens of species have been described, but these may be reduced to some dozen or so major assemblages which in turn fall into three fairly clear categories.

First, there is a group of interblending species in Central and northern South America, north and west of the Andes. These have light fronts and faces, a simple "hair-do," and comparatively hairless faces. They range in color from jet-black above and pure white below to various browns above and cream below, or they may be sandy all over especially in the south and east of this range. All of them tend to carry their tails curled downwards in a spiral and are therefore often called "Ring-tails." Unfortunately, the proper name of the vividly white-fronted ones is capucinus while the plain, brown-colored species, popularly called "Cinnamon Ringtails," are rightfully named albifrons—which means white-fronted.

The second aggregation are found in the Orinoco basin and the Guianas north of the Amazon, and in Trinidad. They are much more heavily built, are clothed

Woolly Monkey

in thick brown coats, darker on the back, limbs, and tails, and they display a wide range of variation in body and facial markings. They are distinguished by having heavy fur around the ears when fully adult; this gives the face a rectangular appearance from the front, and in southern forms it may be developed into tall crests arising on either side of the head above the ears. The typical form is *Cebus nigrivittatus* of the Guianas (Plate 42)—apparently the original organ-grinder monkey.

Farther south in lower Amazonia there are a host of forms in which the fur on the top of the head rises into a peak centrally over the crown (typified by *C. apella*).

Capuchins are diurnal and wander about the forests in small bands, eating fruits, some leaves, insects, snails, and other small animals. They are very agile and appear to be exceptionally aware and intelligent in the human sense. Their distribution appears to be determined by rivers which even when very small form effective barriers between recognizable races or species.

Woolly Monkeys (*Lagothrix*)

Perhaps the most appealing of all monkeys, with faces that always have the ineffably sad expressions of extremely gentle old men, and with slow, deliberate, and carefully unaggressive movements, woolly monkeys make the most gentle of all pets. These rather large and usually corpulent monkeys are found—in sundry color varieties—throughout the upper Orinoco valley and thence south over the divide into the western part of the Amazon drainage basin. They are covered in very dense, woolly fur, so short it appears to have been cropped, and their tails are long and fully prehensile, with a naked pad below the extremity. The skin is black, the face naked, the ears small and almost hidden in the woolly ball of the head-fur, and the eyes rather large, clear brown and pathetically soulful. They are predominantly fruit-eaters, but relish the raw flesh of vertebrates during specific periods of the year. They also chew on flowers and leaves at other times. They move about in family parties in the upper canopy and have a wide vocabulary, including a scream of rage that has to be heard to be believed.

Woolly Spider Monkey (Brachyteles)

These are so far the rarest of South American monkeys, and appear to hold a position intermediate between the Woolly and the Spider Monkeys. They are confined to the *Tupi* forests in the southeast of Brazil beyond the great divide of the drier hinterland. They are very large monkeys with shaggy fur, naked, pinkish faces, long arms and legs, and huge, elongated feet and hands. They are slim like the Spider monkeys, and have immense, fully prehensile tails with a naked "finger-pad." Their thumbs are either minute or completely nonexistent, though specimens with a rudiment on one hand and none on the other have been found.

Spider Monkeys (Ateles)

Here again we are confronted with a bewildering assortment of closely interrelated animals varying in color and color pattern this way and that; further, they display all kinds of individual variations within the tribe

Black Howler Monkey

in some areas, but in others all are just as strikingly alike (Plate 43). In some respects there seems to be no rhyme or reason attached to these gradations in color. They are found from southern Mexico to Uruguay in all forested areas. In build, they are indeed spidery, having immense limbs and long prehensile tails with finger pads. The head is small, the face naked and the hair of the crown is directed forwards. The hands are either entirely or practically thumbless and all the fingers form a combined hook. The fur is thin and scraggly and lacks the under-wool of the previous genus, except in the mountain group typified by a species known to science as *variegatus,* which is parti-colored, is found in a variety of forms all along the eastern slopes of the Andes, and has thick, soft, almost silky fur. For the rest, these monkeys may be divided into a central group in the Orinoco, Guianas, and the Amazonian basin that are jet-black with naked pink or black faces, a southern agglomeration among the foothills of the vast central Brazilian uplands that are silvered or tashed with brown and have white whiskers, and a variegated brownish assemblage north and west of the Andes in Colombia, Panama, and Central America that vary in an almost endless manner according to locality. All, however, are alike in habits, being denizens of the tall forest and never, if possible, coming to the ground. They move about in large troupes feeding on leaves, fruits and green nuts, and when in a hurry they are almost as agile as Gibbons.

Howler Monkeys (Alouatta)

Anyone who has ever been beyond the tourist periphery of South America and slept in the real forest will forever remember the soul-stirring vocal outpourings of these magnificent, great, bearded, leaf-eating, prehensile-tailed monkeys. They sound like the roll of distant thunder preceded by the death-agonies of half a dozen tortured jaguars. From southern Mexico, where they are pure black, to the southern edge of the Amazonian forests where there are a variety of black, brown, gold-and-brown, reddish-brown, and black-and-red forms, these large animals are always to be found in the hinterland, where they move majestically about in large parties led by a colossal male. The prodigious roaring howls are produced by a single individual except on rare occasions when some sort of ceremony is performed by the whole troupe; it is achieved with very little effort by means of a curious, spherical, bony sound-box formed of the hyoid bones which is attached to the throat between the expanded back-projections of the lower jaw. In the northeastern part of South America these animals are clothed in scintillating, metallic coppery pelts, and have deep, full beards. The face is long, protruding, and almost doglike, and the teeth are very large, with pronounced tusks. The tail is fully prehensile and has a large, naked finger-pad below. They are ponderous animals that trundle rather than run, gallop, or leap through the trees; yet, they can travel faster than a man can run on the forest floor below. If South America selected an international emblem it should be this magnificent animal emblazoned in gold on a jungle treetop against the clear blue of the sky.

[76]

SIMIOIDS

The true monkeys are confined to the Old World. One species, the famous Gibraltar Ape, which is a kind of Macaque or Rhesus, barely enters Europe; the rest are found in Africa and Asia, south of the great desert belt and the Himalaya Mountains, but extend northwards up the far eastern side of the continent of Asia to Japan and Amuria. There are well over two hundred different kinds, divided into sixteen genera, and these fall into three major groups, the Colobine, the Long-tailed, and the Dog-faced. The first have long tails and are vegetarians, most of them being leaf-eaters, and they are found in Africa and the Orient. The second are exclusively African, and take a wider diet. The members of the third group have, for the most part, shorter tails and spend more time on the ground. They are all omnivorous. In pre-glacial times monkeys inhabited most of Europe and central Russia and those parts of the Middle East that are now desiccated but were then moister and more fertile. Climatic change does not seem to be the cause of their disappearance since they still live under the most rigorous conditions in Amuria.

COLOBINE MONKEYS (*Colobinae*)

These large, heavy-bodied monkeys with long tails dwell for the most part in trees and forested areas. The Guerezas of Africa and the Proboscis Monkeys of Borneo are leaf-eaters; the Langurs of India and the Oriental Region are leaf- and fruit-eaters. There are six genera, three of which may be lumped together as Langurs, which includes the sacred monkeys of India.

Guerezas (*Colobus*)

These are very large monkeys that grow to three feet in length and have forty-inch tails. They either have no thumbs at all or mere tubercles, sometimes with a tiny horny spike—all that is left of the nail. There are about a dozen recognized species, half of which form a widespread agglomeration, colored black and white, which extends from central Abyssinia in the east, south to Tanganyika and thence west via the Congo to the Cam-

East African Guereza

Capped Langur

Langurs (Presbytis, etc.)

More than twenty species of large, extremely lanky monkeys with very long tails are grouped together under the general heading of Langurs. They are the common arboreal monkeys of India, Ceylon, Burma, Indo-China, Malay, and the greater Indonesian Islands. They come in a wide variety of sizes, colors, and shapes divided among half a dozen basic types, two of which are distinctive enough to be given separate names.

The first group is composed of very large, brownish-grey animals with jet black, naked faces, ears, hands, and feet, which are found all over India from Sind to Bengal and, in a slightly different form, all along the Himalayas from Kashmir to Bhutan, where they range to the upper limit of the trees and may often be seen galavanting about in the snow. Other color variants occur in northern Burma. They are distinguished by having the hair on the top of the head arranged radially so that it pokes out over the eyes like a visor. The young, of which there may be twins, are carried on the mother's belly.

Members of the second group are smaller and have the hair of the head growing smoothly backwards and falling over the neck. One species, with a yellow head and a glossy black body, inhabits southern India. Three related forms live in Ceylon, one of which, the Purple-faced Monkey, is brown with grey haunches but has a striking set of pale cream whiskers. Another form, found in the mountains, has five-inch long, coarse grey-brown fur; still another is a sort of permanent semi-albino, being white with a dusty mid-back and pink palms and soles. There are related species with back-brushed head-fur in Burma, Siam, and Malaya, but some of these have such a topknot that they are called Capped Langurs.

A very distinct species in the same area has the hair on the head raised into a tall, fore-and-aft ridge or crest and still another whole group, also found in Malaya, Sumatra, and Java, have the hair parted in two places so that a thick bang falls over the forehead, and the hair on the back of the head stands up in a crest.

Most outstanding of all Langurs are the Negro and Red Monkeys of Sumatra and Java, the first of which is coal-black all over and has a very fine, silky fur while the brightest colored of all, known as the "Douc" (Pygathrix), is found in Indo-China and the island of Hainan. It has a very short body and limbs, bright yellow face with grey whiskers, a brown head with a vivid chestnut collar, a dark brown body with a white diaper-shaped area over the hind back, and a white tail. The upper arms, legs, hands and feet are black, but the forearms are white, and the lower part of the legs bright red. The Douc is perhaps the most colorful of all mammals.

Snub-nosed Monkeys (Rhinopithecus)

Over the hump between the lowlands of India and Burma on the one hand, and inner China on the other, are found a number of large monkeys of most grotesque appearance. They are very sturdy and covered with long fur that ranges in color from variegated browns and grey to a pure metallic gold, with the naked skin of their faces ranging from black to a bright blue in the golden

eroon in West Africa. Basically, these animals are jet-black but the different species or races have, in varying degrees, developed a white band across the brow, white cheeks and chin, and a capelike structure of very long white hairs starting on either shoulder, extending along the sides of the back, and then joining over the tail. The tail itself, which is club-shaped, is black at the base but white at the end, but the amount of white varies and in one species from Kilimanjaro the tail is pure white throughout, and very bushy. The other half dozen species are West African, two being found on the island of Fernando Po. They lack the white cape, though two are also basically black, and one of these has a white forehead, cheeks, neck mane, and a tail ending in a tassel. Another also has white thighs. The pure black form bears a long bonnet of coarse hair and has a slender tapering tail. One, from the far west, is pale brown with a white head, the hair of which is neatly parted in the middle. These monkeys have stomachs divided into a number of open sacs that aid in leaf digestion. They travel in bands and stay near water.

species. Their faces are like those of the Langurs, except that the nose is rather long and turned straight up, as if the animal had run head-on into the back of a bus. The purpose of this structure is unknown. These monkeys occur all over the mountains from eastern Tibet to northeast China south of the great Gobi desert, and mostly in regions that are covered with snow during a major part of the year. They travel in large tribes and migrate regularly. They are omnivorous and often raid villages, carrying off anything edible or drinkable, especially if bottled, but the local people appear to treat them with much awe and some reverence. They display a high degree of communal organization and discipline.

Proboscis Monkey (*Nasalis*)

The nearest relative of the Snub-nosed Monkeys and the last of the leaf-eating Colobines, is an utterly ridiculous-looking animal of very large size found only in the swamp forests of the lowlands of the island of Borneo. It is of a reddish-orange-brown color, darker on the top of the head and back, and lighter almost to white below. The young and the females have retroussé noses rather like the above species, but the males have enormous, tubular "schnozzles" that hang down over their mouths to below their chins and may measure over three inches. What these are for is unknown except that it gives their voices an absurd nasal twang. They eat only certain leaves, notably bamboo, and live in troupes. They are gentle, calm, and rather slow, and they love to bask in the sun. They also regularly go swimming for pleasure.

LONG-TAILED MONKEYS (*Cercopithecinae*)

The second great group of Old World Monkeys is exclusively African, and all have long tails. There are two major groups (genera) and two others consisting of only one species each. One of these is the only truly ground-living member of the whole lot but it readily ascends trees when they are available. They are confined to the forested zone of the continent south of the Sahara and north of the deserts, treeless veldts, and temperate zones of South Africa.

Guenons (*Cercopithecus*)

One can only be appalled by the task of presenting an adequate description of this vast host of closely related but greatly varied animals in any compass less than a whole volume. The number of species involved is not actually known but probably exceeds a hundred, and in one small area only sixty miles in diameter the author recorded no less than eight species, each belonging to a distinct grouping or sub-genus, while from a hundred miles to the east half a dozen quite other species have been recorded, and still a third assemblage was known to occur across a river to the west. And thus it is, all across Africa from Senegal on the Atlantic to Abyssinia in the east, and thence south to the borders of the Union and the southern limits of the forests in Angola. Almost every area has its own assemblage.

All these monkeys are strikingly similar from an anatomical point of view but they may be separated into groupings on the basis of their over-all color arrangement and other ornamentations. There are about a dozen of these groupings, the commonest and best known of which consists of the so-called Green Monkeys (Plate 46), otherwise called Vervets or Grivets, but which are, for some peculiar reason, known to the Dutch of South Africa as Blue-Monkeys. In point of fact they are all of a yellowish-brown color with brindled fur and sometimes a slight olive tinge. They are long-legged and very long-tailed, and have black, naked faces and hands, a white brow line, and light sideburns neatly brushed back over the ears. Their undersides are lighter. They are exceedingly neat, tidy animals that live in large troupes on the open savannahs and are very quick, intelligent and quite fearless, attacking and driving off even the larger baboons.

The next great grouping may be called the Monas. These are forest-dwellers and have darker, greener coats with blue faces and bright yellow brows, whiskers, and undersides. The limbs and tail are usually dark to black, but there is a great deal of variation in this coloring within various species that are spread all across Africa.

Somewhere near the Monas are the *Cephus* or Moustached group, which range from the Gabun eastward and have yellow cheeks but grey undersides.

The Putty-nosed Monkeys,—often called "Green-Monkeys" because they are a dark, brindled greenish color—are forest types varying in sundry ways, but all having an inverted, heart-shaped, white-haired blob on the nose. They range from Gambia to Central Africa. Close to them are another assemblage with bright red or orange noses typified by *C. erythrotis*, and apparently centered around Nigeria.

Quite different and most distinct are the beautiful Diana Monkeys of the West Coast of Africa (Plate 45), black above, pure white below, with pointed white beards and bright red markings on the flanks and hind limbs, and sometimes a lemon-yellow edging to the white chest.

Even more distinct is a group which ranges from the Cameroon to the eastern Congo, and which lives in gallery forests on the edge of the great jungles. These (*C. pogonias, nigripes, grayi,* etc.) have the hairs on the tops of their heads raised into a tall peak. They are vividly colored, *pogonias* being a really bright green above and brilliant orange on throat, chest and undersides. A mountain assemblage typified by such species are *C. preussi* in the west, and many species in the east are predominantly grey with white fronts, pronounced light whiskers, a parti-colored tail, and a rufous area on the hind back. They are known as the Diadem Monkeys.

But the list is almost endless, and one recent shipment from East Africa to an American importer contained no less than thirteen distinct types, four of which could not be traced from published descriptions. Guenons are easy to recognize as a whole, but to identify the precise species is a matter for an expert and often necessitates exact data as to where it was obtained.

Allen's Swamp Monkey (*Allenopithecus*)

Of distinctive coloration and habitat, and also showing sufficient anatomical differences to be separated from the typical Guenons, this species was not discovered until 1923. It is greenish above and white below.

Military Monkeys (*Erythrocebus*)

The only other Cercopithecids that are sufficiently odd to be classed by themselves are the Patas and Nisnas Monkeys of West and East Africa respectively. These animals are of a reddish brown color, lighter on the cheeks, throat, and underside. Their tails are somewhat shorter than those of the forest monkeys, and they have limbs proportioned more like those of dogs and other running animals. They live out on savannahs and scrublands, and they are predominantly terrestrial, though they can climb well. They are wholly omnivorous, but can apparently subsist on either animal or vegetable food alone, and they eat a great number of insects, scorpions, and lizards. They run in large companies and sometimes mingle with Vervets.

Mangabeys (*Cercocebus*)

These are very large arboreal monkeys found throughout the forest areas of Africa. In many respects they come close to the Dog-faced Monkeys, having noticeably prognathous muzzles, large teeth, and huge tusks. They also have large cheek-pouches like the Macaques,

Red-capped Mangabey

but they lack the noise-producing throat-pouches of the Guenons and others, and make only little mewing and twittering noises. There are about a dozen different kinds that can be divided into those that have the hair of the head raised into a tall, pointed, central peak, and those with normally rounded crowns. Most of both kinds are some form of dull grey in color, with lighter or white undersides and fronts, and often with pronounced backwardly brushed sideburns. All have a prominent ridge of fur across the brow, which is often white, and in the case of the Collared Mangabey this continues round the face in the form of a white ruff. The related Red-capped Mangabey has a bright brick-red skullcap, and both, plus the White-crowned, and the Sooty, have startling white upper eyelids which are believed to be a signaling device, since they blink their eyes when stared at.

DOG-FACED MONKEYS (*Cynopithecinae*)

The current practice is to class these genera with the Long-tailed Monkeys in a single undivided family, but the Dog-faced Monkeys are so different in appearance, habits, and distribution, and form such a closed unit that there is every justification for separating them into a sub-family of their own. Besides, they form a distinct and well-known group that is readily distinguished by the public. It consists of the Macaques (*i.e.*, Rhesus-like monkeys) and the Baboons.

Macaques (*Macaca*)

Any attempt to describe this host of closely related monkeys in a limited space is even more hopeless than

[*continued on page 98*

Patas Monkey

38. *Golden Lion-Marmoset*

PINNEY

41. *Colombian Douroucouli*
PINNEY

→

39. *Black-plumed Marmoset*
PINNEY

40. *White-plumed Marmoset*
MARKHAM

42. *Black-capped Capuchin*
PINNEY

43. *Golden Spider-Monkey*
LA TOUR

44. *Squirrel Monkey*
LA TOUR

45. *Diana Monkey*

LA TOUR

46. *Vervet Monkey*

PINNEY

49. *Silvery Gibbon*
PINNEY

→

47. *Mandrill*
LA TOUR

48.
Anubis Baboon
COMMANDER GATTI
AFRICAN EXPEDITIONS

50. *White-handed Gibbon and Young*
YLLA FROM RAPHO-GUILLUMETTE

51. *Chimpanzee*
LA TOUR

52. *Orangutan*
LA TOUR

53. *Gorilla*
(*immature*)
VAN NOSTRAND
FROM NATIONAL
AUDUBON

54. *Hairy Armadillo*

MARKHAM

55. *Giant Anteater*

56. *Pika*

58. *Arctic Hare*

57. *European Rabbit*

59. *Varying Hare*

60. *Chickaree*

CRUICKSHANK
FROM NATIONAL AUDUBON

61. *North American Grey Squirrel*

MARKHAM

62. *Mantled*
Ground-Squirrel

63. *Thirteen-lined Ground Squirrel*

67. *Young Prairie-Dog*
© WALT DISNEY PRODUCTIONS
→

64. *Flickertail*

HARRISON FROM NATIONAL AUDUBON

65. *Common Ground Squirrel*

HARRISON FROM NATIONAL AUDUBON

66. *Hoary Marmot*

MILLER FROM NATIONAL AUDUBON

68. *Beaver*

69. *Kangaroo-Rat*

Long-tailed Macaque

[*continued from page 80*

trying to describe the various Guenons. Worse still, there is even less agreement, both among zoologists and interested laymen, as to which animals constitute a particular species and what English names should be applied to each. Nonetheless, there are eight very distinctive groups of species.

The first is the famous Barbary Ape of north Africa —a tribe of which lives on the Rock of Gibraltar. This is a rather shaggy-furred, yellowish-brown animal without a tail, that is today practically domesticated, and may have been introduced from the East by the Arabs in their first millennium conquests which carried them to Spain. No other Macaque is found west of what is now Pakistan.

Best known of all Macaques is the Rhesus of Bengal, which has a naked, flesh-colored face and a tail only about half the length of the head and body. Large males grow to almost two feet, and are very powerful, aggressive creatures. They are of a slightly yellowish, sandy-brown color, somewhat lighter below, and are used in enormous numbers for scientific research. They have recently been shot halfway out of our atmosphere in rockets, thus preceding man into virtual space-flight. They live in the forests as well as on cultivated land, and ascend mountains to great heights. There is a related form that inhabits the Himalayas from Nepal to northern Burma.

Beyond the "Hump" another group takes their place. These are shaggy-coated mountain types that culminate in the huge species known as *M. arctoides* or the Bear-like Monkey of the icy mountains of Sze-Chuan in west central China. Others occur in Amuria and Korea. Somewhat closely akin to these are the famous Japanese Apes that have bright red, naked faces and great mops of olive green hair parted in the middle. These very odd animals dig under the snow in winter for food and go swimming in the sea for shellfish in the summer. Most macaques like water and are seldom far from it, bathing regularly, and swimming strongly. There is one in Formosa that lives in caves around the coast, and is semi-aquatic, feeding exclusively on the beach, and diving in shoal waters for shellfish, crabs, and certain seaweeds.

Of the same general color as the Rhesus—that is to say various sandy or reddish to yellowish browns—is still another group of species inhabiting Burma, Siam, Malaya, and the larger Indonesian islands. The first of these is the so-called Crab-eating Macaque, a sturdy animal with a long, thick tail. It inhabits mangrove swamps along the coast and in estuaries, swims strongly and feeds on shellfish, crabs, and some vegetable matter. Inland, its counterpart is the Pig-tailed Macaque, a still larger species, growing to about twenty inches in length, and having an eight-inch semi-naked tail that it can curl up like a pig's.

In the Island of Celebes is found a black species named the Moor-Macaque that has a small stump for a tail, lives in large gangs, and unlike any other monkey actually hunts other animals, though also eating a varied diet of roots, fruits, insects and other animal material.

Of quite a different nature are the Bonnet and Toque Monkeys of southern India and Ceylon respectively.

Both have profuse long hair on the tops of their heads radiating in all directions but parting on the forehead. The former is one of the largest Macaques, males growing to almost thirty inches in length and having two-foot tails. The latter are smaller and lighter in build and have thick, wavy, rather rough hair. The most extreme form of Macaque is the Lion Monkey, now usually called the Wanderoo, which is a magnificient glossy black animal that is two feet long and has a short tail ending in a large, club-shaped tuft. The face is black, but clothed with very short, sparse grey hairs which multiply and increase in length outwards in all directions, forming a tremendous boa, or face-frame, curving forwards around the edge and shaped like the ruff of a Barn-Owl. Down the mid-brow is a divide clothed in the black, short hair of the rest of the head. Wanderoos are forest animals and normally extremely wary, but they make most affectionate, intelligent, and hardy pets, without any of the grossly delinquent habits of the average macaques, which are incorrigible "slobs." Macaques of various kinds have been introduced into many parts of the world, and notably islands such as the Philippines, Mauritius, and even the West Indies.

The Black Ape (*Cynopithecus*)

In the island of Celebes, on Batchian, and some of the southern Philippines, and apparently on some other smaller islands, live bands of very remarkable, large, baboon-like monkeys misleadingly referred to as "apes." They are black all over with short stubby apologies for tails, naked faces with beetling brows and long, doglike muzzles with pronounced cheek swellings. The hair on the top of their heads forms a tall peak curving slightly backwards at the tip. They are exaggerated relatives of the macaques, and have developed a baboon-like appearance. They live mostly in mangrove forests near the coasts, and feed on sea-food as well as fruits and some leaves. They are gentle, retiring creatures, seem to have a high degree of intelligence, and are much respected by the human inhabitants of the countries where they are found.

Baboons (*Papio*)

There are about a dozen recognized forms of true baboons, all found exclusively in Africa. They are the bulkiest of all monkeys, and are essentially ground-living animals though they can climb trees and rocks with great agility. They all have well-developed tails, but of varying length, none being as long as the head and body. The commonest of the species is the Anubis (Plate 48), which ranges all across Africa north of the great forest belt, from west to east. It is a large species with a drab olive coat, a shiny black face, a considerable crest along the neck and fore-back, and a trace of a mane over the shoulders.

In the west there is another species known as the Guinea Baboon; it has ear tufts, is the smallest of known baboons, and inhabits the broken forests and cleared lands right down to the coast. To the south of the rain forests, that is to say, down the east side of the continent and in South Africa, other distinct species take their place. Among these are the largest of all baboons,

Hamadryas Baboon

the Chacma of South Africa, which has a black face and dark cheek-whiskers, ears, and neck. A second smaller species, without trace of a crest, and lighter in general color is known as the Yellow or Thoth Baboon. The face is pink. The northeast African species may be distinct, but there appears to be much blending of forms geographically.

The most exaggerated form is known as the Hamadryas which was the Sacred Baboon of the Ancient Egyptians. It comes from Abyssinia but there is a distinct race in southern Arabia. They are huge, powerfully built animals with a large mane or cape covering the whole fore-body, but their tails lack a terminal tassel. They are of a greyish color—dark in the young and females, but paler in adult males. In some areas these baboons come in contact with the Geladas, and terrific battles, amounting almost to organized warfare—with surprise raids, the taking of prisoners, wide maneuvers, and other grossly human tactics—continue over long periods. These animals were tamed by the Egyptians and were used to pick fruit; they were also sacred, and their bodies were mummified.

Baboons are particularly interesting animals because of their social organization and apparent high degree of ability equivalent if not identical to human powers of reasoning. They live in large troupes of all ages, each bossed by one or more active, full-grown male. They are omnivorous creatures and have to grub for a living by digging and turning over stones for roots, bulbs, insects, snails, lizards and other small morsels. These activities, the movements of the troupe, and its discipline and protection are regulated by the leading males; those of the young ones, by the adult females. Baboons have tremendous jaws, and the males a set of teeth with tusks that match the larger carnivores. Their speed equals that of a horse. They will defend themselves against leopards and men and have become not only a pest, but a very serious menace in some parts of South Africa, where they even invade the suburbs and wreck homes.

The Gelada (*Theropithecus*)

An exceptional type of baboon-like primate known as the Gelada is found in southern Abyssinia. It is a huge animal with a long-nosed and quite hideous face, dark greenish-brown to black fur and an enormous mane covering its forequarters. The chest is naked, and this and all other unhaired parts of the body—which includes pronounced buttock pads—are bright pink. Its tail is

[99]

fairly long and ends in a large, club-shaped tassel. Its fangs are enormous.

Drills (*Mandrillus*)

In the wet forests, where baboons are not found, two species of virtually tailless monkeys of monstrous appearance take their place. One, the Drill, is dark olive green all over with a shiny black face and pale pink buttock pads. It roams about in large bands on the forest floor, eating anything edible, and although it avoids humans, it can be extremely dangerous if molested. The Mandrill (Plate 47), which is found only in the Gaboon, is a grotesque animal. It roams about the deep forests in large gangs, but also makes forays out on to the open savannahs and among rocks in the mountains.

ANTHROPOIDS

The eighth and final major division of the Primate mammals contains about a dozen medium or large animals (including ourselves) without tails and with comparatively large brains. Apart from Men, which are today almost worldwide in range, they are confined to the forested parts of Africa and the Oriental Region from eastern India to the island of Hainan and south to Borneo, Sumatra and Java. Previously, they had a much wider distribution, but a large number of forms have become extinct. They show some possible relation to the Simioids, but appear to have evolved along separate lines during a great period of time, and it is possible that they had a separate origin among Tarsioids. They may be separated into two very distinct groups, the Lesser Apes or Gibbons, and the Great Apes including Men. How-ever, we still bestow upon ourselves separate family status, though this is really quite unwarranted.

LESSER APES (*Hylobatidae*)

There are about half a dozen valid species of Gibbons, one of which is so different as to warrant a separate generic standing. This animal, the Siamang, is confined to Sumatra; the other species are spread from Bhutan in the central Himalayas through Burma and Indo-China to Hainan and south to Java. All Gibbons have long slender tusks and hard callosities on their buttocks. In these respects they are different from the greater apes and more like certain lower primates.

The Siamang (*Symphalangus*)

These are the largest of the lesser apes, adult males measuring three feet when sitting up and having an arm span from finger tip to finger tip across the chest of almost six feet. Like men, their second and third toes are joined together to the second joint by a web. Their final oddity is a pouch connected to the throat; this can be inflated from within and blows up externally like a big red balloon. They are clothed in rather long, shaggy black hair, but the males develop a greyish beard. They move about the high forests up to elevations of several thousand feet, and make the countryside hideous at daybreak and sundown with the most incredible howling imaginable. They are fruit-eaters. They seldom descend to the ground, where they are awkward, but they can swim rather well, using a breast stroke and keeping their heads well above water. They are found in the island of Sumatra.

Siamang

SUSCHITZKY

Silvery Gibbons

Gibbons (Hylobates)

In describing the Gibbons we will start at the north-western end of their range in the mountains of Bhutan in the central Himalayas, with a small, slender, jet-black species named the Hoolock. Its only marking is a pure white bar across the brow, which is still evident even in certain individuals that are grey, grey-brown, or even cream in general coat color. They also set up a tremen- dous whooping racket at dawn, and usually keep it up for two or three hours. Variations of the Hoolock are spread all over Assam, upper Burma and, apparently, east through the mountain forests of upper Thai and Indo-China, for a closely related form is found on the island of Hainan.

Next of kin to the Hoolock is the White-handed Gib- bon (Plate 50) found south of a line drawn through

[101]

middle Burma to Cochin and extending down the Malaya peninsula. This animal is also very varied in general color, from jet-black to pale brown, but it always has a rim of white that completely encircles the face, and white upper surfaces on the hands and feet not unlike gloves and socks. The naked face and palms are black. They come to ground occasionally and can run well on their hind legs, but in the trees they rush along trapeze-fashion using their hands and arms for locomotion and their feet for carrying small supplies of food.

There is another quite distinct Gibbon found in this area and in the Sulu Archipelago and Borneo. This has been given the rather redundant English name of the Agile Gibbon, though it could hardly be more so than any of the others. It varies enormously in color from a very dark brown to pale cream, and may even be particolored with a dark body and light limbs, or vice versa. These are the famous Unka-puti of the Malays, who almost invariably have one in their homes and treat it with as much affection as they do their own children, if not more. There is a related form in Sumatra, and the famous Silvery Gibbon or Wow-Wow of Java (Plate 49) is hardly distinct from some varieties of the Unka-puti.

Gibbons are the cleanest, gentlest, and in many respects by far the most intelligent of the apes. One that lived and traveled with the author all over Malaya and the islands, ate at table, made its own bed, and was responsible for the systematic collection of several thousand species of insects which would never have been found by any human collector.

GREATER APES (*Pongidae*)

Exclusive of Man there are three major types of great apes living today—the Orang-Utan, the Chimpanzees, and the Gorilla—the first in the islands of Sumatra and Borneo, the two latter in Africa. It must be understood, however, that all of them show great individual variation in body build, facial expression, and the amount and color of their hair. Moreover, the racial, or what might be called "national," and even tribal resemblances are often very great and, like men, they even show family likenesses. There are definitely two races of Gorilla, and there may be two kinds of Chimpanzee, though in each case the races can interbreed and both types may have the same parents. The general form of the apes is sufficiently well known already so that we need only point out that the thumbs are small and fully opposed, the big-toes very large and even more widely opposed, the body short and very obese, the hind limbs short, and the arms very long. Their eyes are small, close together, and point forwards, the nose-bridge is sunken but the nostrils are wide and flaring. The muzzle is protuberant and the lips long but very thin and mobile. The Orang and the Chimps are arboreal; the Gorilla is terrestrial.

Gorillas (*Gorilla*)

The gorillas are the largest and the most specialized of the apes (Plate 53). Their heads, at least in adult males, have developed, or perhaps we should say reverted, to a baboon-like or almost doglike form, with long, prognathous muzzle and a tall, bony crest along the ridge of the skull to which huge muscles, used in crushing food, are attached. They also walk on all fours, using the outside of the feet and the knuckles.

They are found in considerable numbers over a wide area of the Cameroons and Gabun north of the Congo, and thence east to the Ituri, and south down the west side of the great mountain divide from Uganda to Katanga. Apart from a large continuous population in the Gabun, they form isolated communities often separated by very considerable distances. One of the distinct "nations" dwells in a remote group of mountains on the border of Nigeria in the far northwest; another on the volcanic peaks of Kivu a thousand miles to the east.

Gorillas are forest creatures and it used to be believed that there were distinct lowland and mountain types, but every described variation in size, shape, and coloration is recorded from both habitats, and it is more than probable that many family parties actually wander from lowlands to the mountain forests when the two types of vegetation are contiguous. Gorillas may be completely black all over, have red topknots, or be silvery grey to various extents. Very often, but not invariably, the whole family, including even the young, may be alike in color, but the females are more often black.

Gorillas are immensely powerful creatures; just how strong is not properly comprehended by any man. Two-inch, tempered steel bars have been bent by frightened gorillas, and tales of twisted double-barreled shotguns are apparently not imagination. Yet, all those who know gorillas, both natives of the countries where they live and the comparatively few white men who have hunted and observed them, agree that they never deliberately attack human beings, will seldom carry through a rush even when defending their family, and if wounded or cornered make every endeavor to get away. Only when a man turns and runs do they carry through an attack, and then they usually swat the fugitive with their great horny hands which can literally knock a shoulder-blade off a man, or rip his thigh from hip to knee. In one area, however, the author inspected over a dozen men with scars on their legs, which they said were the result of bites by male gorillas; strangely, all were said to have been inflicted on the next to the last man in a party traveling single file through the dense mountain forest where these animals dwell. These tribesmen had never had contact with any white men, and are noted for their veracity in matters of animal behavior. They further insisted that the gorilla was not a superior form of monkey like the Chimpanzee but a debased form of wild man, and yet that the males never molested their women. As one proof of the contention that they are almost men, these natives showed the author sleeping platforms constructed by gorillas a few feet off the ground in gnarled dwarf trees made by bending down branches and anchoring them with true knots with a double over-and-under twist—both "grannies" and "reef knots."

Gorillas will climb trees to eat certain leaves and fruits, but the oldest male usually remains on the ground —a custom which has given rise to the belief that he sleeps there. This is not necessarily so, for he may often sleep aloft and especially in the lowland forests. It appears that this greatest of the Primates, is, in fact, not the ravening ogre he has been depicted, but just a great big, easily scared vegetarian, desiring nothing more than to be left alone in his forest fastnesses to raise his solemn, quiet little kids, and be allowed the occasional privilege of marauding a human banana plantation.

The Chimpanzees (Pan)

Just because a gorilla may have the amazing armspread of over nine feet and is otherwise given priority among the apes, it should not be assumed that the Chimpanzee (Plate 51) is a lesser breed. Old male chimps can be of monumental proportions, and possibly just as bulky, and their arms just as long—in fact, longer in proportion to their bodies. They are also of almost inconceivable strength, though more in pulling actions than in bending or striking. There have been many chimpanzees of extraordinary size in captivity; the writer recalls three in particular, all heavily furred, black-faced males, that lived in the Berlin, London, and Rochester zoos at different times. The last was also an object lesson in Chimpanzee appearance in that it was housed alongside a very large female which was almost hairless and had a grey-pink skin and an almost "white" face. Anyone previously unacquainted with chimpanzees would never have imagined that they were of the same species.

The range of facial contour and expression in chimps is certainly as great as that among men, and like us, their skins may be pinkish, yellow, or brown, even to a depth of color that we could call black. Some have profuse hair, others are bald either on the head, the body or both. Their teeth vary in size and texture as much as ours, and like most other animals they suffer from caries, abscesses, and other dental troubles. Their eyesight is subject to the same variants as ours, and they seem to suffer from most of our other afflictions.

Psychologically chimps are demonstrably of quite a different temperament from the Gorillas, being active, excitable, inquisitive, resourceful, and brash. Even in their native forests (they do not live in the mountain forest) they create a tremendous rumpus, roaring about, shouting, brawling among themselves, and taunting any human who approaches them. They are also much bolder than gorillas, and although they prefer to keep to themselves, they will carry through an attack once it is launched. Natives who live in territory where these apes, as well as gorillas, are found show a much greater respect for the chimp in any real encounter.

The "intelligence" of chimps in captivity is proverbial and there is little doubt that they use their brains in the same manner as human beings. Countless observations have established the fact that chimps work out problems just as men do, and there are even cases on record of their having surpassed normal adult humans in reasoning, notably in getting a banana down from a roof in an empty room—the man swatted at the food with a pole until it fell and was smashed; the ape set up the pole, climbed it, and retrieved the fruit whole.

Chimps still live all over a very wide area in Africa, the bounds of which are more or less those of the tall equatorial and high deciduous forests. In some areas they are extremely numerous. A political officer travelling in one area of the central French Cameroon reported counting over two hundred of them between villages only fifteen miles apart.

Orang-Utans (Pongo)

These cousins of ours (Plate 52) are of an entirely different appearance, nature, and temperament. There are two kinds, both clothed in bright, reddish-orange hair; one is found in Borneo, the other in Sumatra. The Orang-Utan—meaning in Malayan simply the "Man of the Woods" or the "Wild Man," has a mongoloid look, a complacent and contemplative temperament, and certain characteristics that are sometimes disturbingly "human" (see photograph on title page).

These creatures develop with age into gross parodies of men, with huge cheek flaps, throat sacks, repulsively obese stomachs, and grotesque little legs. Their arms are even greater in proportion to the body than those of the chimps, since these animals are more wholly arboreal. In youth, they show astonishing intelligence and sometimes traits that are rather embarrassing, for Orangs often behave like completely uninhibited human beings and, what is more, they seem to appreciate the fact that they are doing so. This is difficult to explain unless we accept the fact that they have what we call a "mentality" similar, if not identical, to ours.

Orangs are forest animals, and make every effort to stay out of the way of men, but curiously enough, when captured young but at an age when they can feed and care for themselves, they will settle down in a human household just like any human orphan. It has been noted that they then show strong individual preferences and seem to have an antipathy for certain specific groups of persons. The reason for these strange prejudices may, of course, be the result of nothing more abstruse than the smell of the perspiration of those who customarily eat certain diets.

The Orangs make rough sleeping platforms like those of the gorillas, and appear to be vegetarians, though in captivity they will eat almost anything fit for human consumption, and appear to thrive on it. One owned by the author relished strong cheeses and rare steak cut into manageable pieces. Although of very great weight when adult, they travel rapidly through the forest treetops, often by bending a branch like a crane to reach the next bunch of foliage, even if it be on the other side of a river.

MEN (Hominidae)

Today, there is only one species of Man living; at least, as far as we know. As we have learned from the fossilized bones and stone artifacts found in caves, gravel, and other deposits, there were in the past quite a number of other species, but all have become extinct unless the so-called "Yeti," "Metch Kangmi," or "Snowman," reported by almost all the native inhabitants and many travellers to exist in the Tibetan Himalayas is in fact a remnant of Neanderthal Man. His alleged footprints exactly match those found in a fifty-thousand-year-old mud floor of a cave in Italy, and he dwells in a climate that nearly matches that of Europe in the Ice Age when Neanderthal man lived there. Men, however, still come in a surprising number of variations, and if these belonged to any other order but that of the primates, each might well have been classed as different species, despite the fact that all can, apparently, interbreed. There are four principal types with sundry intertypes. They now inhabit almost the whole earth except Antarctica, parts of some deserts, and the utmost tops of the larger mountain ranges.

Man (*Homo*)

For his size, Man is probably the most numerous of all mammals alive today. Undoubtedly, he is the most adaptable in the variety of climatic conditions under which he manages to exist, though the tiny House-Mouse and one or other of two species of rats have managed to follow him everywhere, and dogs and cats are almost as universally distributed. There are probably more misconceptions about the mammalian species, Man, current today than about any other animal—that is to say, from a purely biological point of view. Apart from his brain and certain habits, he is not especially advanced nor by any means the ultimate in specialization. His strength is modest for his size, but his endurance is extraordinary. A trained human runner can ultimately outrun a horse.

There used to be a widespread belief that the so-called races of men could be arranged in an ascending scale, with the dark-skinned types having flaring nostrils at the bottom nearest the apes, and the light-skinned with narrow nostrils at the top. More factual consideration shows quite other results. The hair of apes is profuse and straight; their hind limbs are short and their torsos long and deep; their lips are thin. The existing men with most hair that is straight are the Caucasoids—including the "Whiteman"; those with the least and most curled hair are the Negroids, who have the longest hind limbs and shortest torsos. The most "primitive" type of man is the Australoid, who seem actually to be linked to fossil Solo Man of Java and thence back to ape-men, yet they have profuse, wavy hair on their heads and almost as much on their bodies as Caucasoids. The Mongoloids, on the other hand, have as little hair as the Negroid but it is straighter than that of the average Caucasoid, and their legs are comparatively short. Thus, the four major types of extant human beings display an altogether contrary set of characters from the physical point of view, and none, with the possible exception of the Negroid, can be said to display non-apelike characteristics.

Man is communal but not truly gregarious like many other mammals; this is to say that, although he lives and goes about in groups, the members of the group are not necessarily always the same individuals. In the past, and in certain areas today where culture is still primitive, man is truly gregarious or what we call tribal.

In feeding habits men are completely omnivorous, eating almost everything digestible. All manner of leaves, roots, fruits, and nuts that are not poisonous, meat and other animal products of all kinds, including even human flesh, is readily digested. Certain desert tribes relish live scorpions; locusts, beetle larvae, and raw earthworms are consumed in vast amounts annually; almost everything in the sea from seaweed to whales and including octopus, sea-urchins, luminous worms, shellfish, and even protozoa and diatoms is eaten; and finally, man uses a great bulk of purely mineral matter—salt, kaolin and other clays, and even such abstruse items as alcohol and sodium-glucomate.

When we turn to the habits of man, we enter a bewildering world of variation, complexity, and extremes that have come to form the separate science of anthropology. Basically, nonetheless, Man is an agriculturist, (like the Leaf-cutter Ants or *Atta*), and a nest-builder, (like the Termites and especially those of the tropics). In fact, Man appears to be fundamentally a tropical mammal and suited best to that climate. His agricultural and nest-building proclivities have led him to an extraordinary development of tools, which in turn has completely changed his original methods of food gathering, transportation, and storage. This in turn has begun to change his methods of reproduction, the care and upbringing of his young, and his individual longevity.

In behavior man, however, is still not greatly different from other mammals, and the higher Primates in particular. He is fundamentally polygamous and females tend to be more numerous than males; he gives almost excessive care to his young, so that they are dependent upon their parents for a very long period, often beyond puberty. Men are co-operative except in their own home territory or "nests," where they may often display most independent and aggressive attitudes to other members of their species. They are essentially terrestrial.

Left-Over Mammals

(*Edentata*)

SCIENTIFIC procedure often mystifies the nonscientist; sometimes it dumbfounds him. The name *Edentata*, which means "without teeth," is frankly quite nonsensical, since the majority of the animals so classed have teeth, and one—the Giant Armadillo—has more than any other mammal, except certain whales. Apart from a few extinct relatives, they could, in fact, be called the *Xenarthra*, meaning the "strange-jointed ones," but this really does not mean very much either. The fact of the matter is that the animals concerned, although related, show very considerable differences of form and habit and they are, collectively, most widely separated from all other mammals. They are very ancient, in many ways truly primitive and a most specialized group, few in numbers and confined to Central and South America.

Until not so long ago, geologically speaking, they

were a fairly successful and quite numerous group of mammals that included some enormous creatures (known as Ground Sloths) as big as elephants. Apparently some of these, and by no means small ones, continued to exist in what is now Patagonia until the arrival of the Europeans in the sixteenth century. In caves in that region dried skins of one such species have been found together with stone walls that had obviously been devised to corral them; and there are several accounts by early explorers of the native Amerinds of the area having caught such shaggy beasts in pits and killed them therein by lighting a fire over them. The author has spoken to two men born on *estanzas* in the west of Patagonia at the base of the Andes who declared they had talked to men who had hunted smaller examples of such beasts in the dense waterless scrub of that country. Whether they still exist may never be known, but similar animals once inhabited some of the West Indies and large parts of southwestern North America.

At the present time, as far as we know, the order is represented by only about a score of species of animals that may be most clearly divided into three very distinct types. These are the Anteaters, the Sloths, and the Armadillos. With the exception of the sloths, all are predominantly insect-eaters, but many of the Armadillos thrive in captivity on a diet of chopped meat, raw eggs, boiled rice, and other substances, and several have been observed eating carrion, and even shellfish under natural conditions. The two sloths are purely vegetarian, the Three-fingered species apparently being unable to subsist on anything but the leaves of a tree called the Wild Pawpaw or *Cecropia*.

One kind of Armadillo, known as the Tatu, or the Nine-banded, is found in parts of North America adjacent to Mexico, and has been spreading rather rapidly northward. Somehow it has crossed the Mississippi—probably aided by men who, strangely, buy quite a number of these animals every year as pets—and has become established in Alabama and parts of northern Florida. This and other armadillos range from there south to Patagonia. Anteaters and Sloths start with the high forest in southern Mexico and extend to the southern limits of that vegetation in the Argentine.

Why these curious animals should be confined to tropical and southern America, is a matter of considerable interest to zoologists, and calls for some explanation of the most recent ideas on the history of the earth since the Dinosaurs disappeared. It is now obvious that about sixty million years ago South America became separated from the other continents by at least a shallow sea somewhere around the present narrow isthmus of Panama. There were already primitive mammals in South America, as there were in Australia, and they not only managed to survive whatever worldwide changes of climate may have contributed to the extinction of reptiles other than crocodiles, tortoises, lizards, and snakes, but multiplied, evolved into many new types and began to fill all the natural niches vacated by the reptiles. Among the original little mammals were apparently primitive marsupials as well as others. There then appear to have been two quite separate times when the

South and North American land masses became joined by a land bridge. On the first occasion many strange creatures including some Edentates that had evolved in South America passed over this bridge to the north, while a few from the north went south. The continents were then separated again and evolution proceeded in isolation in the southern continent, both the original inhabitants and the new arrivals giving rise therein to all manner of strange forms. The last time the continents were joined, a whole host of northern animals flooded into South America overrunning and quickly extinguishing most of the Edentata, and notably all the so-called Ground-Sloths. Only the arboreal tree-sloths, the digging armadillos, and two arboreal and one ground-living anteater being left.

The fact that marsupials, ostrich-like birds, and even certain parasites of the latter, are found in both Australia and South America does not, as was once believed, indicate that the two continents were connected via Antarctica. Rather, it would seem, this only demonstrates that both were isolated at about the same early period from the land masses of the northern hemisphere.

ANTEATERS (*Myrmecophagidae*)

Although closely related, the three living anteaters are of very different appearance and size. Though called anteaters, their food consists mostly of Termites, and they will eat other insects. They have no teeth.

The Giant Anteater (*Myrmecophaga*)

This grey-black animal with icicle-shaped head (Plate 55), and enormous bushy tail is comparatively common throughout most of the forested part of tropical America. It grows to an overall length of more than six feet and, despite its idiotic appearance, minute brain, and apparently lumbering gait, it is one of the most competent and least often molested of all mammals. Even jaguars give the large males of this species a wide berth, for these animals are amazingly quick and deadly infighters. Their forearms are shaped like those of a man, enormously muscled, and armed with stout short fingers bearing gigantic claws. With these they can rip open the concrete-hard nests of termites, and of course any animal. Moreover, they have a habit of waltzing about on their hind legs with unexpected agility and can then strike with their hands in almost every direction. To feed, they stick their pointed muzzles into ripped-up logs, termite or ant nests, and then flick up the insects with a foot-long, wormlike tongue covered with gummy saliva. Along with the food, a great deal of earth and dirt goes into the stomach and is used for digestion. Although it has often been denied, the great, bushy tail apparently *is* used by the animal to sweep insects together, as well as to brush them from its own body. It is also employed as a sort of combined sunshade and umbrella when the animal is resting or sleeping. The young ride on their mother's back for the first few months of their lives.

Lesser Anteater (*Tamandua*)

The Lesser Anteater is quite a different animal, having a shorter muzzle, large ears, and an almost naked

Tamandua

tail which like that of an Opossum is fully prehensile. It is covered in a very close, dense, hard fur, and comes in an astonishing variety of colors and color-patterns. In some areas it is pale yellow all over, in others it has a jet black, diamond-shaped saddle covering its forebody. Although fairly agile on the ground, it spends most of its time in trees feeding on certain termites and ants that build their nests on boles, branches, or creepers. Tamanduas are incredibly strong for their size, and, while hanging from a swaying creeper by their hind feet and tail alone, they can demolish nests that would take a strong man using a sharp, heavy machete considerable time and energy to open. They will defend themselves against all comers by standing on the tripod formed by hind legs and tail, with arms outstretched and claws open. They slash forward with lightning speed with both arms at once, and their large middle claws can rip a man's arm open. They carry their young on their backs but sometimes leave them clinging to a branch all day while they go foraging.

The Pigmy Anteater (Cyclopes)

This little animal—only one foot long—is seldom seen, though found throughout the equatorial forests of South America and in Trinidad. It has a short, curved snout, very small ears, and is clothed in soft, fluffy, reddish-brown fur. Its tail is also furry but strongly prehensile, and either parent carries the single young one on its back till it is almost fully grown. They have been observed feeding their young with regurgitated insects chewed into a sort of mush. They make a pathetic little piping whistle and have a most woebegone appearance, but they can give nasty wounds by pinch-

ing between their largest finger-claw and the fleshy heel-pad on their hands.

SLOTHS (Bradypodidae)

There are only two known types of living tree-sloths, though each has been divided into a number of species. Aggravatingly and quite erroneously, they have been called the two-toed and three-toed, when both have five *toes*. However, one, the Unau, has only two *fingers,* and the other, the Ai, has three *fingers*. Both are wholly arboreal and spend almost all their time hanging upside down in trees, even sleeping, mating, and giving birth to their young in that position. Most surprisingly, however, both can swim very well with a sort of breast stroke, and the body right side up.

The Two-fingered Sloth (Choloepus)

This is a beast with a body about the size of a very large cat, but covered with coarse, shaggy fur that may be over six inches long on the back. It is tailless, and has very long arms with slender wrists and only two immense, permanently recurved claws to each hand. The fingers hardly have to be flexed at all in order for the beast to hang himself up. The mouth contains one of the nastiest sets of teeth found in a mammal, including four sharp, interlocking canines that are triangular in section and have three cutting edges. Unaus appear to be sluggish creatures, wandering about the forest trees, reaching out with their hooked hands to draw leaves and fruits to their mouths, and then munching on them solemnly and methodically. However, they can strike with their claws in an over-arm slashing swing almost faster than the human eye can follow, and can thus inflict ghastly wounds; they can also give a terrible crunching bite which may just go on until the biter itself is killed. Apart from Opossums, Unaus are probably the toughest of all mammals, and can survive a beating or wound that would be fatal to any other animal. Both they and the Ai display another almost unique character. The hairs of their fur are filled with tiny pits, and in these, minute green algal plants often grow parasitically, and may make the whole animal bright green. This renders sloths almost invisible among the mosses and lichens that festoon the branches among which they live. The young are carried first on the mother's chest, and later hanging from her neck.

The Three-fingered Sloth (*Bradypus*)

This is also known as the Ai and is on the whole a smaller animal, clothed in a dense mat of hard, slightly curved hairs. It is also tailless. It is rather mild-looking as opposed to the Unau with its glaring, glassy, and often bright-red eyes. The general color is silvery grey with a white, cream, or even bright yellow, well-defined "face" covered with shorter hairs, but some are sandy brown all over. Usually there is, between the shoulder blades, a most extraordinary marking that is unique among mammals. In its most exaggerated form this consists of two kidney-shaped black spots surrounded by white in a bright yellow sunburst. It somewhat re-

Two-fingered
Sloth
and young

Three-fingered Sloth

The young, usually only one at a birth, are minute—about the size of a man's thumb—and appear to be all arms and legs like spiders. They meander about their mother's body for many weeks, but are weaned to a leaf diet at a very early age. Ais are much more docile than Unaus, and although they will defend themselves by slashing with their arms, their motions are slower and never very vicious, and they seldom bite even if provoked. They move about in companies, systematically stripping one tree after another and eating flowers as well as leaves. These sloths are often infested with swarms of small, cockroach-like moths which scuttle in and out of their dense hair—just about the most surprising parasites found on any mammal. Ais are to be found moving about during the day as well as at night and they appear to possess color vision, since they will not eat artificially-colored Cecropia flowers, nor any food in colored light.

ARMADILLOS (*Dasypodidae*)

Scientists recognize nine genera of armadillos comprising about a score of distinct species. They are spread all over South America from the Argentine pampas to Panama, and one species, as we have said, is found as far north as the southern United States. They are covered in a carapace of little checker-like bones each bear-

sembles the center of the flower-clusters of the Wild Pawpaw tree upon which these animals feed. Although these are "open" trees having comparatively few large leaves and usually grow on river banks or in open places, the Ais resting in them are often very hard to see.

Giant Armadillo

ing a horny plate, separated by soft skin from which sparse hair grows. The top of the head carries another small shield and the tail is either covered in a solid sheath of bony rings or is dotted with separate plates. The main carapace hangs down on either side of the body and is divided across the back into a varying number of bands or rings.

Peludos (Chaetophractus, etc.)

These are circled by six bands of armor and from between these a fairly profuse growth of long hair straggles over the body. Their heads are small and their tails short. The front feet bear five toes with immense claws. They dig burrows in loose soils, or excavate endless little sinks, like miniature shell-holes, in pursuit of subterranean insect food. They move about by day as well as night and seem to be perpetually on the hunt, running about like little dogs with their noses to the ground, snuffling. They eat almost everything, including rotting meat, small birds, snails, any snakes they can kill by scratching and gouging with the saw-edges of their shells, and they are adept and industrious mousers. Closely related to the Peludos are the Weasel-headed (*Euphractus*) and the Pigmy Armadillos (*Zaedyus*) which delightfully means "The very pleasant one." The former is distinguished by having a very pointed little head which is flat on top; the latter by its very small size. The Weasel-headed species is virtually without hairy covering, and is about sixteen inches long; the Pigmy is less than half that size, is more wholly nocturnal, and keeps to places where the ground is composed of soft sand. It is found along the coast, and browses among seaweed and other detritus washed high on the beaches. There is also an extreme form of the Peludo, covered in long, silky hair, which is known as the Fleecy Armadillo. It is found in northern Patagonia.

Giant Armadillo (Priodontes)

Largest and most extraordinary of all armadillos is the animal shown in the photograph on page 108. This lives in the equatorial Amazonian forest belt where it tends to stay by rivers. An adult male weighed in the author's presence topped 120 pounds, and the great claw on its third finger measured eight inches around its curve. It has five conspicuous bands around the body but both fore and aft carapaces are rather soft and pliable so that the animals can curl up into almost a complete circle when resting. Despite their rigid appearance these are very agile animals that spend most of their time balanced on their hind legs with their fore feet held just off the ground. They can walk on their hind legs alone and stand straight upright, and when digging they sit up on their hind legs and the heavy tail. A pair once dug out of a concrete pit overnight, starting from some minor cracks, lifting aside slabs a foot thick, four feet long and as much as three feet wide. These animals have almost a hundred teeth, which is more than twice the normal allotment for mammals.

The Cabassou (Cabassous)

This is the second largest of the armadillos and is also an inhabitant of the wet forests. It is also known as the Eleven-banded but usually has twelve or thirteen separated belts of scales around its middle. It has large rounded ears but a very small head, and a large number of teeth, but never as many as the Giant Armadillo. The third toe is enormous but the second and fourth are long and slender, and the outer two minute. The tail is covered with firm, naked skin dotted with isolated scales especially along the upper side. Cabassous are great diggers and make enormous warrens in the forest floor that may descend as much as fifty feet and come out at water level at the bottom of river cliffs.

Pebas (Tolypeutes)

These little animals, of which there are three species, come from the Argentine pampas country where they are also known as Aparas. They all have three movable bands. Their general appearance may be seen in the photograph at lower left of the title page and their method of defense in the accompanying illustration.

Nine-banded Armadillo

They do not live in holes, though they are adept at digging, a procedure that occupies most of their time in pursuit of insect food. They do, however, eat a certain amount of vegetable material. They are balanced like the last two types, so that only the tips of their front claws reach the ground when running. When rolled into

Three-banded Armadillo in defensive attitude

SUSCHITZKY

a ball, they are more or less impervious to attack by any animal except a large jaguar and the Maned Wolf, which have a wide enough gape to crack them in one bite.

Tatus (*Dasypus*)

These include the Common Nine-banded Armadillo of Texas and Central America, and several similar if not identical animals that are found all over South America both in the wet forests, on open savannahs, and even in stony areas upon mountains to a considerable height. One species found in Peru (Plate 54) has a thick coat of long fur overlaying the normal, horn-covered, bony shields, except for the head and tail. These armadillos are not balanced on the hind legs but on all fours. Although they appear defenseless as they trundle about the surface of the earth in search of insect food, they are, considering their size, very fast when finally alarmed. They can also dig with prodigious speed. Although their ears are large, their hearing is not acute. They always bear four young and all of the same sex because they are really two sets of identical twins *both* derived from a single fertilized egg.

Pichiciegos (*Chlamyphorus*)

The Fairy Armadillos which are found in the drier

parts of western Argentina, and in Bolivia, are in almost all respects unique. The smaller is only about six inches long, is bright pink in color, with long plumes of silver-white hair completely clothing the face and undersides and sprouting out between the back and the rump shields. The muzzle is horny, with the nostrils underneath; the eyes are minute. The back shield is attached only to two bony bumps on the top of the skull and by a narrow strap of flesh along the mid-back; otherwise it hangs loose over the flanks. The backside of the animal is cut off abruptly like the back of a bus and is covered with a solid, fan-shaped shield of plates. This has a notch at the bottom through which protrudes a short rigid tail shaped like a spoon. This can be lowered but not raised, and it trails behind the animal. The Pichiciago can raise itself behind on this tail so that when it digs, the loose earth brought under the body by the front feet can be continuously shot out by the hind feet that churn away with a violent treadmill action causing a continuous stream to jet out behind. The tail can be closed under the body and the flat backside used as a plug to the animal's burrow at times of extreme emergency. They feed on insects for which they burrow continuously, and they stay below throughout the cold weather. The larger species (*Burmeisteria*) has the back shield attached all over.

Scaly Mammals

(*Pholidota*)

IN no way related to the Edentates, but in some respects somewhat similar to them, are the curious Old-World creatures known as Scaly Anteaters, or Pangolins. There are seven species, all of which are currently grouped into a single genus. They are elongated, low-chassied animals with small pointed heads, long tails that are the shape of a half-moon in section, and they are covered all over the back and sides, from the muzzle to the tail tip and beneath that organ, with shell-shaped, horny scales. These are actually composed of agglutinated hairs like the horn of the rhinoceros, and they overlap each other from front to back to about half their extent. They grow from the skin in regular bands, but in a most complicated arrangement. Between them is soft, white skin, and the underside, including the cheeks, throat, chest and belly, is also naked but sparsely sprinkled with hard hairs. Both hands and feet are armed with large curved claws.

The most extraordinary part of a pangolin is its tongue, which is wormlike, comes to a fine point, but is so long that it has to be housed in a sheath supported by pliable rods of cartilage, which reaches from the back of the mouth to the pelvic region at the hinder end of the lower belly. If you open up a dead pangolin for dissection, the first thing that pops out of the stomach cavity is this hoop of cartilage, for it is actually longer than the torso of the animal. Thus the tongue is about half as long as the combined head and body and it can be extruded to almost that length.

The method of feeding is, as with the Edentates and ant-eating members of other orders of mammals, to break open logs containing communal termites or ants, or the clay, woody, or fibrous nests of such animals whether underground, on the surface, or in trees, and to lick up the inmates by shooting the tongue into the galleries so exposed. Pangolins' tongues are laved in a very sticky saliva to which the insects adhere along with quantities of earth or other material, but unlike the Edentates and others, the pangolin manages to rid the tongue of most of the latter before it swallows its food. An old theory that the insects attack the tongue with their jaws and thus become attached thereto, appears to be partially correct, but the speed with which the pangolin flicks its tongue in and out makes it obvious that some other method of separating the food from the unwanted dirt is employed. This uncanny process may take place in the throat, which performs a rhythmic action of its own during feeding. If sawdust be mixed with chopped meat, raw eggs and milk, and fed to captive specimens, almost all of the indigestible sawdust will be left in the bowl after the meal while everything else has disappeared!

Pangolins live in the moist, equatorial zones of Africa, India, Ceylon, Burma, southern China, Siam, Indo-China, the Malay Peninsula, and the greater Indonesian Islands as far east as the Celebes. There are three fairly distinct types in Asia and four in Africa, but there is much dispute about this among systematic zoologists. There is one simple way of telling one lot from the other: the row of scales extending down the middle of the upper side of the tails of the Asiatic species is continuous right to the tip, but that on the African doubles up about two thirds of the way down. The species found in India and Ceylon, and another ranging from Nepal through Assam to China, have very long front claws and comparatively short tails. The third Asiatic form, which ranges from upper Burma to the Celebes, is much lighter in build, with a long tail and rather short claws. All these have a few hairs growing between the scales and little, ridgelike external ears. Those of Africa have no external ears, merely an ear-hole, and the scales sort of peter out on their hands and feet. One has a short tail, another, one of the longest tails found among mammals—twice the length of the head and body. A third species is distinguished, at least when young, by having the outer edge of all the scales trifurcate like a trident, but as these animals grow older the points wear off and the scales become lozenge-shaped. The fourth species is rather remarkable, mostly for its huge size. It may grow to more than six feet in length.

All pangolins are surprisingly alike in habits. They can dig shallow burrows for themselves, but usually inhabit the holes of other animals, after enlarging them somewhat. The author has found many coiled up therein with large pythons. They trundle about the forest floor by night, sniffing at insect nests but the long-tailed species, both in the East and in Africa, also climb

Young Long-tailed Pangolin

with agility, and are sometimes almost wholly arboreal, sleeping in hollow trees.

The tails of all of them are prehensile in that they can be curled downwards, and they are thus used to envelop and lock around the rolled up body for sleeping or defense. In fact, there is a small, naked pad under the tip of the tail that hooks over the farthest scale it can reach on the lower back when the animal is thus rolled up, and this makes the resulting ball almost impossible to open. The tails of the arboreal species are, however, fully prehensile in that they can also be turned around, corkscrew fashion, either way. At the slightest indication of danger, even a low hiss given several paces distant, these strange creatures throw themselves on their sides and snap together into a tight ball.

The Giant Pangolin, although in all manners performing like the rest of its kind, has a few additional

Giant Pangolin asleep

defensive tricks. First, it usually lives near water, bathes regularly, and will, if molested, plunge into large rivers if they are nearby, and disappear below, rather than roll itself up. They can stay below for a considerable time, and can either swim below the surface or walk along the bottom. This species also may rise on its hind legs and even attempt to defend itself much like a Giant Anteater by waving its immense fore-claws at its adversary. However, being singularly myopic and of rather low general sensitivity—though having acute hearing and smell—it seldom knows accurately where the attacker is. If it does get a grip on an adversary it never appears to know what to do, and after giving it a terrific hugging it lets it go. This treatment is, however, usually enough to kill anything smaller than a large dog, and the great fore claws have been known to remove most of the muscles from a man's leg, though apparently more by accident than design.

We have on several occasions had cause to mention the strength of certain mammals—strength that is remarkable either because of the large or small size of the animal concerned. The Giant Pangolin is neither a monstrosity nor a small animal but, while all pangolins are very powerful for their size, the strength of this species is truly prodigious. An animal collector showed the author a photograph of a full-grown male attached to a rope tied around its middle hauling ten stalwart natives across an open grassy plot towards a river and explained that the animal finally moved them all to the water's edge by maintaining a continuous steady pull. The identical animal, when weak from a five weeks trip to America and when it had coiled itself up, could not be opened by as many men as could get their hands on it at one time.

There is an old tale that pangolins obtain their food by rolling in an ants' nest until the insects swarm under their scales and start biting the soft skin, whereupon the animals enter the water and float the insects off. There is no evidence to support such an idea, and many pangolins live in areas where there is not even water to drink. However, it is now known that foxes and other animals do slowly submerge themselves backwards so that all their fleas are finally concentrated on the tip of their nose and are then either floated off or made to retreat onto a tuft of grass held in the mouth, which is finally dropped. Some collectors assert the Giant Pangolin does likewise when too many insects get under its scales.

Leaping Mammals

(*Lagomorpha*)

BECAUSE the Hares and Rabbits have two very prominent front teeth in both the upper and lower jaws, and because their heads, if divested of the large ears, are rather like rats', these animals were until recently regarded as a group of Rodents or Gnawing Mammals. However, the fossilized remains of both rabbits and true Rodents have been found together in rocks that were formed sixty million years ago or in the earliest days of the age of mammals, and although both were more primitive than their present-day representatives, they were just as distinct then as they are now. Thus, if they ever had a common ancestor it must have been an exceedingly ancient type that lived in the Age of Reptiles. Further, the Lagomorphs, or Hare-shaped-Ones, are anatomically quite different from the Rodents, apart from a superficial resemblance between their teeth. They are today almost universal in range and have been introduced into Australia and New Zealand, where they have multiplied out of all control, on the one hand causing devastation to crops, but on the other providing the basis for the felt industry with the fur plucked from their pelts. The order is clearly divided into two very unequal groups.

PIKAS (*Ochotonidae*)

In mountainous districts throughout Eurasia and North America—but not in all such localities by any means—colonies of fluffy little egg-shaped animals (Plate 56) with small, rounded ears may be encountered among the loose rock of screes. These are variously know as Pikas, Conies, or Whistling-Hares. These animals are communal within the colonies and keep up a ceaseless, high-pitched to supersonic whistling among each other and between colonies night and day, a noise that seems to get right inside your head. They also have the curious habit of collecting masses of grass and other vegetable matter and making sort of hayricks of it outside the holes between the rocks in which they dwell. This is their food and they spend half their lives caring for these stores, carrying it all below if rain threatens and then bringing it all out to dry again as soon as the bad weather has passed. They bear four or five young at a time but are not nearly so prolific as the Leporids, or hares and rabbits.

LEPORIDS (*Leporidae*)

Of all mammals these are, as a whole, probably the most easily identified by even the least interested non-specialist, yet they are all so very much alike in appearance and habits that even the greatest expert can not put the proper name on an individual without knowing where it came from. Literally hundreds of species and varieties have been described from all over Asia, Europe, parts of Africa, and North and South America, but apart from about half a dozen types, they all fall into one or other of two great genera. Moreover, the reasons for separating even these is so obscure that they are not worth describing in a book of this intent.

In brief, they include the Rock-Hares of Asia (*Pentalagus* and *Pronolagus*), the Bristly Rabbit of India (*Caprolagus*), which has small ears and tiny eyes and which digs true burrows, the Short-eared Rabbit of Sumatra (*Nesolagus*), and the Pigmy Rabbit (*Brachylagus*), which lives under dense, semi-desert scrub in certain areas in the western parts of North America.

The only well-known type that warrants a special name is the Common Rabbit of Europe and North Africa (*Oryctolagus*) which (Plate 57), although almost indistinguishable from some cottontails, has the almost unique distinction among Leporids of being a true burrowing animal. It is communal and its vast burrows interlock, forming what are called warrens. Although excellent eating, it has always been regarded as somewhat of a pest in Europe, but it now appears about to be brought under control, if not actually doomed, by the discovery of a deadly disease that can be artificially introduced into its communities. All the rest of the vast hosts of Leporids are divided between the genera *Lepus* and *Sylvilagus,* which simply means Hares and Wood-Hares. There is no true distinction between hares and rabbits; these are simply two names that can and have been used to denote various types of both genera, and often quite indiscriminately. On the whole, the name hare is preferred for the larger, long-legged, large-eared types that leap (the genus *Lepus*), while rabbit is reserved for the smaller, short-legged species that run (*Sylvilagus*). The popular American name Jack-Rabbit for various hares greatly muddles the issue, but the title Cottontails for all the Sylvilagids is most useful and appropriate.

The hares vary considerably in size, color and general appearance, so that anyone can spot quite a number of them. First, in the extreme north beyond the pine forests and thus out on the open tundras and even among the perpetual glacier-fields right up to the northern tip of Greenland, are to be found very large hares with huge hind feet (Plate 58); these are pure white except for black ear-tips. These animals are very numerous and form the principal food of the hosts of carnivorous mammals and birds that dwell in those regions. From time to time and on a fairly regular cycle these animals greatly increase in numbers until extreme swarm conditions pertain throughout enormous areas; then they suddenly die away in one season and become everywhere rare. It has now been shown that the numbers of those animals which feed upon these hares has also to follow this rhythmical variation throughout the years. In Eurasia there are numerous kinds of hares that are hardly distinguishable from the Jack-Rabbits of north and eastern North America. It is in the latter continent, moreover, that Leporids reach their zenith of diversity.

Here, there are the so-called Snowshoe-Rabbits, or Varying Hares (Plate 59), which in many areas turn white in winter and which have absurdly large hind feet clothed in winter in long, fluffy fur. Then there are the Jack-Rabbits or Antelope-Hares of the western prairies and deserts. These are the most exaggerated of all Leporids, having small heads, positively immense ears, and very long, narrow bodies. Their forelegs are slender and long, but their hind legs are monstrous. They are prodigious jumpers, covering over twenty feet at a leap in some cases, and their sustained speed is such that nothing short of a Cheetah can catch them.

Apart from the European or True Rabbit mentioned above, all the rabbit-like animals of the Old World are really small hares. In the New World, however, the Cot-

TAYLOR FROM FISH AND WILDLIFE SERVICE

Antelope Jack Rabbit

tontails take their place. These are compact little "bunnies" that live in meadows and other natural or artifical open places in wooded areas, and in swamps. They seek cover under dense vegetation and in the holes of other animals, and some, like the Marsh and Swamp Rabbits of the southern United States, and sundry species in Central and South America, nest in mounds of reeds and are semi-aquatic, diving into water to escape danger, and in some cases swimming about regularly in search of food. Neither hares nor rabbits live in woods or forests, though they may be found on open areas within them many miles from the nearest open country. In parts of South America, they may be found on every open savannah, but they are never seen so much as a few feet within the surrounding forests.

A very great deal is known about these animals, both from the age-old domestication of the European Hare which has resulted in the production of an incredible

NEW YORK ZOOLOGICAL SOCIETY

Cottontail

variety of artificial forms of all manner of colors and color combinations, from more recent studies on the North American hares and cottontails, and finally from much research on the little *Oryctolagus* both in Europe and Australia. Although these animals would at first appear to be singularly uninteresting, doing nothing besides eat and breed, they actually have many strange habits. One of the most extraordinary, to our way of thinking, is their method of digestion. This is not unique to Leporids and is now known to occur in many Rodents. When fresh green food, as opposed to desiccated winter forage, is available, the animals gobble it up voraciously and then excrete it around their home lairs in a semi-digested form. After some time this is then re-eaten, and the process may be repeated more than once. In the Common Rabbit, it appears that only the fully grown adults indulge this practice. The reproductive rate of "rabbits" is notorious, and it may indeed be phenomenal. While that of the Arctic species is rhythmical, that of temperate and tropical types appears to be sustained and dependent upon the available food supply. In Australia and New Zealand it ran riot.

Gnawing Mammals

(*Rodentia*)

BEFORE entering upon any description of the Rodents it is necessary to take a deep breath, metaphorically speaking, for this is the largest order, containing over a third of all the genera and over half the total species of living mammals. This gives us a minimum of 5,000 distinct animals that must be split up into over three hundred genera to be dealt with. Moreover, the vast majority of these are very small animals, less in size than the ordinary Brown Rat, and whole slews of these look almost exactly alike, regardless of whether they be truly related or not. In bulk of actual individual animals alive at any one time, they so far exceed all other mammals put together that any attempted computation of their numbers becomes worthless. Many of them swarm either periodically or from time to time, and on one occasion in western North America as many as 12,000 specimens of one species were estimated *per acre* over an enormous area. One permanent "city" of Prairie-Dogs once covered an area of 25,000 square miles in the same part of the world, and was estimated to have been inhabited by over four hundred million individuals.

Thus, it would at first seem to be a hopeless task to try to arrange all this host of small creatures in any semblance of order, or to describe them adequately without resorting to a plethora of technicalities. Nonetheless, the whole can be reduced to a fairly simple pattern. The basis of this is to make a primary division of all Rodents into three great groups, called the Squirrel-like (Sciuromorph), the Mouse-like (Myomorph) and the Porcupine-like (Hystricomorph). Then, it is merely a matter of taking one well-defined family after another and, doing this, it will be found that there are seven families in the first, ten in the second, and sixteen in the third group. However, one or two families constitute real conundrums, since their true position is not yet finally settled.

Despite their great variety of form and habit, the Rodents constitute the most clear-cut and readily definable order of mammals, both as they exist today and as found in a fossilized condition. Since the dawn of the age of mammals they have been quite distinct, and there is nothing known, dead or alive, that in any way links them to any other mammals living or extinct. Only in the fact that they are neither Monotremes nor Marsupials do they agree with other mammals. The most primitive living rodent appears to belong in the Sciuromorph group, but the Porcupine-like ones undoubtedly branched off from the main original stock, whatever that may have been, at an earlier period, and thus represent the more direct descendants of the primitive type.

Rodents are today practically universal in range. There are even quite a large number of truly indigenous Australian types, the only mammals apart from the Marsupials, Monotremes, and a few Bats that ever reached that continent. They are found to the ultimate of the northern tundras right to the beaches of the Arctic Ocean. They reach higher up mountains than any other mammals; they dwell in trees and some glide in the air; they live in all types of forest, woodland, scrub, swamp, farmland, and grassland; they inhabit deserts even where there is no visible vegetation; and they burrow under the ground everywhere. They have taken to living in the habitations of men and under his modern cities. In some instances they can survive on less oxygen and at lower and higher temperatures than any other mammal but man. The little, naked, wrinkled-skinned Sand-Puppies of the deserts of East Africa burrow blithely along just under the surface of loose sand that is so hot it will blister human skin and would certainly fry the proverbial egg. The Arctic Vole (*Dicrostonyx*) burrows in or runs about on the snow in sub-zero temperatures for nine months of the year. Finally, there are hosts of different rodents that have taken to the water.

The largest Rodent is a Hystricomorph called the Capybara which inhabits streams and river banks in South America, grows to a length of three feet, and may weigh as much as 250 pounds. The smallest is a kind of mouse that lives in tall grass and is only about twice the size of the smallest shrew.

Rodents are by far the most economically important of all mammals, not because we depend on them—though without them we would be swamped by the insects in a matter of weeks—but because we must forever fight them for survival. Just after the second World War, rodents in the United States were found to be destroying more of the crops in the fields annually than the total allotted to rehabilitate all other countries. Some further hundreds of millions of dollars worth of food was then consumed by them after the harvest had been gathered and stored. Their destruction of other property, from lead-covered power-lines to priceless art treasures, is inestimable. They are, in fact, the most versatile, sturdy, and efficient of all mammals, and some display terrifying intelligence and powers of learning.

SCIUROMORPHS

This, the first great assemblage of the rodents, contains many surprises for the non-specialist. Outstanding among these are the Beavers, the Pocket-Gophers, and a group of little mouselike creatures. The reasons for associating these and sundry less-known animals with the squirrels is mostly anatomical and need not detain us. There are seven families thus assembled, one of which far transcends all the others in diversity of species and distribution.

SEWELLELS (*Aplodontidae*)

Perhaps the most primitive living rodent is a small, heavy-bodied, harsh-furred animal with tiny eyes and small ears, and rather resembling a tailless muskrat, that is found only in certain limited mountainous areas on the Pacific coast of North America between Puget Sound and central California, and along a narrow belt of territory running down the Cascades and the Sierra Nevada to the east. They are called aplodonts from the Greek *aplöos* meaning simple-toothed, since their cheek teeth have no roots at all. They are about a foot long and live in communal burrows, usually near water, and they swim very well. They eat succulent herbage.

SQUIRRELS (*Sciuridae*)

As opposed to the Sewellels, the Squirrels are perhaps the most advanced and successful of all Rodents, although they are still somewhat generalized. They are found all over Eurasia, Africa, and the Americas, and they come in hundreds of distinct forms that are divided into about forty genera. In order to reduce this to manageable proportions, however, it is feasible to describe them under seven heads, all of which have popular names and all of which are quite widely appreciated as constituting distinct assemblages of animals.

Typical Tree-Squirrels (*Sciurus, etc.*)

The general form of the typical tree-squirrel is sufficiently well shown in Plate 61 and does not need any detailed description. There are well over two hundred described forms found wherever there are woods or forests all over Asia, Europe, Africa, and North and South America. These have at one time or another been broken down into a score of genera, but apart from two obscure forms in central Asia, one in Borneo and one in South America, which display demonstrable anatomical differences, they are all as one, structurally. In size, color, and color pattern, however, they vary greatly. In habits they are surprisingly alike. They eat nuts, fruits and insects. They live in trees but descend to the ground fairly frequently, especially in temperate

Sewellel

climates or where the trees are stunted or stand apart, and they all make nests either in holes in trees or, in the manner of birds, out on the open branches. In colder climates some may even go into a state of semihibernation and in the drier parts of the tropics some of them aestivate. Most of them have several broods of a number of young during the year, but the colder the climate the more numerous the offspring and the more concentrated the breeding season. All are distinguished by having long and bushy tails, though the bushiness varies from the slightest to an extreme of several inches, giving the owner a regular sunshade—*sciurus* means "shade-tail." They are mostly sleek-furred and rather brightly colored, and many, like the common European Red Squirrel and certain species in the west of North America, bear pronounced plumes on their ears. They are agile and use their hands to handle food and raise it to their mouths. They descend trees head first.

The Chickaree (*Tamiasciurus*)

The little Red Squirrel of the North American pine or boreal forests (Plate 60) forms a quite distinct tribe, and is truly unlike the above. Not only is it smaller, but its general build is more compact, its teeth are different, and its habits are somewhat dissimilar. It eats a variety of foods but relishes the seeds of pines.

Palm-Squirrels (*Funambulus, etc.*)

It is virtually impossible for the non-specialist to identify or even define this tribe as opposed to the typical tree-squirrels on the one hand, or the following group on the other. They are much more varied in size and appearance, ranging from the great, gaudily-colored, plume-eared Ratufas of the Orient, and the Giant Booming Squirrels of Africa with their vividly black and white ringed tails, to a tiny, delicate creature, one of the smallest of the whole family—the African Dwarf Squirrel (*Myosciurus*). The Ratufas have contrasting upper and undersides, and parti-colored facial markings; some are wine-red above and bright orange below. Most of the African types are a rich reddish-brown, with ringed tails, but some are longitudinally striped on the body—the Bush Squirrels (*Paraxerus*)—and there are species that are vivid green above and yellow below. They are arboreal, and few except the Bush Squirrels spend much time on the ground. As their common name implies they make the fruits and nuts of palms a considerable part of their diet but eat insects and leaves.

Oriental Tree-Squirrels (*Callosciurus, etc.*)

As a tribe, these are even less easy of definition than the last, but they nevertheless form a distinct assemblage. They are all Asiatic and most come from the southeastern part of that continent. *Callo* or *kallo* means "the beautifully adorned," and this is just what many of these animals are. All manner of rich reds, golds, browns, and yellows enter their coloring, often set off with vivid black or white markings, and sometimes the individual hairs may have rings of as many as six different colors between their base and tip. There is one tiny species (*Nannosciurus*), the smallest of all squirrels—it is only about the size of a small shrew—that is greenish-brown in color, and lives in the dense matted heads of equatorial forest trees. The common brown squirrels that associate with the Tupaias in the Indonesian forests belong to this group. They do not appear to make nests, but dwell in hollow trees or sleep out on open branches. None is known to aestivate.

African Ground-Squirrels (*Xerus, etc.*)

With this tribe we come to the first true squirrels that are really very different in color, pelage, structure, and habits. There are several types in Africa, mostly distinguished by size alone, and one odd kind found in Central Asia that looks similar, but the relationship of which is uncertain (*Spermophilopsis*). They are all long-bodied and rather short-legged animals with very short, rather sparse, but hard, coarse, straight fur composed of hairs that are individually ringed with black and some light shade from white to yellow, gold, or reddish. The undersides and often the greater part of the limbs and the throats, cheeks and behind the ears are usually pure white. Their tails are rather short in most species, but exceptionally long in at least one West African type, and are often banded with dark and lighter belts that tend to form a herringbone pattern. They live on the ground on the savannahs and in scrub and semi-desert areas, eating insects, roots, and grass seeds, and some of them dig. Others rest in the holes of other animals or under grass tussocks. Some appear never to drink at all and most of them aestivate, but erratically.

Northern Ground-Squirrels (*Marmota, etc.*)

The larger of these are better known to Americans as Groundhogs or Woodchucks, but the tribe actually includes also the Prairie-Dog, the vast host of true Ground-Squirrels or Gophers, and the Chipmunks. They are distributed all around the northern hemisphere from above the Arctic Circle to the southern limits of the temperate belt. The largest are the so-called Hoary Marmots of Alaska which, like the Groundhogs and the Marmots of Europe and Asia (Plate 66), have short tails, great heavy bodies and rather short limbs, and which live in holes that they themselves excavate in earth or among the rocks of mountain screes. They have thick, coarse fur, and all hibernate during the cold months of winter.

The little Prairie-Dog (*Cynomys*) (Plate 67) is a truly gregarious animal, resembling a very fat rat with a short tail, and living in enormous communities on the open prairies of western North America. They are sandy brown in color and spend much of their time sitting up on their haunches. They dig burrows with vertical entrance-shafts in the middle of a sort of small volcano of earth on which they sit. At the slightest sign of danger they stand up on their hind legs and give a high-pitched cheer and then duck down into their burrows. Their hole has a sort of ante-chamber near the top and in this the animals turn around and then emerge to peep over the rim. They maintain their families on the small plot of grass growing around their mound.

All over the treeless areas of northern Asia and North America and in the latter as far south as central Mexico, dwell countless hosts of little ground-living squirrels to which the names Susliks or Gophers (Plate 65) have been given on the two continents respectively. They come in more than a dozen basic forms, and although most of them are of a dull, greyish-brown color, some show the most vivid markings. The Mantled Ground Squirrel (Plate 62) of western North America may be of a combined olive brown, orange, reddish, black and white pattern, and the Thirteen-lined Ground Squirrel (Plate 63) of the deserts is quite unique, being light below but reddish above, with seven bright white longitudinal lines over the back and flanks with six lines of white spots between them. Most of these little animals take refuge in holes of other animals but some of them dig their own. They eat grass, seeds and roots, but all of them will take insects and many eat snails and other animals. The most engaging is the Flickertail (Plate 64).

The Chipmunks are, in a manner of speaking, only an extension of the Gophers that dwell in the forests and woodlands rather than out on the plains, and they are linked to them via such types as the Mantled Ground Squirrel. They are lighter-bodied, on an average much smaller and more active, and their tails are more bushy and squirrel-like. They vary enormously in color, but are mostly greenish-brown, and all have on the flanks more or less of a light stripe bordered by black, and similar marks enclosing the eyes. They are extremely busy little animals throughout the warmer months, raising families and collecting nuts and other food that they stash away in hidden hoards under tree stumps and elsewhere. Fortunately they forget where most of these hoards are and never make use of them in the lean months during which they hibernate fitfully. As a result, their activities do much to foster the re-afforestation of the land and to promote the distribution of nut-bearing bushes and other plants.

Flying-Squirrels (*Petaurista, etc.*)

There are a dozen different sub-groups of squirrel-like animals that have furry flaps of skin stretched between their fore and hind limbs like parachutes and tails with long, plumelike hairs arranged along either side to make horizontally flattened rudders. They are found all over the forested areas of North America, Europe south of Scandinavia and Britain, and throughout Asia south to the Himalayas, India, Ceylon and the Indo-Chinese area, and the Indonesian islands. Most of them are Asiatic, but one (*Sciuropterus*) is found in Europe and another (*Glaucomys*) in North America.

Apart from the large Taguans (*Petaurista*) of the Indian region they are all much alike, having very soft, thick, silky to woolly fur, short, rather rounded heads, and large eyes. They are nocturnal, and get about among the treetops by taking prodigious leaps. These usually start almost straight down, with the limbs held spread-eagle fashion and the tail held straight out behind. Then the tail is brought up sharply and the animal becomes horizontal. At the last moment the arms are raised and the tail is again raised and the little creature swoops almost upright to land on the branch or tree-trunk. In such leaps these animals can even turn to either side to a limited extent by using their limbs and/or tail. They are nut- and insect-eaters and live in holes in trees. The great Indian Taguans are of similar habit but are at least partly diurnal and while they also sleep rolled up in a ball with their tail curled over their heads, they may often be observed basking in the sun on their backs with their parachutes spread wide. There is one type in the higher Himalayas that has woolly fur and tolerates snow.

BEAVERS (*Castoridae*)

An animal that has intrigued men since prehistoric times and has greatly influenced his life in several respects is the Beaver (Plate 68). Once found all over Eurasia and North America from the retreating front of the glacial ice-cap to the southern limits of the temperate woodlands, it is now almost extinct in Europe, much restricted in range in Asia, and greatly reduced in numbers in North America. There was even a time when it was considered threatened with extinction in the New World, but new ideas of conservation and some public enlightenment have saved it. It has now been rehabilitated throughout much of its original range and even in localities that have been substantially taken over by man and his works.

The appearance of the beaver is well known and its behavior and habits are just as widely known but are sorely misunderstood. At first sight the Beaver and his works appear to indicate the exercise of an intelligence akin or identical to our own, and this is particularly intriguing to men, since the results are technological and we are basically technologists. On closer investigation, however, the activities of these animals lose much of their practical wonder, but in doing so they become even more mysterious. Although, like men, they work together in family communities with full co-operation and considerable discipline under the guidance of experienced individuals, and thereby accomplish engineering works of prodigious extent and great accuracy, they appear to do the whole thing entirely mechanically. Captive beavers will work ceaselessly at these same ac-

DERMID

American Flying Squirrel

tivities when there is no need for them to do so and even when they can not achieve anything by so doing.

The whole business of cutting timber, hauling stones, building dams across running streams, digging and transporting special mud to plaster them, excavating long canals to bring additional water, erecting vast houses with complex entrances, internal flooring and other devices, and spending months of labor cutting special sticks and anchoring them in the mud at the bottom of the artifical ponds, is directed at but a single objective—namely to provide a safe winter home with ample food supply under the winter ice in which to raise young.

How the Beavers do this brings up the everlasting question of the difference between instinct and reason, and it is here that we come upon the real mystery. Despite their purely automatic and apparently mechanistic

[117]

activties, and their lack of practical forethought, beavers appear to draw upon sources of information that are beyond our ken. How else can they so accurately judge the height of floods that are not to come for several months; how do they know so exactly how far water is going to be backed up by a dam and just what stresses and pressures it will exert? Can such knowledge be the result of cumulative racial experience, and if so how is it transmitted from generation to generation? Is it, too, entirely mechanical, or does it indicate a form of intelligence other than our own? Whatever be the cause, the effects have been far-reaching throughout the whole area where beavers have dwelt and labored since the retreat of the ice-cap.

By damming streams beavers create ponds, but since these rapidly silt up, their work is unending, and throughout the millennia, millions of acres of pasture land have thus been created where only sterile rocky river courses would have otherwise been. Further, by raising the water table all around their ponds the plant growth of much larger areas is completely altered; the conifers are pushed back and broad-leafed trees allowed to take hold. Thus, enormous areas of the best soil and pasture in the homelands of the white man and in those countries which he has colonized—northern Asia and North America—would never have existed had it not been for the Beaver.

Two to eight kittens are born in the early spring in the lodge which stands in water but has a dry platform above water level within. The parents care for the young for a year. Their food is the bark of certain trees and bushes, and great stores of these are laid in for the winter by pulling them down to the pond bottom and then anchoring them in the mud.

POCKET-GOPHERS (*Geomyidae*)

There is another family of Sciuromorph rodents known as the Pocket-Gophers that has throughout a somewhat smaller area done almost as much to alter the landscape as have the Beavers. These are found over the western part of North America from British Columbia, Alberta, and Saskatchewan in the north, to Honduras in Central America in the south, and from the Pacific in the west to the western foothills of the Appalachians in the east, and down the gulf coast of Alabama to Florida. Although the numerous species have been divided into about a dozen different genera, they are all surprisingly alike in structure but for their pelts, which tend to be thick and fluffy in the colder and damper areas, and very sparse, short and hard in the dry southern areas.

They are fossorial animals—i.e., professional diggers—that seldom come to the surface, and they are closely adapted to their life of tunnelling which results in a general appearance that is quite atrocious to our sight. The forelimbs are immensely sturdy and the spatu-

Pocket-Gopher

E. P. WALKER

late hands are armed with huge claws. The short tail is enclosed in a skin half a size too big, so that the vertebral column may be slipped up and down within it. The mouth is hard to explain, since the two upper front teeth have moved out along the underside of the projecting upper jaw so that the skin of the upper lips joins behind them. The real mouth is a small round hole in a furred membrane just above the lower front teeth. To either side are slitlike openings to very large, fur-lined pockets that extend backwards past the cheeks and on into the sides of the neck.

Pocket-Gophers live on roots to obtain which they burrow along just below the surface, using their upper front teeth as picks, and their forefeet as combined shovels and grabs. The loosened soil is then moved back under the body by the hind feet, and when too much has accumulated behind, the animal turns around in its burrow, places its hands side by side with the palms facing forwards against the pile, and then pushes it back to the nearest point where a vent has been opened to the surface; out of this it is then erupted. The animal can then, and often does, rush back to work backwards using its sensitive tail as a feeler. Larger living and nesting chambers are excavated at greater depths and around them vast stores of food are accumulated. To fill these, the animals attack roots with their teeth, excavating large hunks. These they then chop up into smaller pieces by twiddling them around against their juddering teeth as we do wood or metal in a lathe. When a piece is thus cut it is sideswiped into one or other pocket with a flick of the wrists. When the pockets are full the animal buzzes off to its store and tips out the load by squeezing the pockets from back to front.

Their activities have been conducted on a mass scale for tens of thousands of years and this has resulted in a gigantic ploughing up of the whole land surface, plus the continued cropping and transplanting of herbs and other vegetation. Had there never been any Pocket-Gophers half of North and considerable parts of Central America would today have an altogether different vegetation, appearance, and possibly climate.

POCKET-MICE (Heteromyidae)

Another family of North and Central American rodents appears to be of the Sciuromorph group though all its members are of small size and either mouse-, rat-, or Jerboa-like (see below) in appearance. They have fur-lined cheek pouches like the Pocket-Gophers, and they show some other characters in common with those animals. They are divided into three sub-groups but in external appearance they form a more or less continuous series from small, soft-furred mice to larger long-legged hopping rats covered in flattened spines.

Pocket-Mice (Perognathus, etc.)

These are very delicate little mice about two to four inches long, with sharp-pointed heads and smallish ears. They are found all down the western side of North America from British Columbia and the Dakotas to Central America. There are more than two dozen recognized species, all of which vary greatly. The fur may be soft and silky, coarse and hard, or even intermingled with small spines. The tail may be naked, lightly furred, crested, or bear a terminal tuft. They are light-colored, and live in dry areas, being nocturnal, and plugging up their holes by day. They are seed-eaters but take insects, and they appear never to drink even if water is available.

Kangaroo-Rats (Dipodomys)

In the drier parts of Mexico and thence north to California, Idaho, Wyoming and Oklahoma, there dwell vast hosts of small rat-shaped animals with large heads and huge eyes, tiny front legs and immense, stilt-like hind limbs (Plate 69). Their fur is silky and long, so that the arms may be completely concealed underneath it when the animal is in motion. The hind feet have a brush of long, stiff hairs all over the soles, and the tail is often longer than the head and body and has a big terminal tuft. This is used as a rudder to make sudden turns when the animal is going flat out by a series of terrific leaps. In color they are all greyish-brown above and pure white below, with black facial markings and a white line around the rump. Some grow to an over-all length of eighteen inches. They are nocturnal and live in holes by day.

Spiny-Rats (Heteromys, etc.)

In Central and Northern South America, one of the commonest rodents on the deserts, farm and grasslands, and even in the woodlands and damper forests, is a form of sleek little rats with pointed heads, long tails and hind legs, and short forelimbs. They have very clean, white undersides, and are otherwise usually some shade of orange or reddish brown. The Maya Indians call them "Poot-em-poot" and they do a great deal of damage to corn crops. Their pelage is composed entirely of flattened and soft but sharp-pointed spines and spinelike bristles. Their habits vary greatly, the same species digging burrows in one area, nesting in dry grass tussocks in another, and using hollow logs or holes in tree stumps in another. They are very agile leapers and some of them can climb bushes and trees to a considerable height, using their teeth and forepaws to heave themselves up on to branches, and then hopping and balancing thereon on their tall hind legs with dizzying agility. One small kind that has fur as well as spines spreads north into Texas.

SCALE-TAILS (Anomaluridae)

The old name for these animals—the African Flying Squirrels—is really very misleading because, of the three kinds of animals included in the family, none can actually fly, one cannot even glide, and another looks more like a mouse than a squirrel. They are really very common and numerous animals throughout the forested territory of Africa. There is considerable doubt as to whether they should be included in the Sciuromorphs or be placed in a separate group of their own. They are, however, most squirrel-like in habits and to a certain extent in both external appearance and internal anatomy. Together, the three types take the place in Africa of the Flying-Squirrels of other continents. The Zenker-

ella provides one of zoology's greatest conundrums for several reasons. Also, it has only once been caught.

Gliding Scale-Tails (*Anomalurus*)

No photograph, unless taken at extremely close range and from below when the animal was in mid-air on one of its long glides, would contribute anything to a proper understanding of the structure of these strange creatures. The head is squirrel-like, but the body is elongated, very narrow and shaped rather like that of a fish. The limbs are long and slender and can never be wholly straightened out. The hands bear four and the feet five sharp claws held close together, and have complex pads on palms and soles. From the hinder point of the elbows a long, thin, pliable, cartilaginous rod, as long as the forearm and at an angle of forty degrees to it, sticks straight out in line with the upper arm. From the sides of the neck to the wrists, thence to the tip of this rod, then to the first toe, and finally from the fifth toe to a point about a quarter of the way down the tail is stretched a double skin-membrane, clothed above in the same very soft fur as the back, and below sprinkled with a fine down colored like that of the belly. The tail bears a bulbous terminal brush of firm hairs, and on the underside at its base there is a double line of tough, triangular scales with sharp points directed backwards and arranged *en echelon*.

They come in a bewildering variety of colors but all appear to be only regional phases, so that one type blends into another. In many areas, however, there may be two or three distinct species, one perhaps being brindled grey above and red below, another green above and yellow below, and a third reddish-brown above and white below. Only two are very distinctly different; a very small type from the eastern Congo, and a giant species from the Gold Coast that measures almost two feet and has an eighteen-inch tail. This last is jet black above and pearl gray below, with vivid white edges to the parachute membranes.

These animals are, of course, wholly arboreal and move about only by night. They are common in the tall deciduous and rain forests, but species are found in the isolated copses of savannahs and in the gallery forests bordering streams in scrub lands. They range all over Africa from Gambia to Kenya and south to Nyasaland.

Anomalures climb tree-trunks like giant caterpillars; first, hooking on with the claws of both back feet, they jam the points of the scales at the base of their tail into the bark to form a firm tripod, then they let go with their hands, and reach as far up as possible. As soon as they have a good grip, they let go at the back and hump their bodies up. In order to volplane they turn head down on the tree-trunk and just jump out into the air. Borne by their parachute and pushed by gravity, they swoop away, but the tail soon comes into action, and, using this as a rudder and by tilting or lowering their arms and legs, they can twist in and out among the trees and branches. The distance they can thus travel is improperly documented and has not been filmed, but it is stated to be well over a quarter of a mile across gorges. They invariably land upright on a vertical sur-face. Anomalures are primarily leaf-and-flower-eaters but will take green nuts and unripe fruit.

Non-gliding Scale-Tail (*Zenkerella*)

This "English" name is silly but nonetheless warranted in that it points up the fact that the creature is simply an *Anomalurus* without a parachute. Otherwise it differs only in having a close, rather woolly, bluish-grey fur similar to that of some of the Giant Arboreal Dormice. The scales under the tail are like those of the Anomalures. The only specimen known was taken from a hollow tree on the Benito River in French Gabun.

Gliding Mice (*Idiurus*)

This title also may prompt facetious comment but it too serves a specific purpose, for these animals are just about the size of mice and greatly resemble them in shape except for parachute membranes. Unlike those of the Anomalures, these organs are not attached to the hind toes, but stretch from wrist to ankle and leave both hands and feet free. The tail, moreover, is completely free right from its base, and the membrane behind the leg is attached to the thigh. There are small scales under the tail-base but behind these to the tip of the tail there are two lines of tiny, stiff bristles set at a slight angle backwards and outwards. These are used by the animal to cling to tree faces. In addition, the tail, which is otherwise naked and mouselike, is sprinkled with very long individual hairs almost like whiskers, the purpose of which becomes plain when the animals are viewed in flight.

Idiuri live communally in the tops of the largest hollow forest trees, and the author has witnessed several hundred of two quite different species come out of a single tree. They run about like ordinary animals and do not "hump" like the Anomalures. Their flight is leisurely and almost casual. They simply launch out into space and go sailing away rather slowly, twisting and turning, rising and falling on small air currents, or by using their tails as rudders, and they are so light that they can be turned by the air-resistance offered by the fine long hairs on those tails. These animals appear to be insect-eaters rather than leaf-eaters and like the Kobegos they carry their minute young clinging to their bellies.

SPRING-HAAS (*Pedetidae*)

Nobody has ever decided quite what to do with this odd animal. It stands all alone in the scheme of things and is not at all like any other rodent. It has been put into all three of the great divisions of the Rodents, but has come to rest in the Squirrel-like division mostly because of some alleged similarities to the Anomalurids. Spring-haas means "Jumping-Hare" in Afrikaans; the scientific name is *Pedetes caffer*, the first meaning a dancer in Greek. They are found all over the open plains and mountains of South Africa, and thence northwest to Angola and northeast to Kenya. They are grey-brown above and white below, and the immense bushy tail has a black end. The front limbs are very short, the hind ones enormous, and the feet have only four toes. Most odd are their teeth, of which there are the standard two above and two below in front, but which otherwise consist of only four on each side of either jaw. These latter lack roots and grow all the time to keep pace with the wear and tear of their crowns resulting from the chewing of coarse material.

Spring-haases are entirely noctural animals and sleep in holes that they dig for themselves. While foraging, or just mucking about, they go on all fours but with a loping gait like a kangaroo; when alarmed they leap but, although making a great display and taking long bounds, they are not really swift. What is more, they make much better time going uphill than down. They make extremely amiable pets and can be turned into daytime animals in about three weeks, when they will follow their owner about, making an assortment of odd little noises. They eat a well-balanced diet of roots, leafage, some fruit, insects and some flesh. Like the porcupines, to which some specialists believe they are related, they show a passion for salts of all kinds and must be protected from starch or directly poisonous powders, which they will eat without discrimination.

MYOMORPHS

The Mouse-shaped rodents form the core of the great order of Rodents and far surpass the other two groups both in numbers of species and genera. The biggest is only a little over a foot long. We may well ask of what possible use or even interest they can be to anybody and, at first sight, they do appear to constitute an almost endless and most dreary confusion. However, as in almost all aspects of nature, the more one studies them and the closer one looks into their lives, the more remarkable, fascinating, and important they become. The group comprises ten quite separate kinds of animals, two of which (the Cricetids, or "Squeaking-Ones," and the Murids, or "Mousy-Ones") are closely related and far transcend all the others in numbers and importance. Once again, the status of some groups, notably the Blesmols, is not certain.

ANCIENT MICE (Cricetidae)

This family includes no less than one hundred recognizable genera, more than half of which belong to the first batch described below as the New World Mice. The total number of species represented is probably near two thousand. Therefore, only certain exceptional or well-known types will be mentioned. The Cricetids appear to be the original rats and mice, as it were, and to have evolved in the Northern Hemisphere some forty millions of years ago at the dawn of the Oligocene period. Thence, they spread southwards into the continental peninsulas of the Orient, Africa, and South America. The Murids or Modern Rats seem to have been developed much later in the same northern area and then also to have pushed south, driving the Cricetids before them until, in the Old World, the latter came to occupy isolated pockets of territory far to the south. The Murids, however, never reached the New World until brought by the Whiteman in the form of the Black and Brown Rats and the House-Mouse.

New World Mice (Peromyscus, etc.)

Of this seething mass of tiny, nondescript animals, a few stand out as being at least recognizable to the layman. Most typical of the whole lot, and one of the first to be discovered, is the delicate little White-footed or Deer-Mouse of North American woodlands (Plate 72). This is just a tiny mouse with big ears and light undersides and a moderate tail. It swarms throughout the country and does a great deal of good by keeping down the insect population. Smallest of all is our little Harvest

Spring-haas

Mouse that builds nests at the tops of grass stems and in wheat and other grain fields. Out West we find the Grasshopper Mice, small fat-bodied, short-tailed and light-colored insect-eaters that will attack and kill other small animals, and show cannibal tendencies.

In South America there are literally hundreds of different kinds, most notable being the tiny *ratonitos de campos,* the first type to be accurately described (as *Akodon*) in scientific literature. Then, in North, Central, and parts of South America, there are the Cotton Rats (*Sigmodon*) which swarm in countless millions, and do immense damage to men's crops, and the hardy, agile Rice Rats (*Oryzomys*) that somehow got to the distant volcanic Galapagos Islands in the Pacific before the Whiteman arrived. Finally, there are the Woodrats, Packrats, and Rockrats with their bushy, squirrel-tails, and mysterious and engaging habits. These animals collect great masses of sticks and other material to build huge nests in caves, at the base of trees or in the branches of bushes, but why they should pilfer tin cans, jewelry, bits of glass, and all manner of other objects, and often replace them with shiny stones or other even more worthless (to us) items is quite beyond man's understanding.

Hamsters (*Cricetus, etc.*)

Because of its introduction as an experimental laboratory animal during World War II, one species of Hamster is today fairly well known to the public. This is a small, richly colored, reddish-orange and yellow species with dark facial markings, that comes from Syria and Palestine. Until now, Hamsters have been of interest only to people in Middle Europe and Russia, where certain much larger species have, throughout the centuries, caused immense damage to standing crops. Periodically, these large rodents (Plate 70), half the size of a rabbit, become extremely numerous over large areas, and they may then literally mow down a field of wheat, oats,

barley, or other grain crop in swathes, chopping every stem off to the ground. They build complex underground dwellings. There are three other genera in Asia and one aberrant form in Africa. Several of the larger species have scent glands on the sides of their backs.

Sokhors (*Myospalax*)

These curious little-lemming-like rodents lead a completely subterranean life, excavating endless tunnels in search of root and insect food, and throwing up small mounds of earth along the way. The best-known species is found in northwest India and is known as the Quetta Mole. Other species live all over Central Asia, from Kurdistan in the west to the edge of the Gobi desert in Sinkiang and north to Kasakstan. In all localities they are found up to heights of many thousand feet in the mountains. They are about the size of large mice, with very short, dense fur.

Malagasy Voles (*Nesomys, etc.*)

In the great island of Madagascar off the southeastern coast of Africa all the mammals are odd. The only rodents are Cricetids or Ancient Mice, which seem to have some relationship to the Hamsters. They form one of those pockets of isolated leftovers mentioned previously, whose ancestors probably entered the island from Africa in extremely ancient times. Since being isolated there, they have diversified into seven quite distinct types, each of which has come to look strangely like some other rodent found elsewhere. Only a specialist can identify these animals, and then only by the pattern of their teeth.

The Crested Hamster (*Lophiomys*)

Also known as the Maned Rat, this unique animal is found only in certain areas of East Africa about Uganda, where there are half a dozen distinct kinds. These animals can not by the widest stretch of the imagination be regarded as rats. They are like no other animal in

North American Rock-Rat (Neotoma)

the world. Their bodies are more than twice the size of a big rat, with a tail about half that length, but they appear to be much larger because they are covered from the tops of their heads to the tips of their tails with very long, coarse, rather wavy hair, the roots of which are bulbous and spongy like an old-fashioned hairbrush that has been soaked too long in hot water.

In color their fur is grey, being composed of a mixture of black and silver hairs. The face and undersides are black, and on the former there is a triangular white mark on the forehead and a white line under each eye. Odder still, long hairs along the ridge of the back form a crest that continues right down the tail. This can not only be elevated like the head feathers of a cockatoo but can be opened like a flower. Then, on either side of this and running along the flanks from the neck to the tail-base, are two belts of very short fur that look exactly as if the long hair had been cropped with an electric razor. These bands are glandular and are completely naked in the young. They live in pairs in hollow trees, and are more or less arboreal, though they will come to earth often and some of them live among the rocks and tangled scrub on mountains. They are leaf-eaters, and are easily kept in captivity, but are most irascible creatures, raising their crests, turning their backs, and making a "churring" noise with their teeth if even looked at.

Voles (*Microtus*, etc.)

We now come to the second largest conglomeration of the Ancient Mice, and one that is almost universal in distribution all over the Northern Hemisphere. The hundreds of species are divided among some twenty genera, and these in turn can be grouped as Lemmings, ordinary Voles, Water-Rats, and Mole-Voles. Among these are several well-known, important, and truly popular animals. It is a sort of tradition among zoologists to start off any discussion of the Voles with the Lemmings (Plate 73), and it is just as well to dispose of this troublesome matter as soon as possible. There are many different kinds of lemmings, and they are grouped into four lots—namely, the Common Lemmings, the Snow-Lemmings, the Bog-Lemmings, and a rare Eurasian type. All are small (about five inches long), fat, compact little animals with thick, rather fluffy fur, small ears almost completely concealed under the fur, long front claws and very short, haired tails. They all live in or around the Arctic circle.

Almost everyone has heard the blood-curdling, heart-rending story of these pathetic little creatures periodically, as *The New Yorker* once put it, "pattering to their doom" from the mountains of Norway by casting themselves *en masse* into the heaving Atlantic Ocean. This makes a nice story, pointing a poignant moral, but unfortunately it is almost complete rubbish, though there is just enough truth in it to keep it alive.

The real truth is that all lemmings show a marked periodicity in their rate of breeding, resulting, it is believed, from the cumulative effect of either "E" or other vitamins derived from certain lichens which grow under the snow and upon which they feed during the spring. Over a number of years, the population increases in certain areas and, commensurate with this increase, the fertility of the animals and to a certain extent their individual appetites and even size follow suit. In time, as litters of young come faster and faster and for longer periods of the year, things get out of hand, the food supply becomes inadequate, and the animals have to start emigrating outwards *in all directions.* This brings them down from the mountains in Norway and, on the western side of the country, to the attention of the human populace in the fjords. As they come they keep breeding and multiplying, but at the same time they keep dying by myriads because of disease, trampling, drowning in rivers and fjords, or from exhaustion.

None of these emigrants (not "migrants," please note) ever returns, for, although some attain new locations highly suitable to their perpetual maintenance, they sooner or later cease to breed, and die away. In some very rare cases some of them may find their way to the shores of the ocean, though this is extremely hard to do in Norway because of the indented precipitous coastline, and may then enter the waves in the mistaken notion that they are rolling down a fjord; and of course they perish. However, the vast majority of these emigrant swarms never reach a coast, and most of them never even get as far as the nearest large river.

Much more interesting than these are the Snow-Lemmings (*Dicrostonyx*) which turn pure white in winter and burrow under the snow or even meander about on its surface in the dead of the Arctic winter. In the fall each year they grow an enormous second claw under the permanent one on the third and fourth fingers, a unique characteristic among mammals. Its purpose is not definitely known, but would appear to have something to do with their snow-shovelling activities. In summer they are referred to as Collared Lemmings. They can close their ears with a comb of stiff hairs controlled by special muscles in the hind cheeks. The Bog-Lemmings are rather dull little creatures that live in moist places south of the range of the two former types.

The main body of the Voles is represented principally by two great genera known to science as *Microtus* (Plate 74) and *Clethrionomys,* meaning the "small-eared" and the "bolt-toothed" respectively and called, popularly, the Field-Voles and the Red-backed Voles. Despite their enormous numbers, there is little to be said about these small rodents. They live on the surface of the earth or slightly under it, either in their own burrows or those of other animals. They eat all manner of seeds, small nuts, insects, and some herbage. Some make nests and store food, others live in hollow trees and climb well, while many of them make great complexes of permanent runways under grass covering about the area of a tennis-court. They breed very rapidly, some having up to eight young at three-week intervals, and the youngsters begin breeding in another five weeks. As a result, some of these animals also swarm from time to time and it was a *Microtus* that once inundated vast areas of the West to the estimated number of 12,000 per acre. There are half a dozen other genera scattered throughout Europe and Asia.

The Water-Rats come in two very distinct forms—

those of the Old World (*Arvicola*), and those of the New (*Ondatra* the Muskrat (Plate 75), and *Neofiber* the Florida Water Rat). The former are the size of rats but have long, rather fluffy fur and shortish tails. They live in the banks of rivers and ponds, and are semi-aquatic, feeding on lush herbage and doing little damage. Their cousins, the larger Muskrats, which have long tails somewhat compressed from side to side so that they form sculls, have been introduced into Europe during the past century, and there they have on the contrary proved to be a great pest. The reason for this is their habit of making entrances both above and below the water level which is all very well under natural conditions, but becomes a terrible menace in lands where the water level is often above that of the land because of ancient and extensive dyking. The Muskrats being, of course, unacquainted with such conditions, solemnly burrow through to the water, often horizontally, whereupon the dyke begins to leak and even a small trickle in an earth wall can rapidly have disastrous effects.

Muskrats make large, dome-shaped nests above water level, using a most odd method of construction. They take sodden reeds and other water plants and roll them into little oblong bundles like a roll of string, and fit them together between growing reeds or other supports, leaving a single concealed entrance. In these they rear their young. They are harmless animals living in swamps and keeping out of men's way, though they are still so common that they live and nest within the limits of many of the greatest cities in America.

Economically, muskrats are of the utmost importance since their pelts form the basis of the modern medium-price fur market, being durable, waterproof, abundant, and rather beautiful. They are also sturdy enough to be "dropped" so that the small rectangular skin can be finished as a long narrow strip suitable for modern coat designs. The Florida Water-Rat is a close relative, found only in the coastal swamps of eastern Florida north to the great Okefenokee Swamp of Georgia. It is somewhat smaller than the Muskrat, and the tail is thick, but round in section. These animals make large platforms in the swamps by dragging their food plants to the same place for months on end, so that all they do not eat accumulates there. In these they then burrow and construct nests. They live in brackish as well as fresh water.

The remaining Voles, of which there are half a dozen genera, may be arranged in a series which displays progressive deterioration from an active cursorial animal to an apparently moribund, lozenge-shaped lump of fur known as *Ellobius*, or the Mole-Vole. Starting with the very common little Pine-Vole of North America (*Pitmys*), which unfortunately burrows under the soil of all types of grass, farm and woodland, *except* specifically pine stands, all these remaining moles are fossorial and spend more or less of their time below ground. In the case of *Ellobius,* the little animals are completely adapted to perpetual burrowing and darkness, having minute eyes, and practically no ears. The fur of many digging voles is short and silky and, like that of the moles, able to be brushed both ways, which aids the animals in shunting backwards along their holes.

Sand-Rats (*Gerbillus, etc.*)

The last great division of the Ancient Mice are desert animals (Plate 76) found in eastern Europe and the drier parts of Asia and Africa. They are divided into about a dozen genera, often on grounds that do not seem warranted but need not concern us here. They are all small animals, ranging in size somewhere between mice and rats. They are root- and seed-eaters, and burrow industriously under desert grass and scrub. The Indian Gerbil takes the place of the voles in other countries, being enormously prolific, having sometimes as many as fourteen young at a time, and occasionally swarming. In the year 1878 such a swarm totally destroyed all crops over an area of about eight thousand square miles in one part of India by literally mowing down the standing grain stalks.

MODERN MICE (*Muridae*)

If it is considered advisable to take a deep breath before embarking upon a description of rodents as a whole, it may be permitted to give a deep sigh on concluding the Ancient Mice and before tackling the Modern Mice, and more especially the first group of them, which we call the Old World Mice. This may be an advantageous point at which to clarify the words "rat" and "mouse." There is really no distinction except when a particular animal is being defined that is popularly known as either one or the other—thus, a Brown Rat and a House Mouse. Otherwise, the distinction is basically one of size, the smaller types being called mice, the larger rats. Even this is a comparatively modern development, since they all appear to have been called mice previously. There is, however, one way in which House Mice (genus *Mus*) differ from the rats of the genus *Rattus:* The former stink—at least to us—while the latter do not. Included in the modern mice alone there are more than seventy genera, one of them containing several score of apparently valid and distinct species. We have not so far mentioned sub-species in this book, but nearly all animals may be thus subdivided into perfectly distinct races. The number of animals involved in the next two pages, therefore, amounts to about a third of all known mammals. Nevertheless, the reader may take heart since the members of this great group are even more alike than are the members of the Ancient Mice, and only those of special note or popularity will be mentioned, while, for those who may be interested in further details, there are available several monumental tomes, which incorporate the most recent findings of systematists in this field.

As was noted on page 121, these are apparently the most recently evolved mammals on the surface of our earth, and their evolution seems to have been very rapid because of their great adaptability and a terrific birth rate which increases the chance for natural mutations and provides a rapid means of fixing regional variations that may arise in conformity with specialized environmental factors or by simple natural selection. Human generations march forward in time with about twenty-year paces, but those of some tropical mice may scamper along in only six-week steps.

Old World Mice (Mus, etc.)

Although Man is undeniably "top-mammal" in certain ways, and the Elephant may be regarded as the most highy "evolved," there is little doubt that some rat, and probably the Brown Rat (*Rattus norvegicus*) (Plate 81) is actually the finest—in every sense of that word and especially in efficiency—product that Nature has managed to create on this planet to date. Further, there are sound reasons for stating that, even if man eliminates himself entirely from the earth by the undue release of long-term radioactive materials or by other means, certain rats could survive: their "efficiency" in the basic sense of *that* word is not yet fully appreciated, for we tend to judge behavior on emotional grounds rather than by its results. Rats preserve a much more practical balance between compassion for and indifference to their own kind than we do. While weaklings or cripples among their numbers may be left alone, "fools" and "criminals" seem often to be deliberately eliminated or killed outright. All of this results in much sounder eugenics than we practice. That there are more individual Brown Rats in North America than there are people, is not the result of man's carelessness, indifference, or wasteful and dirty habits; it is the result of the greater stamina and, frankly, commonsense of the rats. What is more, these animals can learn as we can, as well as by that other "instinctive" manner employed by the Beaver.

Of the more than seventy different kinds of Modern Mice, the members of the genus *Rattus* therefore call for first notice. There are, according to Dr. Simpson, more than 550 currently recognized forms of this genus. All were originally Asiatic animals but two, and probably more, have been carried by man, have followed man to, or have simply colonized on their own account, almost every land area on the earth exclusive of the Antarctic and some islands in the Polar regions. The two most concerned in this worldwide occupation are the Black Rat (*Rattus rattus*) (Plate 80), which appears originally to have been an arboreal species from Indonesia, and the Brown Rat (*Rattus norvegicus*) (Plate 81), which was a fossorial inhabitant of the treeless steppes of Central Asia. The former started its mass emigration first, probably in Roman times or earlier, and probably did so primarily in ships and in cargoes of tropical fruits and other edible produce. It then came ashore along ropes and cables and took to dwelling in the tops of houses and granaries. The Brown Rat came much later, probably first moving westward with the slavonic and mongoloid hordes that swept into Europe. Being a terrestrial animal and a digger, it took refuge in the basements of buildings, ousting the Black Rat.

Since their arrival in foreign lands the fortunes of the two animals have depended on man's changing ways. When sewage systems were recreated in Europe the Brown Rat benefited; when concrete buildings, overhead electrical and other cables, and kitchens on top floors were developed, the Black Rat came back. That the Black Rat harbors the fleas that carry bubonic plague is well known, and that both species contaminate food, spread disease, and destroy billions of dollars worth of food and property annually is at last becoming common knowledge. The part played by the domesticated varieties of both these rats in laboratory experimentation, as test animals for medical analysis, and as subjects for study in a wide variety of researches from psychology to space-flight, is not so readily appreciated.

Finally, there is the research on the rats themselves. This now warrants not a volume but an encyclopedia to itself, and is beyond our scope, but may perhaps be summed up by one example. Rats behind bars offered a large unbreakable dog-biscuit lying horizontally on the floor outside the cage immediately reached through the bars with their hands, tilted it on edge, and pulled it between the bars into their cage. The more one considers this apparently simple act the more appalled one becomes and the better one understands the great group of animals called the *Muridae*.

Next to the typical rats both in numbers, versatility, and breadth of distribution come the tiny mice (genus *Mus*) (Plate 78). These delicate little animals, of which there are probably as many recognizable forms as there are of *Rattus*, are indigenous to Europe as well as most parts of Asia and of Africa, in one form or another. Today, they have for so long been carried back and forth all over the surface of the earth by men, and they are such adept interbreeders that it is quite impossible properly to unravel them taxonomically. Although the so-called House Mouse does a certain amount of damage to human property, makes a mess, and may transport filth and some diseases, it is really a very pretty, delicate, and cleanly little animal, and it probably does more good than harm in the over-all picture by clearing away "crumbs" and by keeping down cockroaches and other more dangerous pests. The habits of mice are too well known to warrant elaboration, but it is not generally appreciated that they can survive and breed under many extreme conditions. They turned up in a winter camp on the Antarctic Continent and have been known to breed in deep-freeze plants and in parts of chemical plants that are lethal to humans. A pair were found nesting in General Rommel's personal possessions in his tank in North Africa during his World War II campaign against Egypt.

Mice—in the pure sense—come in an incredible variety of forms, even under natural conditions. The smallest is the tiny European Harvest-Mouse (Plate 83). There are silky, woolly, hairy, spiny, and even prickly-furred species, and even in the genus *Mus;* then there are typically spiny forms (*Acomys*) (Plate 79), and others (*Leggada*) that have skins not much more substantial than wet toilet-paper and sparsely covered with wiry little hairs. Rather alarming to behold are such species as the Barbary Mice (*Arvicanthis*) and certain West African species (*Lemniscomys*) that are vividly striped or covered with white spots respectively. Then there are swarms of plain, ordinary mice of various dull browns, grey-browns, and greys, known by all sorts of popular names and typified by the European Field-Mice (*Apodemus*), that inhabit almost every square mile of the Old World, including the remotest islands, but each of which seems to display some distinctive habit or feature.

In addition, there is a large group of rather more exotic genera found in New Guinea and the remoter Indonesian islands to which such popular names as Fruit-Rats have been given, and still another peculiar group that is confined to Australia, where they have mostly developed into forms resembling Jerboas (see page 129) or the little Pouched Mice and Rat-Kangaroos of the Marsupialian fauna of that continent.

Even then, the catalogue is far from complete; there are pouched mice (*Saccostomus*) in Africa (Plate 71), and also bulky and obese forms like the Bandicoot-Rats of southern India (*Bandicota*) and the gigantic specimens of typically ratlike form from West Africa (*Cricetomys*) which for some foolish reason have been called Hamster-Rats. Nothing could be further from reality, for they are less like a hamster than almost any other rodent. These animals have all the native cunning and educational ability of the Brown Rat combined with much greater size and strength. They also have the added ability of the Cricetine Pack-rats to build houses, collect, transport and store material, and they employ a sort of communal nursery system by which they can save orphans.

Wading-Rats (*Deomys, etc.*)

When all those genera of rats and mice which we have classed collectively as the Old World Mice (known technically as the sub-family *Murinae*) have been disposed of, there still remain over a score of other genera of Modern Mice that are in one way or another sufficiently different to be separated into several other subfamilies. The first of these (known technically as the *Dendromyinae*) is typified by a stilt-legged African rat known as *Deomys*, which for some extraordinary reason has become known popularly as the Tree-mouse. The animal concerned is actually a small rat with a very long, pointed head, and extremely long legs that lives in swamps and other moist places in the high equatorial forests and spends most of its time wading in shallows in search of water insects, snails, and other small animal food. It is current practice (see Simpson, *The Principles of Classification*, etc.) to associate five other genera with *Deomys* in a special group. We suggest another arrangement wherein these five and several other genera, including such types as *Hybomys* and *Malacomys*, which are normally classed with the *Murinae*, as well as the Veldt-Rats (*Otomys*) be brought together in one subfamily. This is all rather technical but is necessary to keep our record straight. At the same time, it should be appreciated by the non-specialist that, however obscure and unimportant this group of little African rodents may seem, they actually have a status equal to that of, for instance, the Great Apes or perhaps even of all apes, and that they are therefore just as important, probably just as interesting, and possibly of even greater significance.

The Shrew-Rat (*Rhynchomys*)

The Philippine Islands, and notably Luzon, are notable for harboring a number of odd rodents some of which are comparatively very primitive, and some of which have relatives only in Australia. It would seem that the break-up of the land-mass which once connected Asia to Australia was neither sudden nor complete, but was progressive and entailed periods when one or other of the present-day Indonesian islands remained connected to either one. Thus, relics of earlier faunas have become trapped in some islands and not in others, and the famous "Wallace Line," which for a century was believed to divide the two continents of Asia and Australia, is not so definite and final as it was once supposed. A case in point is a very curious little rat of the Philippines. It has rather short legs and an immensely long snout and looks for all the world like a Shrew. Nothing is known of its habits and it remains one of the zoological collectors' prizes.

Cloud-Rats (*Phloeomys*)

We here meet a group of some half-dozen genera of odd rats that are distributed as follows: two in New Guinea, two in the Philippines, one in the East Indies, and one on the mainland of Asia. As will be seen below, the next group spreads from the Philippines to New Guinea and thence to Australia, so that the famous zoological gap, mentioned above, between the continents is more or less bridged. *Phloeomys,* the so-called Rind-Rat, is a large, long-haired species with very odd teeth; it lives in the lowlands. Others (Plate 82) with strange, parti-colored black and white hairs not unlike the Crested Hamster, live in the mountain mist-forests; others are arboreal.

Australian Water-Rats (*Hydromys, etc.*)

The most typical and outstanding member of this group is a large rat, about two feet in over-all length, black above and bright orange below, that has the distinction of having only eight back teeth. Its large hind feet are fully webbed and it leads much the same life as the Muskrat. Its relatives come in all sizes, some live by water, but there are others that live out near the deserts, and there are even arboreal types. In New Guinea there are several relatives of the typical Australian Water-Rat and also a very odd creature which is more wholly adapted to an aquatic life than any other rodent and is known as Monckton's Rat (*Crossomys*). This has completely webbed and enlarged, paddle-like feet, small eyes, no apparent external ear at all, very close fur like a Water-Shrew, and a tail bearing wide fringes of thick, stiff hairs on either side that transform it into an adequate paddle.

MOLE-RATS (*Spalacidae*)

We now proceed to the second family of the Modern Mice, creatures so entirely different that it is hard to believe that they are related to any mice. This and the following two families are in fact of doubtful affinities because of their extreme specialization for a subterranean existence. The Great Mole-Rat (*Spalax*) is a unique animal found in eastern Europe to the Caucasus, in Asia Minor, Syria, Palestine, Mesopotamia to Persia and south to upper Egypt. It is a rather unpleasant-looking furry lump, without a tail, with short limbs, rather small hands, and large feet. Its minute eyes are covered with skin, and neither they nor the wartlike little ears are visible, being covered with a thick brush of fur that arises from the cheeks. There is no neck, and the head may be recognized only by the small, naked muzzle and a round hole from which protrude two pairs of large rodential front teeth. The fur is thick, dense, and woolly to silky, and like that of a mole can be brushed any way. They lead the lives of moles also, burrowing continuously underground for bulbs and roots, and throwing up occasional mounds of earth. Their main tunnels are dug about eighteen inches below the surface, but they construct sleeping quarters and make huge storerooms that they fill with food about four feet below this level. There is only one species, which is a dirty, yellowish grey-brown color, often with white spots below.

ROOT-RATS (*Rhizomyidae, etc.*)

In Tibet, northern India and thence east to China and south to Malaya and the large Indonesian Islands there are to be found a number of similar animals which, however, belong to quite another family. They are popularly known as Bamboo-Rats (Plate 77) and they are not so wholly subterranean in habits, some living under grass, among tree roots, or in the piles of rubbish among the stems of Giant bamboo clusters. The Indian species are about a foot long, but in the islands there is a giant race almost twice the size. They also have minute eyes, but these are not covered by skin, and there are small naked ears and a short tail. Another kind is found in Abyssinia in East Africa. They are nocturnal animals and prodigious eaters; moreover, they make the most terrible rumpus while feeding, groaning, mumbling, and chewing so loudly that they may be heard from a hundred yards away.

BLESMOLS (*Bathyergidae*)

There is a third group of lowly, but even more profoundly specialized fossorial rodents found all over the eastern side of Africa south of the Sahara. These are in no way related to either of the two above families, nor is it even certain whether they should be included in the great tribe of Myomorphs, for they have no known relatives either living or extinct and they are anatomically unique. Nonetheless, we place them here lest they be forgotten or lost in the welter of other rodential types. There are five recognized genera, four of which have much in common but the fifth of which is altogether bizarre.

Strand-Rats (*Bathyergus, etc.*)

Over a wide area in Africa between the Cape and Kenya there may be found very numerously various kinds of subterranean-dwelling rodents of positively appalling mien and various size, to which the popular names of Blesmols, Strand-Moles, or Sand-Rats have been given by the European colonists. The largest (*Bathyergus*) is found on the coastal flats, where it appears to live and behave much like the Mole-Rat of Europe. This animal is clothed in a thick, soft, molelike fur, appears to have no limbs at all, but bears sharp claws on the small front feet, and has just an apology for an eye, about the size of a large pinhead. The front of the animal is most singular, there being, first of all, a circular hole in the fur, lined with naked, dark skin in which the small nostrils may be seen. Second, below this, two enormous pairs of chisel-shaped teeth jut out at an acute angle, arising apparently from the fur. No sign of the tiny mouth is visible.

The animals do all their digging with these teeth, using their front feet only for scraping away the loosened earth and their hind feet for passing this out behind the body and then tamping it firmly into the hole to make a solid and continuous plug. This whole procedure can be carried on in a vertical position when the animal is going straight down. Thus the holes of these animals are impermanent while the animals actually "live" in the soil but encased in a small capsule of air. Smaller species which are only about six inches long and thus only half the size of the Strand-Rat live on higher ground.

Sand-Puppies (*Heterocephalus*)

Of altogether different, and absolutely unique appearance, are the mouse-sized creatures found in the fiery

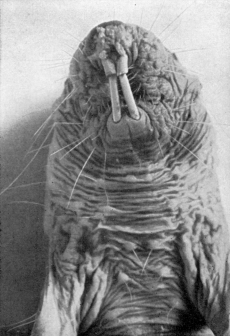

Sand-Puppy (Heterocephalus)
PARKER FROM LONDON ELECTROTYPE AGENCY

Blesmol (Cryptomys)
ZOOLOGICAL SOCIETY OF LONDON

sands of the Somaliland deserts; they look like newly born puppies or small, uncooked cocktail sausages. They are naked but for a scattering of fine, stiff hairs and their skin is wrinkled all over. They are long-bodied with rather large, lizard-shaped heads without external ears and with tiny, lidless eyes. The limbs are small and somewhat vestigial, and the tail is short. They live entirely below ground and never appear on the surface, and are insect- as well as root-eaters. They make rather large nests of the interwoven hair roots of special plants, and instead of raising mounds, make small volcanos from the bottom of which they eject streams of earth in spurts and with considerable force. The most astonishing thing about them is that, despite their nakedness, they can burrow along in the topmost layer of sand at midday when the air temperature above may rise to over 130° in the shade and when the sand is literally at furnace heat.

DORMICE (Gliridae)

Of the ten families of the Myomorphic Rodents only the first two, namely the Ancient and Modern Mice, are in any way similar, and then only in certain of their forms —notably those which we would call simply rats and mice. The three fossorial families—the Mole-Rats, Root-Rats, and Blesmols—may appear superficially alike. Each group is, however, on a quite separate branch of the tree of life. Similarly, most members of the present group look more like squirrels than anything else.

The Dormice are divided into seven genera, one of which is African and typically sciuromorphic; and others vary greatly from small, mousy creatures to truly arboreal animals with long, bushy tails. They nonetheless form a closely-knit little group that is very ancient; the bones of typical dormice have been found in rocks at least fifty million years old. The name "dormouse" is derived from the French "dormir," to sleep, in reference to their profound hibernation for long periods in northern climes.

The Hazelmouse (Muscardinus)

The oddest form of Dormouse happens to be the best known, and is one of the commonest smaller mammals of western Europe (Plate 86). It has particularly interesting habits and engaging ways and has, for centuries, been a pet of German, British, and French nature-writers, and small boys, so that it has been presented to the world at large as "The Dormouse." It is about twice the size of a house mouse, three inches long not counting a three-inch tail, has large, bulbous, jet-black eyes, a wide head, and soft, close, silky fur dark at the base but a beautiful shade of rich brown at the tip. The tail is slightly compressed from top to bottom and is clothed in short fur the same color as the body. Underneath, the animal is lighter, ranging from pale buff to pure white. Hazelmice are found all over Europe from Scandinavia and Britain to North Italy and Transylvania.

They live among bushes, not trees, and make two kinds of nests: during the summer, small spherical ones, low down; in winter, much larger ones in the center of dense, evergreen vegetation, and in these they store a mass of food. In the fall, they store up fat in their bodies and then go into a profound form of hibernation, from which they can be aroused only with difficulty. When they are up, however, they have a meal from their food store and then go right back to sleep with their tails curled over their heads. The young are born in the large nests, in spring in the north and in fall in the south. As their name implies, they are particularly fond of hazelnuts, which they can empty through a single tiny hole bored in one side with their front teeth.

Squirrel-tailed Dormice (Glis, etc.)

There are five genera in this group, typified by the so-called Common Dormouse of continental Europe (Plate 85). This is a six-inch animal with a five-inch tail, clothed in thick, soft, ashy grey fur above, and pure white below. The tip of its snout is white and it has dark "spectacles." It is found from northern Germany south to Spain, Italy and Greece, and east to Russia in the north and Syria in the south. It nests in holes, either in trees or in the ground, and in the north hibernates for about seven months out of every year. It eats insects and other small animals, as well as nuts, and is itself extremely good eating, especially when fat and ready for hibernation, as the Romans long ago discovered. The young grow at an incredible speed—they have to, when the parents have only five working months out of the year. The Asiatic tree Dormouse (Dryomys) is smaller, reddish above and white below.

The Garden Dormouse (Eliomys) is about five and a half inches long with a four-inch tail that is short-furred except for the end which bears a large tuft. It has very much larger ears, is grey above and white below and has dark marks round the eyes. It is almost entirely carnivorous and makes all kinds of nests in trees, bushes, or on the ground. It is found from Belgium to Baltic Russia and Bulgaria. There is another genus in Japan, and there are squirrel-like dormice in central Siberia.

African Dormice (Graphiurus)

These, the largest of the family, are distributed all over the forested part of Africa, even to the semi-temperate woodlands of the Cape. Some may be as much as two feet in length, and there are a very large number of described forms. They all have exceedingly thick woolly fur that displays the most beautiful soft shades of coloring—pale greys, all manner of smoky greys, beiges, and mauves. Some are white below, others a steely blue all over. They have very bushy tails, usually of a darker hue, and always flattened from above to below. Like their northern Eurasian cousins they have a mere stump in place of a first finger. Most of them sleep in hollow forest trees, and they are one of the commonest night animals of Africa.

Little is known about the majority of them, but some have remarkable habits. They are spotlessly clean animals that make large homes, not mere nests, in hollow branches, in which huge larders of food are maintained —all manner of unripe fruits and nuts, decapitated insects, and small birds and other animals neatly killed

[continued on page 145

70. *Common Hamster*

MARKHAM

71. *Cape Pouched Mouse* (*Saccostomus*)

72. *Deer Mouse*

73. *Snow Lemming (summer phase)*

74. *European Field Vole*

75. *Muskrat*

76. *Sand Rat*

MARKHAM

77. *Root Rat*

MARKHAM

78. *House Mouse*

MARKHAM

79. *Spiny Mouse*

MARKHAM

80. *Black Rat*

MARKHAM

81. *Brown Rat*

YLLA FROM RAPHO-GUILLUMETTE.

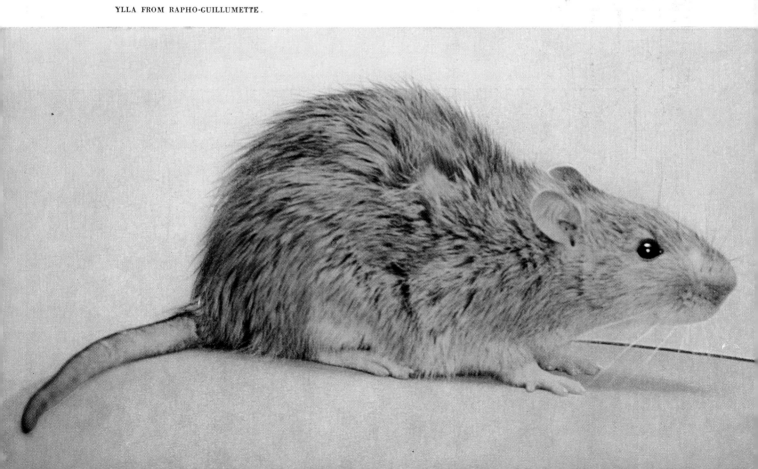

82. *Cloud Rat*

PINNEY

83. *European Harvest Mouse*

MARKHAM

84. *Common African Jerboa*

MARKHAM

85. *Common European Dormouse (semi-albino)*
MARKHAM

86. *Hazelmouse*
MARKHAM

87. *East African Porcupine*
MARKHAM

88. *Mediterranean Porcupine*
PINNEY

91. *Pacarana*

LA TOUR

→

89. *Capybara*

MARKHAM

90. *Mara*

PINNEY

92. *Cutting-Grass*

MARKHAM

93. *Acuchi*

PINNEY

94. *Coypu*

MARKHAM

95. *Californian Sea-Lion*

LA TOUR

96. *Northern Fur-seal*

KENYON FROM NATIONAL AUDUBON

97. *Harbor Seal*

LA TOUR: MARINELAND OF THE PACIFIC

[*continued from page 128*

and laid out in piles. The strange thing is that decomposed meat is never found in these stores, although even a small corpse will get very smelly in a few hours in the tropics.

SPINY DORMICE (*Platacanthomyidae*)

Placed near the true Dormice are a few very remarkable little rodents found in India, and there only in the tallest trees of forests in certain areas, notably the Anamalai Hills of Malabar. They are reddish-brown above and pure white below with spines interspersed through the dorsal fur. They are about five inches long and have bushy tails of equal length. The amazing thing is that they bore their own holes in the top of trees, using their teeth. In these they construct nests of soft material gathered from the epiphytic plants growing on the treetops. There is a published statement anent these animals—to wit, "They eat hot peppers, damage Jack-fruit, and drink fermented Palm juice"—that is interesting but not strictly accurate. Almost any animal will eat the hottest peppers with relish; Jack-fruit may be acceptable at certain seasons; and "palm-mimbo"—the semi-fermented juice that drips from the heads of certain palms—is available to all arboreal animals in open calabashes suspended from hollow bamboos driven into the heads of those plants by the natives. It is a sort of beer that intoxicates even goats.

SELEVINIDS (*Seleviniidae*)

A special family had to be created to receive a remarkable little rodent discovered in Russia in 1939 and named *Selevinia paradoxa*.

Any animal that waited till 1938 to be discovered and which then necessitated the formation of an entirely new family for its reception must, indeed, be very singular. They are small mouselike animals with long tails and most peculiar teeth. If they should be placed near any other rodent family it would be that of the Spiny Dormice or *Platacanthomyidae,* of Asia south of the Himalayas. They were first found as bones in refuse left by vultures in Kasakhstan by a collector named Selevin. Later, live specimens were obtained. They have the smallest molar teeth of any rodent.

JUMPING-MICE (*Zapodidae*)

It may seem that there is no end to the list of mice and, indeed, we are still far from completion of the catalogue, but it must be appreciated that the term "mouse" is a very general one, like the name "fly" which is applied indiscriminately to many tens of thousands of small flying insects of innumerable different origins. There are still two quite separate families of truly mouse-formed rodents to be mentioned; the first, the *Zapodidae,* is few in kinds but has a very wide and curious distribution. From their teeth and certain other points of their anatomy they appear to be very ancient

E. P. WALKER

and rather primitive rodents. There are three little lots of Jumping-Mice, the first found in Europe and Asia, the second in Asia and North America, and the third in North America alone.

Striped Mice (*Sicista*)

Throughout a considerable area centered around the Caspian Sea, but stretching from northeast to southeast Europe and thence east to central Asia, there are to be found a few forms of little mice about two and a half inches long with naked tails of the same length; they live in small tunnels under grass, and generally behave like Field Mice. In color they are fairly striking, being white below, having greyish yellow flanks, and above this a pale grey-brown band on either side enclosing a black stripe down the mid-back. Very little is known of their habits except that they make small nests and have about four or five young at a time.

Jumping-Mice (*Zapus and Neozapus*)

The appearance of these tiny, fragile creatures may be seen in the accompanying photograph. Although there are numerous described species, they are all very much alike and are all distinguished by their immensely long tails. There are some species in Asia, whose headquarters are in the Chinese Province of Sze-Chwan but which spread thence into eastern Siberia. The Woodland Jumping-Mice (*Neozapus*) which are distinguished only by having fewer teeth, are wholly North American.

These tiny creatures live under and amid grass and woodland litter, and make lined nests in deep holes in which they sleep and raise a couple of litters of half a dozen young ones each year. In these nests they also hibernate profoundly for almost half the year, rolled into a tight ball with their tails curled around them. They eat insects as well as seeds and berries. Since they have very long hind legs, they are tremendous jumpers and are the flying things often seen when hay is being cut, shooting out of the grass a few inches ahead of the mower or scythe like corks from an old-fashioned pop-gun. Their place in the scheme of life appears to link the next family, the Jerboas, on the one hand, to, of all unexpected creatures, the Old World Porcupines, on the other.

JERBOAS (*Dipodidae*)

Herewith we come to a group of most engaging little animals that used to be quite well known as pets not

American Jumping Mouse (Zapus)

only in Europe but even in America. They have intrigued civilized men since the time of the earlier dynasties of the Ancient Egyptian Empire, whose royal artists sometimes included them in their murals in temples and tombs. From those depictions there once arose, by simplification of lines, a hieroglyph meaning "swiftness." There are a very large number of different species of a dozen distinct genera and these may be divided into three major groups on fairly technical grounds.

They are desert animals and are found all over the drier parts of far eastern Europe, Russia and central Asia, western Asia, and north Africa. The identification of any particular Jerboa is a matter for specialists only, but for others who may be interested there are certain clues that may help to some extent and which are interesting in themselves. First, in Eurasia there are jerboas with five toes and others with only three; also there are in Asia those with short ears, and some in Yarkand that have very large, long ears shaped like those of tiny rabbits. There is also one rare Asiatic kind that has a curiously flattened tail like a horizontal paddle. In Africa also there are those with small rounded ears and others (*Scarturus*) with enormous, batlike ears that can be opened out like two huge, loudspeaker horns, or folded up and laid back alongside the shoulders when the animal is going in top gear. And Jerboas have a regular set of gears too—they can walk backwards or forwards, pace along on their stiltlike hind legs, or change suddenly into full-flight by leaping with both hind legs together. When they do this, they go off over the sand at such speed the human eye can hardly follow them.

The form of the common African type (*Jaculus*) (Plate 84) is, except for the shape of its ears, typical of them all. The fur of Jerboas is rather long, extremely soft and silky, and is always some light beige color above and white below. There is almost always a dark facial mask and the tuft on the tip of the tail is usually black, black and white, or white. Jerboas range in size from about three inches, with a six-inch tail, to ten inches, with a foot-long tail. They dig with their teeth, and live in communal burrows, and in the northern part of their range they hibernate, while some in the true desert regions of the south aestivate. They come out in millions after rain to feast on insects and the herbage that springs up with miraculous speed on the deserts as soon as any moisture is available. Crossing some deserts at night in a car at such times may disclose an amazing sight in the headlights, at the extreme range of which the little Jerboas will be seen flying off in countless droves. In captivity they have a strange love of all kinds of brooms and often make nests in them.

HYSTRICOMORPHS

All the remaining Rodents belong to the Porcupine-shaped tribe and there are vast numbers of them of much more varied appearance than either the Squirrel- or Mouse-like ones. They are currently divided into fifteen families, eleven of which are exclusively American, and four of which are exclusively of the Old World, while three of the latter are found in Africa only. However, there are valid arguments for reducing the American families to half this number by combining the five short-tailed guinea-piglike types in one, and the three long-tailed kinds into another compound family group. The trouble is that the Hystricomorph tribe is probably a compound assemblage itself, while the animals thus grouped together are mostly remnants of vast hosts of creatures that have become extinct and left no remains that we have so far unearthed. Most of these animals are fairly well known, have popular names, and may often be seen in zoos.

OLD WORLD PORCUPINES (*Hystricidae*)

Everybody thinks he knows a porcupine when he sees one, but it is highly improbable that anyone but a mammalogist would class certain kinds from the Oriental Region with the typical spine-covered creatures. In fact, the typical spiny monsters in color plates 87 and 88 are really extreme forms; at the other end of the series is a ratlike affair. The family may be divided into five very clear-cut genera as follows:

Crested Porcupines (*Hystrix*)

These are the largest of the Porcupines, and the four species are distributed over Africa, west and southern Asia, and throughout parts of southern Europe. The Great Crested Porcupine of the Mediterranean periphery and West Africa (Plate 88) grows to a length of twenty-eight inches, but has a tail only four inches long. The quills on the tail are hollow; those on the body are of two kinds—long, thin, and flexible, alternating with short, rigid, and sharp-pointed. The great plume on the head and the mane are composed of pliable bristles. The long body quills are white at base and tip and banded black and white between; those of the tail are white, but on the hind body the short spines are all black.

They are nocturnal animals which sleep by day in caves, in holes of other animals, between rocks, or in burrows of their own making. They eat all manner of roots, bark, fallen fruits, and a certain amount of coarse green-stuff, and are enormously powerful brutes with gnawing and grinding teeth, jaws and the appropriate muscles to work them that surpass all but the carnivores in strength. They are not aggressive but they are extremely arrogant, wandering about at night mumbling, grunting, and rattling their quills. This is part of their defensive mechanism, a proclamation to all and sundry that they are dangerous customers. However, these animals have extremely thin and delicate skins and are actually most vulnerable to attack by any animal that knows them, for their heads are almost entirely unprotected and they will succumb to a single blow across the muzzle. Nonetheless, if cornered they will put on a great show which can be most unpleasant or even disastrous to an attacker. They are really very quick-moving animals, can waltz around with remarkable speed and then suddenly rush backwards at an attacker, with quills and spines bristling, and if they make contact numbers of the latter become detached from their skin and lodge in their opponent. Meantime, they stamp with their back feet and growl, which is most unnerving and distracting.

South of the Sahara and ranging from the Gabun to

Kenya and thence south to the Cape, this genus is represented by the largest of all porcupines (Plate 87), which is often over thirty inches long including an eight-inch tail covered with long, white spines that the animal rattles in rodential rage at the approach of anything from a locust to a lion. Throughout the Near East from the Caucasus and upper Iraq, through Persia to India and north to Turkestan, another large species with a pronounced white collar is to be found in rocky areas, though it is seldom seen since it is strictly nocturnal and very wary and rather silent. Closely related is still another species which ranges to the east through India from Kashmir to Ceylon and east to Bengal. This is about thirty inches long and has a four-inch tail when full grown. In this species the tips of the hairs, spines, and quills of the cheeks, and a collar round the neck are white; those of the body are black below and white throughout the terminal half, and those on the tail are pure white. They are very common animals in India and do immense damage to root crops, standing grain, and fruit trees.

Noncrested Porcupines (Acanthion)

East of the range of the Crested Porcupines, and often ascending to considerable heights, from Nepal to Yunnan and thence south through Assam, Burma, and the whole Malayan Peninsula to Sumatra, Java, and Borneo, there are to be found a number of large animals of very similar shape and appearance but lacking crests. Their quills are much shorter and cover only the hinder back and rump. Their heads are clothed in rounded bristles, somewhat longer on the neck; the forequarters are covered with flattened spines, each with a groove on its upper surface. The underside is covered with similar but smaller spines with grooves on both sides. They have short tails, the hairs on which form strange little cylindrical capsules, hollow within and usually open at the end. The tails make a whirring noise when shaken. Their habits are similar to those of their crested allies but they are forest-dwellers.

Sumatran Porcupine (Thecurus)

Externally, this looks like a half-size edition of the former type but internally it is a very distinct animal, having close affinities with the Brush-tailed species to be described below. Instead of being hairy, their heads are clothed in short flattened spines, grooved on both sides. These also cover the rest of the body and even the base of the tail and rump, except for a limited, oblong area on the hind back. From this grows a mixture of short quills and some stiff long hairs. Their tails, which are very short, are also covered with strange hairs having a thin solid stem about half an inch long and then swelling out into hollow capsules but pointed at both ends and seldom open at their tips. They are more carnivorous than other porcupines.

Brush-tailed Porcupines (Atherura)

There are two species of Brush-tails, the distribution of which is very odd. One is found in West and west central Africa, the other in Burma, Thailand, Malaya, Sumatra, Borneo, and Java. They are astonishingly alike, being about two feet long with a tail half as long. The body spines of the Asiatic species are flattened and bear a deep groove from base to tip; those of the African are round and smooth. The face, underside, forelimbs and neck are clothed in stiff fur blending into bristles and, thence backwards, into the spines which get longer and longer towards the back until they form a domed shieldlike mass that can be erected all over the rump. The "naked" tail is clothed in hard scales and, in the Asiatic species, terminates in a brush formed of alternating broad and narrow hollow spines, in the African, in a mass of extraordinary, dry, pliable structures composed of alternating solid constricted portions and hollow inflated blisters. These are dirty white in color and when agitated by a rapid side-to-side vibration of the upturned tail produce a rustling noise like that made by a rattlesnake. There is in Africa either a quite separate species, a genus, a geographical race, or a mutation that may crop up anywhere, in which the tail is reduced to about three inches in length and is composed solely of the tasselled terminal portion. Brush-tails dig extensive burrows but many of them live in the large cavities under giant termite towers, sometimes in association with pythons, large hairy spiders, and other gruesome types, none of which appears to molest them. They are capable of moving with unexpected speed despite the shortness of their limbs, but when fully alarmed they are quite insane, rushing hysterically through the undergrowth without regard for obstacles and, as a result, often knocking themselves cold on a log or stone or sailing out into some pond or stream. Luckily they swim well.

The Rat-Porcupine (Trichys)

The end of the Old World Porcupine line is reached in mountains in Borneo, Sumatra, and Malacca, where lives a rat-shaped animal clothed in stiff fur mixed with bristles and spines; this, despite its appearance, is a close relative of the Brush-tails. Nothing that is unusual has so far been discovered about its habits, and it appears to lead a life much like that of the Atheruras, though it can apparently climb very well and may collect some of its food in the tops of low trees. It is of a dull brownish color, slightly lighter below and has a prodigious set of whiskers of various kinds sprouting from sundry glandular areas about the head as well as from the muzzle or rhinarium. Its tail is long and bears bristles that are flattened from side to side instead of from top to bottom relative to the surface of the skin. This animal may represent the kind of rodents from which all the more exaggerated porcupines sprang in ancient times, or it may simply be a degenerate form that has been able to get along in an isolated area where the necessity for furious defense did not exist.

NEW WORLD PORCUPINES (Erethizontidae)

The porcupines of the New World are quite different from those of the Old, and had an entirely separate origin. By developing spines and quills, they have come to look somewhat like the latter and their habits are in certain respects similar but they display curious traits quite their own. They are found from Alaska to the Ar-

North American Porcupine

gentine though almost entirely in wooded and forested areas. They come in four distinct forms, two of which are very well known.

North American Porcupines (*Erethizon*)

There are two clearly recognizable forms in North America, one inhabiting the areas east of the great central plains—see accompanying photograph—and the other the areas to the west, from Alaska in the north to New Mexico in the south. The former is smaller, with comparatively short fur from which black-tipped yellow spines protrude; the latter is much larger and is covered in an immense fuzz of plumelike, firm hairs. The western species is much more at ease on the ground than the eastern.

These animals are really much more remarkable than the average person realizes. Although slow-moving and defenseless against a man wielding only a stick, they have managed to survive amidst the sudden welter of land-clearing and general civilization-building to which the whole continent has been subjected in recent years, and they are even making a comeback in the more

thickly populated areas of the east. Unlike their Old World counterparts, they usually make no noise, but if they are discovered they put up a much more serious defense with far less rumpus. Erecting their short, sharp, and barb-tipped spines (they end in miniature harpoons), they maneuver deliberately about so that they keep their rumps facing the enemy as far as possible. If the aggressor rushes in he will get a paw, chest, or face full of spines that pierce his skin and pull out of the porcupine. The spines are never "shot" by their owners. Once inserted into another animal's skin they can seldom be extracted except by a man, and then only with some ingenuity or a very hard jerk. As a result, they normally work on inwards until they finally set up gangrene or penetrate some vital organ.

These porcupines have tremendous jaws and front teeth and gnaw away at anything, especially if it bears the slightest trace of salt, of which they are inordinately fond. The author has a photograph of a large, thick, glass bottle gnawed right through by one of these animals! They will kill trees by ringing them, and collapse the largest log-cabins by cutting through the underpinning. They usually bear two, but sometimes three or four young at a time once a year in the spring, but the babies are the largest, in proportion to the size of the parent, of any mammal known; a single baby may be twelve times as big, comparatively, as a normal human baby of seven pounds, and would thus be equivalent to an 84-pound human child. These porcupines do not hibernate but wander about the woods all winter.

The Bristly Porcupine (*Chaetomys*)

The next animal in what is apparently a series that joins the North American Porcupines to the typical prehensile-tailed tropical forms (*see below*) is a curious Brazilian form found among the bushes and tangled growth around open savannahs and in copses isolated upon these. It is a large form clothed in short spines (as opposed to quills) of various lengths and having a long, non-prehensile tail that trails behind it. This is circular in section and is covered with regular rows of large square scales, from between each pair of which there sprouts a single stiff bristle. Although this animal has been known scientifically for over a century, information upon its habits appears to be nonexistent. In several respects this animal bridges the last gap to the Coendous.

Coendous (*Coendou*)

A considerable number of species of these prehensile-tailed, tree-dwelling Porcupines of Central and South America have been described, but most of these types seem to blend one into the other. However, there are two very distinct groups; the first centered in Mexico, and the second in the Amazonian forests. The former is distinguished by appearing to be covered with thick hair usually colored black which conceals the short spines, and has a long, naked, scaly tail. The latter is covered in short, light-colored spines, is almost naked below, and has a pale, flesh-colored tail. The habits of both are identical, though the Mexican form is often to be found

Coendou

in rocky as well as forested areas. They amble about at night in search of fruits, leaves, unripe nuts, and bark, and sleep in holes in trees or in logs. They are amiable beasts, but if molested will stand up on the tripod formed by their hind legs and their tail, and growl at you over clenched fists in a pugilistic stance. However, they make very docile and rather intelligent pets, showing unexpected affection for friends who call them by some sharp-sounding name. They are most ingenious creatures that can gnaw or pry their way out of almost any cage.

Mountain Porcupine (*Echinoprocta*)

The equatorial forests north and west of the Andes Mountains in Ecuador and Colombia contain a distinct fauna among which is a large arboreal porcupine very like the North American forms, though separated from the nearest of them by the whole of Central America. This animal was first caught in 1860 but was not seen again till 1920 yet does not appear to be too rare around Bogotá, at altitudes of two to four thousand feet. It is clothed in long fur and four-inch quills which are white on the head and neck and banded with black on the body. It has a short, five-inch, spine-covered, non-prehensile tail, short spine-covered ears, and long black whiskers. Most odd, are the hands and feet, which only have four digits bearing long needle-sharp claws: the feet have huge pads where the big-toe should be and this is supported within by both the stub of that digit and by a fan-shaped bone growing below it like a sixth toe.

CAVIES (*Caviidae*)

We would prefer to call these "Guinea-pigs" despite the fact that they are not, of course, pigs and do not come from either of the Guineas. The family clearly includes the Mara, an animal of altogether different appearance from the modest, mumbling little creatures which we all know so well. There are five genera of Cavies, four of which are alike; the fifth, the Mara, is much larger and looks like a small deer. They are exclusively South American rodents but are not found in the Amazonian forest area nor to the north or east thereof.

Guinea-pigs (*Cavia, etc.*)

Our domestic guinea-pigs which now come in a wide variety of colors—white, black, brown, red, yellow, and in combinations of these, speckled, blotched or with long, plumelike hair—are descendants of a species known in the wild state as *Cavia cutleri* which was originally domesticated by the ancestors of the Incas and related peoples in the Andean region in very ancient times. Early Spanish records note that they were brown or reddish in color then, that they swarmed in the dark houses of the local peasantry, and were their only domesticated animal apart from the llamas. They are gentle little animals of enormous fertility and considerable stamina. They make good eating and excellent laboratory animals. In the wild they live in small communities, making runways in dense mountain grass and other

SUSCHITZKY

Wild Guinea-pigs

groundcover, and are nocturnal. There is a much smaller kind in Bolivia that lives in enormous townships like the Prairie-dogs, and tunnels the whole earth just below the surface so that it collapses as one walks over it. Another kind lives in damp and marshy districts around the periphery of the forests on the lowlands of southern Brazil, Paraguay, Uruguay, and parts of the Argentine, and is known in the old natural-history books as the Restless Cavy, though where it got the name nobody, especially in South America, knows. The fourth kind cannot dig at all and has very short claws. It inhabits the open dry areas of the same general region as the last. Despite their apparent lack of notable habits, these little animals could provide material for a large book of information on rodential ways.

The Mara (*Dolichotis*)

This animal is an inhabitant of the open, dry pampas plains and other treeless semi-deserts of the Argentine and Patagonia. It is a comparatively large animal (Plate 90), as tall as a terrier, and is extremely swift. Like most other members of the long-legged, cursorial rodents of South America, moreover, it is very nervous and hysterical, leaping away without regard to anything if suddenly disturbed, and often meeting with serious accidents and even sudden death as a result. Like many deer and other hoofed animals, Maras use the white rump-patch as a signaling device among themselves, its display being a warning to its own kind, who then follow this "flag" in their headlong flight. They are pure vegetarians and are diurnal, often basking in the brightest sunlight all day. They have long, languorous eyelashes to protect their eyes from the sun's glare. They dig their own burrows and sleep in them.

CAPYBARAS (*Hydrochoeridae*)

This animal (Plate 89) is rapidly becoming the best-known gnawing mammal through its appearance in many zoos, and also because it is the largest living rodent. Old males sometimes measure almost four feet over the curve of their backs, from their somewhat horselike muzzles to their button-sized tails, and, in at least one case, weighed as much as 220 pounds. Their bodies are covered in very coarse bristles like those of a hog, and they are without doubt the most mild-mannered, pompous and, from our point of view, dignified of all animals. In the wild they live mostly near water, and although they are cautious beasts, they will sometimes stand in disdainful silence and well-ordered ranks on the bank of a river, and watch a rowdy motorboat chug by; in captivity, however, they quickly adapt themselves to the ways of the household and take up a position of gentle determination with regard to all its members, including dogs. The author encountered one in Suriname trained as a reliable "seeing-eye" by a blind Boer farmer. They apparently have an IQ equivalent to that of the pig, which comes second only to apes and ourselves. They are grazers and browsers, and spend a lot of time just sitting complacently on river banks and blinking. The females bear half a dozen young a year, in one, two, or three litters. They all follow the mother about for many months. Males usually have small harems of females of various ages and take many years to reach full size.

PACARANAS (*Dinomyidae*)

A very unexpected creature was found wandering about in the backyard of a house in a small township in Peru in the year 1872. Nothing like it had ever been seen before by anybody except the natives of a few remote areas in South America, and nothing more was seen of it for some decades. It is still a great rarity in collections, but has now begun to turn up alive from time to time. As Plate 91 shows, it is rather an appalling-looking beast, covered with large spots and having a long tail. Nothing much is known of its habits in the wild state except that it digs, is a vegetarian, and can put up an unexpectedly stolid defense when cornered. In many ways it links the Cavies, Capybaras, and the Pacagoutis on the one hand, to the long-tailed Hutias and Degus, to be described later, on the other. It has a cleft upper lip, and only four fingers and toes. Its tail is clothed in coarse hair.

PACAGOUTIS (*Dasyproctidae*)

This combination title, coined from Paca and Agouti, covers four distinct animals, two of which are among the best known of all tropical American animals, and the other two of which are very obscure and rare. They are forest animals and feed primarily on fruits and nuts, though it must be understood that such things, falling from jungle trees at all seasons of the year, contain all manner of strange vegetable objects from immense three-foot beans to tiny blobs of vermilion jelly that might not be recognized as either a fruit or a "nut" by any northerner. Pacas have beautifully white and tasty meat and are a regular item of diet in many parts of South America. Agoutis are a little more gamy but so delicious and nutritious that they were once imported into the West Indies and released as food for the slaves.

The Paca (*Cuniculus*)

Though exceedingly common from Mexico to the Argentine, and in some places living to as many as fifty per acre, this large rodent is little understood even by most zoologists. When young or partly grown it is a common exhibit in zoos, fairs, and circuses, often under such titles as "What-is-it" or "The largest Rat in the World," and it sometimes appears on the dinner tables of tourists in the countries where it lives under various innocent or high-sounding titles. Yet, the size to which it can grow in the wild is seldom realized. A length of well over two feet is not uncommon and old males de-

Paca

velop grotesque heads of huge size with the bones of the cheeks forming enormous, blister-like swellings. Their bodies may become bulbously fat yet they remain one of the fastest things on four feet.

They are rugged fighters, suddenly jumping opponents or imagined aggressors, and delivering frightful wounds with their enormous upper front teeth with which they can rip their way through planks of almost any thickness. Despite this, they make very amiable pets, will follow you around like a dog, and act as a substitute for a large garbage pail, eating almost anything and in prodigious quantities. In the wild, they dig large holes and rummage about the forest floor at night, gnawing on fruits, especially those of various trees of the fig family. Their skins are extremely thin and fragile, and large strips may be ripped off in their headlong rushes through spiky undergrowth. However, it grows back again in a quite miraculous way, complete with the hard bristly hair. Pacas' eyes are enormous, and bulging, and they have a tremendous array of whiskers, yet, as experiments have shown, their senses of smell and hearing are also uncannily acute. They dig with all four feet and their teeth, and even large roots are no obstacle. The young are little larger than a mouse when born and take many years to reach full size.

The Mountain Paca (Stictomys)

High in the mountains of Ecuador and Colombia there is a small relative of the Paca, brown in color, that makes extensive burrows. It stores up a lot of fat in its body at one season of the year, and appears to go into a semi-hibernation during the cold, wet season following. Though these animals are really quite unlike the typical lowland Pacas (Cuniculus) the differences are mostly internal. Apart from their smaller size and more slender muzzles the only noticeable distinction is the very long, slender digging claws and the strangely granulated appearance of the soles of the hind feet. The eyes, in addition, are less bulbous.

Agoutis (Dasyprocta)

Even commoner than the Pacas are the numerous, closely interrelated species of delicately-built, long-legged Agoutis. They too range from Mexico to the limits of perennial green vegetation in the Argentine, and they are just as common in captivity but are even more commonly ignored. Admittedly, there is nothing much about an Agouti that catches the attention of the public; yet they play a considerable role in Nature and on closer investigation prove to have remarkable habits.

They are one of the most "nervous" mammals known, having every sense triggered for instantaneous action. The author once saw a party leap off a hundred-foot cliff for no reason other than the snapping of a twig on the forest floor. Despite their excessive timidity they are terrifyingly violent among themselves, individuals sometimes stamping other members of their family party to death systematically, relentlessly, and with no cause apparent to us. They are also predominantly refuse-eaters, but they sometimes browse. They take their food in their forepaws and eat it as they sit on their haunches and in doing this they display much finger-

SAN DIEGO ZOO

Agouti

dexterity despite their long claws. Like the Paca, they are covered with glands from many of which spring sensitive bristles that can be moved by special muscles in the skin. They do not dig, but scratch industriously and thereby make "foxholes" which they often roof with crisscross twigs, and then cover with leaves. They are communal animals for the most part but their habits vary widely over their vast range. They are known even to climb sloping trees to collect green fruits; they swim well, and some have been seen grubbing about on oceanic foreshores eating shellfish. A twenty-foot leap from a sitting start is recorded, and they can jump chasms in a manner that defies understanding.

The Acouchi (Myoprocta)

Much smaller than the Agouti and having a tiny, slender tail, bearing a little plume of white hairs, the Acouchi (Plate 93) of Guiana is an altogether different animal. Although even more slender and delicate and supersensitive, it is altogether less hysterical and can be tamed, whereupon it shows remarkable intelligence and even apparent affection for those it trusts. It is a very rare animal and little is known about it. It appears to live among marsh herbage in certain very limited areas. The coarse but beautifully colored hairs on its hinder back may be six inches long and can be raised at will. This is done when the animal is alarmed, when it also drums with its back feet and grinds its teeth.

HUTIAS (Capromyidae)

We now come to a remarkable collection of "living fossils" with a very strange distribution. As a group they are hard to define, but their general form may be seen in the accompanying photograph of the Hutiacouga of Cuba and the Coypu of South America. To most people they look like enormous, fat rats; to anatomists they present quite another picture. Five of the living types are isolated on various of the West Indian Islands, where they represent the only indigenous mammalian land fauna apart from bats and the Solenodons. A sixth is isolated in the mountains of northern Venezuela, and the seventh, a rather different animal with the delightful name of *Myopotamus coypu*, is widely distributed in southern South America. It has been introduced to North America and is known to all discriminating women as the fur "nutria."

[151]

Hutiacouga

Long-tailed Hutias (*Capromys*)

There are two species groups, one with and the other without prehensile tails, found in limited areas of Cuba and, in the case of the latter, the Isle of Pines. The best known is the Hutiacouga, a 22-inch rodent with coarse, long fur, a rat-shaped body, and a short, very thick tail covered with sparse bristles. It lives in trees and appears to be a leaf-eater though there is a belief that it also takes a lot of the little lizards (*Anolis*) whose relatives are sold in America as "chameleons." The Hutiacarabali is smaller and has a slightly longer tail that is prehensile throughout its terminal half, and is more wholly arboreal.

Short-tailed Hutias (*Geocapromys*)

It might be better to call these Tailless Hutias or even West Indian Coneys, for the tail is a mere stump concealed in the fur while the animals are astonishingly like the African Rock Hyraxes or Coneys in general form. There are three species with a very odd distribution. One is found on the easternmost Plana Cuays of the Bahamas only, the second in the Blue Mountains and Cockpit Country of Jamaica, and the third upon an absurd little half-mile long area, known as Swan Island, in the Bight of Honduras which is claimed by three countries but inhabited only by a weatherman for a U.S. airline. How this animal ever became isolated there is a challenge to geologists who have for long been endeavoring to reconstruct the past geography of the Caribbean. It must have survived there since very ancient times. All three species look much alike and lead much the lives of rabbits among tangled vegetation, hiding in holes and among rocks. They are nocturnal and, from the lay point of view, quite useless, yet they are living fossils that, by their odd distribution, tell us much of the past history both of the area where they live and of the Rodents as a group.

Zagoutis (*Plagiodontia*)

About a hundred years ago a French Jesuit priest interested in palaeontology unearthed some semi-fossil-

ized bones and teeth of a fair-sized rodent from the clay covering the floors of caves in Haiti, and surmised that the animal was either only recently extinct or still living. Subsequently, a few fresh specimens of the bones and some alleged skins of the same animal reached museums variously labelled as coming from Jamaica, the island of Dominica, or Hispaniola. It was not until this century, however, that any definite evidence of the existence of this animal or its true origin came to light. Since then a few live specimens have been brought to the United States. It appears that this large guinea-pig-shaped rodent with a naked prehensile tail lives in the tops of trees in the recesses of the mountains of the northern part of Haiti only, and that it feeds exclusively on leaves and flowers of certain trees. Its distribution is as much a mystery as its habits, and its present location indicates an ancient land-bridge between the Greater Antilles and South America.

The Venezuelan Hutia (*Procapromys*)

In 1899 a small hutia-like animal, with a tail half the length of its head and body, was collected in the bare rocky mountains facing the Caribbean between the port of La Guira and Caracas, which is situated above the port. On examination, this proved to be anatomically of great interest, and virtually a living remnant of the ancestors of the other Hutias (hence its name *Pro* meaning "coming-before"), the *Capromys*. A few other specimens have since come to hand but nothing whatever is known of the animal's habits. One is led to wonder what other small missing-links may remain to be found in areas only a few miles from the capitals and principal ports of other countries.

The Coypu (*Myopotamus*)

This large, aquatic rodent (Plate 94) is a native of South America, from Peru almost to the lower extremity of the continent and is found on both sides of the Andes but outside of the forests. The heavy-set body reaches almost two feet in length, and the tail is about half that length. It is covered in a very fine underfur—the "nutria" of the fur-trade—and this is, in turn, covered by a long coat of coarse over-hair. They have five toes and fingers, the former completely webbed. They make shallow burrows in pond and river banks with an enlarged nesting chamber at the back, and they produce up to eight young at a time; these have the happy habit of riding on their mother's back when she swims. They are vegetarians but of rather a special nature, as has been proved since they were introduced into North America. This was first done in 1899 without results, and then again, unsuccessfully, in the early 1920's. The animals were then released and found a niche for themselves in much the same areas as the indigenous muskrats, but they never have interfered with those animals and have established themselves over large tracts of suitable marshy territory in Mississippi, Louisiana, Texas, and in Washington state.

TUCOTUCOS (*Ctenomyidae*)

As we have seen from the start of this survey of the mammals, almost every group of ground-dwelling form

[152]

Tucotuco

has given rise in one area or another to a burrowing, molelike form. Sometimes closely related families have fossorial types with reduced external ears, lozenge-shaped bodies, short, trailing tails, and stout limbs with large claws. The Tucotucos are the fossorial form of the Hystricomorphs and they are as well, if not better known, to all country people of the Argentine as are the Pocket-Gophers of our own western states. There are several species varying in size from about six inches to a foot in length, exclusive of the tail. They live in colonies in drier and particularly in sandy areas, where they spend their time tunneling along just beneath the surface like moles, and throwing up mounds of earth. They make an extraordinary noise night and day that has been aptly described as the hammers of the Nibelungen of German folklore, tapping away forever in unison upon tiny anvils to the accompaniment of some unheard chant.

OCTODONTS (*Octodontidae*)

Another group of ratlike rodents from South America contains five quite distinct genera that, with one exception, are of interest only to specialists—though, if more were known about them in their natural habitat they would probably prove to be just as interesting as any other animals that we know. Two are worthy of mention; the first, known as the Cururo (*Spalacopus*), looks much like a Tucotuco but has small ears and spends more time above ground; the second, *Aconaemys*, is found only in the higher parts of the southern Andes which are covered with snow for more than half the year. The best-known type, the Degu, looks like a large rat and has a tufted tail that it carries arched over its back in a squirrel-like manner. It is very common in many parts of Peru and Chile, where it lives in large communities with interlocking burrows among tangled, low vegetation. It makes large stores of food and generally behaves like a squirrel but has no reason for hibernating or aestivating, since the climate where it lives is too equable. They are best known as garden pests.

THE RAT-CHINCHILLA (*Abrocomidae*)

Closely related to the Degus but found in Bolivia are a few species of rat-sized rodents with excessively soft, fine, and silky fur. They have large ears and plain, naked, ratlike tails and live in small family groups on the open valleys of the Alto Plano. In some respects they bridge the gap between Octodonts and Chinchillas.

CHINCHILLAS (*Chinchillidae*)

Almost everybody has heard of the Chinchilla because of the value once placed upon its fur and, in the United States at least, because of the volume of publicity and advertising that has been devoted to this animal for almost thirty years. What is not commonly known is that the animal of the fur trade is only one of several related rodents and that it is not extinct as was once supposed. Fifty years ago the pelts of these little rodents commanded an enormous price because their fur is so fine that you cannot even feel it with a fingertip if your eyes are closed. After its discovery, in the high Andes, it was ruthlessly trapped and became so scarce that its

Mountain Chinchilla

Chinchilla

pursuit was finally abandoned, and it was pronounced extinct. In 1922 an American engineer named M. F. Chapman, after a five-year search, managed to bring fourteen live specimens down from the mountains of Chile and by the exercise of extraordinary effort and ingenuity got them safely to the United States in a refrigerated cage. There are now estimated to be about 70,000 in this country. The early concern for the temperature at which the animals had to be kept is now over, for it has been discovered that this sometimes rises to a hundred degrees in their native habitat. They have proved to be very hardy little animals, but they do require enough heat to keep their drinking water from freezing in sub-zero weather. In the wild, they are nocturnal and live in communal burrows on the mountains above the tree-line in Chile and Bolivia. They have rather weak claws and a big-toe that bears a small, flat nail. The natives used to flush them from their burrows with trained weasels of a local species just as Europeans use the Ferret to hunt rabbits. Another, short-tailed species is found in Peru.

Mountain-Chinchillas (*Lagidium*)

Of much larger size, even than the short-tailed species, is Cuvier's Mountain Chinchilla, which grows to a

Viscacha

length of twenty inches and has a twelve-inch tail. It is of a beautiful pearl grey color above, and yellow below, while the tail is black and carries a crest. The multiple whiskers are black and may be eight inches long. They exist on moss, grass, and some roots, and live on the highest ranges of Peru, Bolivia, and Chile up to 16,000 feet. Like the common Chinchillas they are wonderful rock-climbers, and although not too shy, are adept at doing an instantaneous disappearing act into their holes. Moreover, these holes—beyond an arm's length from their entrances—always remain at a constant temperature of about fifty-four degrees despite the excessive heat of day and the intense cold of night at the altitude where these animals live.

Viscachas (*Lagostomus*)

The commonest of the family and one of the best known of all South American animals are the large rodents, known as Viscachas, which once swarmed all over the plains of southern Brazil, Uruguay, Paraguay, and the Argentine pampas. They measure almost two feet when stretched out and have short, eight-inch tails. They live in large communal warrens called "viscacherias" that may cover 2,000 square feet of territory, and contain twenty to thirty individuals residing in a dozen or so burrows. The entrances of these are huge, being a yard in diameter, and in the chambers below all manner of other animals may live with the Viscachas— burrowing owls, a bird called a *Geositta*, and snakes, while a kind of swallow often lives in holes in the sides of the entrances. The little animals clear the ground all around their warrens and they collect things as do Pack-rats. They are considered a pest and are being rapidly eradicated.

PORCUPINE-RATS (*Echimyidae*)

This purely descriptive English name is adopted to cover an enormous host of small, long-legged, rat-shaped members of the Hystricomorph rodents found throughout South America, and is chosen to distinguish them from the Sciuromorph Spiny-Rats which have cheek-pouches and belong to the Pocket-Mice. There are no less than fourteen described genera of Porcupine-Rats representing an unknown number of species, for these animals come in a seemingly endless number of variations, while comparatively very little is known about them. They are identified technically by differences in their teeth, measurements, fur-texture, and other characters that are quite beyond our scope herein. Most, but by no means all, have bodies clothed in small, slick, flattened spines; others wear a mixture of spines and bristles or fur; still others are furry or even silky. Most of them jump about on stiltlike hind legs. They are primarily forest animals, taking the place there in South America of the Old World Mice and the Wading-Rats of Africa. They have long tails, but several species throw out a completely tailless mutation.

AFRICAN ROCK-RATS (*Petromyidae*)

This and the following large African rodent appear to belong in the Porcupine-formed tribe but there is still some element of doubt about this. They certainly do not belong in either of the other two great tribes, but they may be early offshoots from the main rodent stock, and of an intermediate position in the scheme of life. The Rock-Rat (*Petromys*) is found only in South Africa and most commonly about Namaqualand, where it is fairly common in dry rocky areas. It feeds on a strange diet of the flowers of certain plants and is abroad both by day and night. It is very like the South American Degu, is about six inches long and has a thick-based, tapering tail of the same length. It is covered in rather long coarse hair and has five toes, four fingers, and a stump of a fifth. It sleeps in holes in rocks.

[154]

CUTTING-GRASS (*Thyronomyidae*)

This large, cumbersome rodent (Plate 92), found all over Africa south of the Sahara in woods and other vegetation outside the forests, but not on open savannahs, dry scrub or desert zones, is commonly called the "Ground-Rat" or the "Cane-Rat." However, it is not a rat, being more like a large, dry, Coypu, and although it lives on the ground, it does not seem to have any special predilection for *canes* of any kind even if they are available. It is a garbage or dropped-fruit eater, as well as a general vegetable browser, and grazer—hence its most descriptive West African pidgin-English name of *Cutting-grass*.

It grows to a length of almost two feet, is very heavy-bodied and has a short, seven-inch and very thick tail. It makes large, strongly woven grass nests in depressions under the densest vegetation and bumbles about at night making a most remarkable noise. This sound is like that of a heavy pebble dropped on a frozen lake and has the same metallic ring as the note of the Indonesian "Brain-fever Bird." The noise made by both these animals can indeed drive one insane since there is no logical sequence to the issuance of the "boings." The natives of the countries where the bird and the Cutting-grass are found bet on both for hours every night on the number of "boings" to come next. The Cutting-grass also makes a curious thumping noise, but whether this is done with the feet or emanates from the throat is not known. It never comes with the "boings" but goes on intermittently most of the night. The teeth of the Cutting-grass are very sturdy and the two upper front ones have three deep grooves down their outer surfaces; the lower front ones are smooth chisels. They have been known to tear a hole in corrugated iron fences.

GUNDIS (*Ctenodactylidae*)

Nobody quite knows what to do with these curious African rodents. They have been bandied about between all three of the great tribes but it is, and should always have been obvious that they are not of the Squirrel-like group. There are four recognized genera, all African, and all confined to the scrub and semi-desert belts that lie around the true deserts and between them and the savannahs. They are mostly pale-colored with rather coarse hair, and are about the size of water-voles. They live among rocks and are predominantly nocturnal. The Common Gundis (*Ctenodactylus*) of the southern border of the Sahara are virtually tailless; *Pectinator* of the Somaliland desert periphery has a long, rather bushy tail. All, however, agree in having only four fingers and toes, and in that the inner two of the latter bear most curious horny combs on their inner sides, covered with a line of stiff, curving bristles. These structures—unique among rodents—are used for currying the fur, a practice which the little animals carry on most industriously and very frequently, sitting up on their haunches and going over their whole bodies systematically. They live in communities of varying sizes and sometimes of huge extent, and some of them dig extensive burrows.

Fin-Footed Mammals

(*Pinnipedia*)

THE seals are universally recognizable and form a compact little group of aquatic mammals that cannot be mistaken for any other. There are only about a score of recognizable species and they are all very much alike in external form though they display a considerable range in size. They are usually made a sub-order of the Carnivores and they almost certainly sprang from some animals that were common ancestors of both. However, they are today so different from all living Carnivores that they must be treated quite separately. Also, apart from one fossil form, which may have been a seal-like otter, or an otter-like seal but is not a missing link and therefore does not indicate any true relationship between the two, no animal living or extinct gives us any real clue to their origin. At most they may have sprung from the very earliest doglike Carnivores; otherwise, they had their origin farther back in time and down the genealogical tree. In either case this sets them apart from all other mammals.

The classification of this homogeneous little group of animals presents several difficulties. It is primarily obvious that there are two major kinds of fin-footed mammals, and these may have quite separate origins. The first we call here the Sea-Lions though this contains the eared Fur-Seals and the mighty Walrus, in addition to the long-necked, small-eared Sea-Lions proper. In purely popular parlance these might well be called Sea-Lions, Sea-Bears, and Sea-Sabertooths, for the Walrus bears just such a relationship to the others that the extinct Sabertooths held to the Cats and Bears. The second group contains all other seals, including the enormous Sea-Elephants.

Seals as a whole are worldwide in distribution, being found in almost all seas and oceans and even in the great landlocked Lake Baikal in Siberia. However, they are by preference cold-water animals. There are distinct seal populations in the northern and southern hemispheres usually represented by different genera or at least species. In between is found only one kind of true seal—the Monk Seal—of the Mediterranean, tropical Atlantic and the Hawaiian Islands. There are no seals, as far as is known, in the central Pacific tropical belt, but there are Sea-Lions in the temperate belts.

Seals are wholly carnivorous but they are by no means invariably nor entirely fish-eaters. On the contrary most of them eat seabirds, other seals, shellfish and a wide range of smaller marine life. The Walrus apparently munches up a certain amount of seaweed along with its normal shellfish diet.

Under their skins, seals have a thick layer of fibrous

tissue impregnated at most times with oil, which may be called blubber. The amount of oil therein, however, varies more from place to place and from month to month than it does in the whales. They are not so wholly aquatic as the whales since all of them can and do flop out on to ice or rocks and some of them, like the Sea-Elephants, spend a great deal of time on shore. The Sea-Lions can bend their hind limbs forward and raise their forebodies on the front limbs so that they can hump along at a considerable speed by arching their backs like enormous caterpillars. Most seals are gregarious though the two sexes may only congregate once a year to breed. The Sea-Lions and the Sea-Elephants are commuters; other seals usually stay around a limited sea territory all their lives but some are true nomads.

SEA-LIONS

The three genera included in this division undoubtedly have a common ancestry and appear to be on the whole more primitive than the other fin-footed ones. The Walrus is a highly specialized beast but it still displays certain characteristics that point towards the terrestrial ancestry of the order as a whole. Sea-Lions have longer necks than Seals and can turn their heads all about. Their limbs are longer and have a greater amount of freedom.

EARED SEALS (Otariidae)

These animals, of which there are four distinct genera, are the "seals" of common parlance. All trained performing seals are of the first genus, and the "sealskin" of the fur trade comes from the third. They are spread all over the Pacific and a part of the South Atlantic. The foreflippers have five fingers, each bearing a rudimentary claw: the hind feet have the first and fifth toes enlarged and clawless but the three small middle toes bear well-developed claws. The webbing extends beyond the tips of the digits and the soles of both front and back flippers are naked.

True Sea-Lions (Zalophus and Eumatopias)

The smaller of the two kinds of True Sea-Lions (Plate 95) is spread all across the Pacific from the west coast of North America to Japan, and south to and around Cape Horn into the South Atlantic, and west to New Zealand and Australia. One species is found around the coast of South America from the equator on the west to the mouth of the La Plata on the east and on the Falkland Islands. The males of this animal bear a fairly long mane. Hooker's Sealion is found on the Auckland Islands south of New Zealand, and another species inhabits the Bass Straits between Australia and Tasmania. This last species (Z. lobata) has a thick underwool when young and is in other respects more like the Fur-Seals or Sea-Bears. The largest species belong to the genus *Eumatopias* and are North Pacific forms, which range north from California. Large males may weigh as much as three-quarters of a ton. All Sea-Lions are gregarious though the South American species wander extensively during most of the year and only assemble to breed. The males arrive at the breeding places first and about the beginning of the hot season according to the hemisphere. Only one pup is born a year. Sea-Lions do comparatively little damage to commercially worthwhile fish, eating a wide diet of squids, seabirds, shellfish and other small animals. They are very inquisitive and extremely intelligent. Just why they should be so universally adept at balancing things on their noses is not known and seems beyond our ken though they can often be seen playing catch with a fish or other object while standing straight up in the water. They tame easily and then like nothing better than to get into a swimming pool with human occupants. They can, however, be very dangerous in their "rookeries," and one should never pat even a trained one on the head. They are consummate divers, can stay under water a fairly long time and may attain a swimming speed of twenty knots in short bursts. A Sea-Lion can project itself clear out of the water on to the top of a five-foot wall.

Sea-Bears (Arctocephalus and Callorhinus)

There are two distinct genera of Sea-Bears or Fur-Seals, one (*Arctocephalus*) being found on islands all round the Antarctic and on the southern extremities of South America, Africa, and Australia, and about New Zealand. The other (*Callorhinus*) is confined to some islands in the Bering Sea in the North Pacific (Plate 96). Like the Sea-Lion, the males of these animals are much larger than the females and occasionally veritable giants may be found among them. The males of the Northern species measure over six feet, the females under four. Then again, the males develop enormous domed heads and have a habit of raising the front part of their bodies straight into the air, giving them a most menacing appearance. These animals roam all over the ocean throughout the year but all assemble at a set time for a brief breeding fiesta on the Pribiloff Islands in the North Pacific. The males gather large harems of females and guard them with the utmost vigor while an assortment of young, weak, and old males are forced to live in bachelor circles.

Their pelts are of great value as these are clothed in a dense yellowish underwool and a firm shiny overcoat. To prepare the skins the inner layer of fat is scraped away because the long hairs of the overcoat grow from this and are thus loosened and fall out. The skins of young males and the females are the best quality. The Northern Fur-Seal was reduced to a bare hundred thousand by 1910 as a result of hunting but has now passed the three-million mark again because of stern protective measures. The Fur-Seals of South America, South Africa, and the New Zealand–Australia region are each distinct species. All of the Fur-Seals appear to be principally squid-eaters though they take fish.

WALRUS (Odobaenidae)

This very singular and vast Arctic mammal was once, long ago, common all around the periphery of the Arctic Ocean but is now found only in a few rather limited areas about the north of Greenland and the Bering Straits. It has always been much prized both for the ivory in its upper tusks which may grow to over two feet in length, and also, in early historic times, for its hide, from which the strongest known ropes for ships were made. They are huge creatures, large males measuring over twelve feet and are so rotund that they may weigh over a ton and a half. Females never reach more than two-thirds this size or bulk. They have little claws on their fingers and large ones only on the outer two toes. Like the Eared-Seals, they can turn their back legs forward and raise the back part of their bodies to hump along; but their scientific name means literally "Those who walk (*baino*) with their teeth (*odos*)" and it is true that the males may pull themselves on to, over, or off ice floes with their tusks. One or two pups are born in the late spring and are suckled for two years. The food of the walrus is almost entirely composed of shellfish, starfish, and sea-urchins for which they dive to considerable depths. They have only sixteen teeth besides the tusks, and all of these are reduced to rather small rounded cones admirably suited for crushing their hard food. They are naked and their skins are rough and warty but they bear enormous "Old Bill" moustaches of huge translucent bristles that curve forwards and downwards.

Walrus

SEALS

The remaining Fin-footed Mammals are commonly and properly referred to as Seals, obviously have a common origin, and are normally grouped in one large family. Nevertheless they fall into four separate groups and one of these—the Crested Seals—is very distinct. There are northern, central, and southern groups of genera and the Crested Seals are represented in both hemispheres.

NORTHERN SEALS (Phocinae)

There are three exclusively northern genera of small seals—the Common Seal, the Grey Seal, and the rare and odd Bearded Seal. There are four very distinct and two additional species in the first genus but only one in each of the latter.

Common Seals (Phoca)

The best-known species is usually called the Harbor Seal and is distributed all over the northern Atlantic and Pacific Oceans ranging from the ice-front (which the animals do not make use of), to the Mediterranean and southern New Jersey in the Atlantic, and to California and Kamchatka in the Pacific. They are small animals, averaging about six feet long, that live in small colonies of a few families about fixed locations all along the coasts and especially where there are rocks, and about islands. Nonetheless, these seals often come swimming by sandy public bathing beaches and they are occasionally met with far out to sea. Generation after generation will be born about the same cove or rocky promontory and none appears to leave though they live to a considerable age. There may be miles of suitable coast for seals without colonies, and then half a dozen will be found close together in a small area. They prefer still water and bays and they ascend rivers to completely fresh water, and even enter the Great Lakes. They spend some time ashore at every tide and the young are born in late May at night on the rocks. They are fish-eaters but take other food and they are inordinately inquisitive and friendly. They tame easily, are very gentle and sometimes quite pathetic in their attachment to human beings. They are a variable yellowish-grey in color and lighter below. The young are yellow but lose their baby underwool within a few hours of birth.

The second species of this genus is the "hair-seal" of the fur trade, known as the Greenland, Harp, or Saddlebacked Seal. This is found farther north and both in the Atlantic and the Pacific. They are migratory to the extent that they all assemble at a few points to breed on the ice floes in spring before these convenient floating rafts melt away. Notable points are off the Jan Mayen Islands, and at the mouth of the St. Lawrence River. The young, born on the ice, are clothed in a pure-white fluffy coat and it is for this that they are butchered in the most atrocious manner, being clubbed and peeled, often before they are dead. That anyone could kill these tiny helpless creatures with their enormous, pleading, liquid, black eyes even for the profit motive is clear proof that Man's patiently developed and much vaunted moral concepts are a complete failure. Left alone, the young keep this coat for some weeks but they then pass through a whole series of color changes lasting five years, and end up a pale creamy yellow with a large black mask and a large, dark brown, lozenge-shaped ring around their backs. Where these seals go throughout the rest of the year is unknown but they scatter far and wide over the oceans and sometimes appear off coasts in enormous loose companies.

The third species—the Ringed Seal—is the smallest of the genus and is dark grey above with numerous oval white rings, and light grey below. It is more slender than the Harbor Seal and has proportionately much longer limbs. Its habitat is the true Arctic waters of both east and west hemispheres, and since it stays in one place from year to year while these places freeze up every fall, the animals have developed the habit of making potholes in the ice as it freezes and descending and ascending through these to fish throughout the winter. Closely related to the Ringed Seal are two very similar

Harbor Seal

species found in, of all unexpected places, the half-salt Caspian Sea and the entirely fresh-water Lake Baikal in Siberia. How they got there is a mystery and more especially in the case of the Caspian animals, for it is a very long time even geologically speaking since that sea was connected to any northern ocean. The last species of this genus is perhaps the most extraordinarily colored and marked of all mammals. It is a small, chocolate-brown animal found only in the Bering Sea and adjacent parts of the Arctic Ocean but it has vivid yellow rings right round its neck, around the front-flippers where they join the body, and some distance in front of the tail. It is known as the Ribbon Seal and is usually taken off Alaska and the Aleutians.

The Grey Seal (Halichoerus)

This is a much larger animal, males reaching well over twelve feet, and is somewhat different in habits, being more sluggish and preferring open, exposed rocks where there is a constant swell and surf. It is a comparatively rare animal, confined to the North Atlantic, where it stays in a belt south of the ice and north of temperate latitudes. It is commoner on the European than the American side of the Atlantic, ranging from northern Norway to south Scotland and Ireland. It apparently shuns oceanic islands but appears again about Greenland and is sometimes found on Nova Scotia. It is of a shiny steel-grey color, dappled with lighter and darker diffuse marks. Unlike all other seals it gives birth in the fall. It is a shy lone creature endowed with incredibly acute hearing. (Plate 97)

Bearded Seal (Erignathus)

This is the largest, rarest, and oddest of the northern Seals, found only occasionally from Newfoundland to the Arctic, and in some areas along the Alaskan coast. Specimens are still found around Iceland and they occasionally appear off the Scandinavian coast; there are also reputed to be colonies in the Sea of Okhotsk. It is a very thick-skinned animal that spends most of its time alone on the high seas and stays in the coldest water near the ice. It appears to be a squid-eater and has very small teeth most of which drop out before the animal is in any way old. It is of a dull grey color but is notable for its exaggerated moustache of enormous, flattened, and slightly curved whiskers. Bearded Seals are hunted by Eskimos for their tough hide and tender flesh and they display a most singular trait when shot, leaping into the air and turning a complete back somersault

from the ice into the water, so that one never knows if they are dead or alive.

TROPICAL, OR MONK SEALS (Monachinae)

One, two, or possibly three species of small dark brown seals are found in warmer waters and constitute a distinct sub-family. There is one species that occurs in the Mediterranean, about the Straits of Gibraltar, and down the west coast of Africa to the Bissagos Islands, as well as on the Canaries and Madeira. Another species used to be rather common in the Caribbean, where it was first recorded by Christopheren Colon on his first voyage. It was subsequently hunted almost to extinction but is still found on some isolated islands off Cuba, Jamaica, and south Hispaniola. There appears also to be a small True Seal in the Indian Ocean centering on the Chagos Archipelago. Monk seals also occur around the Hawaiian Islands. Monks have small claws on all the fingers and toes but these may be missing from the two outer toes which are much longer than the others. They are open water fish-eaters and appear to live in colonies about fixed rookeries.

SOUTHERN SEALS (Lobodontinae)

There are four genera of seals exclusive to the Antarctic. They are obviously of common origin and they differ markedly from the Northern Seals in several respects. There is only one species in each genus.

The Crab-eating Seal (Lobodon)

These are small seals, colored a uniform olive brown above but with yellow cheeks and undersides. They stay by the Antarctic coasts along the ice-front, and also around islands off the coast, and appear to be, as their popular name implies, crab-eaters. However, their teeth are very remarkable, there being small front teeth, long slender tusks, and then five cheek-teeth on each side of each jaw, all divided into three sharp points arranged fore-to-aft. This mechanism looks much more like a device for catching slippery items like pelagic fish or squids rather than for crushing crabs. These animals have no claws at all on the back feet.

The Leopard-Seal (Hydrurga)

Of much wider range is the somewhat fearsome Leopard-Seal or Sea-Leopard. This ranges from the periphery of the Antarctic Continent northwards in all directions to the southern coasts of South America, Australia, and New Zealand, and is seen off the Marie Islands south of the Cape of Good Hope. The animal is about twelve feet long when fully grown and has the longest, most formidable jaws of any seal. The front teeth are long and sharp, with big slender tusks or canines but the cheek-teeth are shaped like tridents. The skin is usually dingy brown with some dark brown and many lighter spots but specimens colored very much like leopards are sometimes taken, the whole pelt being yellow with quite vivid black spots that are sometimes hollow and contain white areas. These seals do not migrate but they do sometimes foregather in enormous crowds, when they may travel long distances in company and then vanish. This species will attack animals that land on or fall into water and they will feed on scraps from whaling operations on the high seas.

Ross's Seal (*Ommatophoca*)

This little known but apparently not uncommon Antarctic seal is obviously specialized in certain respects for a particular life in some as yet unexplained niche. Its eyes are proportionately very large—in fact enormous—and its teeth are small and weak. Its coloring is remarkable in that the back is greenish and the belly yellow but the sides are marked with bright yellow stripes running obliquely from top-front to back-bottom. The animal is a bottom feeder taking seaweed as well as the smaller soft-bodied animals. It ranges north to the same extent as the Sea-Leopard. When on the ice it has a strange habit of flattening itself out until it is almost disc-shaped.

Weddell's Seal (*Leptonychotes*)

This seal is rather like the Sea-Leopard, being pale grey above spotted with yellow or white, and pure yellow below. It is found on the pack ice and occasionally on the coasts of extreme southern South America, the South Orkneys, Shetlands and Georgia, and on the Falklands. It is a shore animal and the commonest seal of the Antarctic. Curiously, it also has rather excessively large eyes and weak teeth.

CRESTED SEALS (*Cystophorinae*)

These remarkable animals come in two very distinct lines—one, a fairly normal seal-like creature except for its head; the other, an altogether unique form known commonly as the Sea-Elephant. The former is confined to the upper North Atlantic from a line drawn from Newfoundland to Norway and north to the ice-front. The Sea-Elephants were formerly found in vast herds on almost all the islands around the Antarctic Ocean, namely in the southern South Atlantic, the Indian, and perhaps the South Pacific Oceans. They also had colonies up the west coast of South America almost to the tropics. Then, isolated colonies also dwelt on islands off the coast of California far to the north.

The Crested Seal (*Cystophora*)

When fully grown these seals are about eight feet long. In color they are some shade of bluish-grey to black, with small irregular and diffuse white spots all over the back. They have small, rather weak, peglike teeth, very small foreflippers and extraordinary hind feet shaped like fans, the two outer toes being enormous, clawless and bearing long flaps of flesh continuous with the webbing. The males gradually develop over some years a most remarkable bladder-like excrescence on the top of the muzzle that is partly supported by a median ridge of bone, is hollow and is connected with the nasal passages. This can be inflated at will and, with it, horrible watery noises can be produced. The young are pure white and fluffy. Crested Seals live in large family parties on the open sea and migrate north and south with the seasons, the young being born on the ice in March far out to sea. The males stage terrific battles at the mating season, when they make a most awful

SORENSON

Leopard Seal

uproar that can be heard for miles. They are aggressive creatures and may attack small boats.

Sea-Elephants (*Mirounga*)

Even a photograph cannot give a true picture of these fabulous and ridiculous creatures. Not only are they immense, males growing to eighteen feet in length and as much as fifteen feet in girth, but this sex is adorned with an eighteen-inch trunk that normally flops down over the mouth but which is also connected with the

SORENSON

Subadult male and Pup Sea-Elephant

nasal passages and can be inflated and raised almost straight up. Worse still, these animals are clothed in very short sparse greyish brown hair, which they moult once a year and in doing so not only lose their fur but also their whole outer skin; they are then bright pink and present the most grotesque and revolting appearance, especially when they lounge around on shore in great misshapen, heaving masses under a hot sun, moaning, groaning, gurgling, and roaring. They live on cuttlefish, seaweed, and shellfish and are fairly agile in the water but spend a lot of time on land. The great bulls heave their immense bulk up gently sloping beaches and into the tussocky tall grass of the islands they most prefer and then go to sleep. Nothing is quite so alarming as to stumble up against one of these animals at such a time since they come "unstuck" with a veritable explosion and rise to full height, blowing and snorting. They assemble in special places once a year to breed and the old males maintain harems. At these times they were formerly slaughtered almost to the point of extinction for their oil, but in recent years they have reappeared even along the coast of California.

Flesh-Eating Mammals

(*Carnivora*)

THE Mammals grouped under this heading form a natural Order, all the members of which are quite definitely related to one another and all of which are at the present day quite distinct from all other mammals. In actual numbers of living types (*species*) that can be clearly distinguished they stand fourth, being greatly surpassed by the Rodents, the Even-Hoofed mammals, and the Bats, but in number of families they stand third, while in variety of form they far surpass even the rodents. Not all of them by any means are exclusively flesh-eaters, and some of them are complete vegetarians. They are, on the whole, the most astonishing collection of mammals, presenting the greatest number of surprises to the non-specialist, and causing the utmost confusion in the minds of the public.

On preliminary investigation this Order would appear to be very simple since it is clearly divided into seven great groups all of which are typified by an animal so common that youngsters in the lower school grades either know it or have heard of it, even if they are not at all interested in animal life *per se*. The major divisions are the Cats, the Dogs, the Civets, the Hyaenas, the Raccoons, the Bears, and the Weasels.

The Carnivores are comparatively—and this word is used in its strictest sense—a modern development, having arisen about the dawn of the age of mammals. To estimate just when they did arrive upon the scene is particularly hard, and for a reason that comes as a surprise to most people. The ancestors of all the animals we call *Carnivora* today have been named Creodonts, and these appear to merge with, of all groups, the Hoofed Mammals. The term modern as used above therefore denotes merely more recent than the extremely ancient time when such orders as the Insectivores and the Marsupials became differentiated. Once the Carnivores as we know them started, moreover, they diversified in a manner that is difficult to follow and which still produces much debate.

The consensus today seems to be that certain of the Civet group (the Viverrines) retain the greatest number of characteristics of the most ancient Carnivores. In fact, certain most peculiar little insect-eating creatures, sometimes classed with the Mongooses but found only on Madagascar (where almost all animals are odd and almost prehistoric) may actually represent remnants of the common carnivore ancestors. Next to the Viverrines, the Weasel family (Mustelines) seem to be the oldest. The Cats (Felines) probably arose from the Viverrines. The Bears (Ursines) and Dogs (Canines) form a close pair, the former being nothing more than vast, tailless dogs, and both seem to have sprung from the Raccoons (Procyonines) or from some common ancestor. What to do with the Hyaenas is really much more puzzling than even some experts assert. It is generally accepted that they arose from some form of Viverrine, and there are fossil forms that have been bandied about between the two groups. Most notable in this argument is an animal still living in Africa and known as the Aard-Wolf (see page 175) which really seems to be half Civet and half Hyaena. Carnivora are today of worldwide distribution, with the exception of Australia and the oceanic islands, but even the former has the Dingo, and most of the latter have domestic dogs.

FELINES

It is not determined as yet with any degree of accuracy just how many distinct kinds of cats may be living today. As many as fifty species have been described, but there is much confusion about many of them, principally because of a simple lack of specimens either dead or alive for study. Further, many cats vary widely in the color and pattern of their coats and may then change from one to another in a single moult or over a brief period of time. Many of them will interbreed, even species that appear very distinct like the Lion and the Tiger. It is now customary to class them all as members of a single family (*Felidae*) but there are valid reasons for separating several especially odd types like the Serval, the Jaguarondis, and the Cheetahs. As there are no allowable family divisions for the Felines, we break them down here into arbitrary groups which may possibly be regarded as representing genera or sub-genera.

GREAT CATS (*Panthera*)

There is no doubt that the first five cats form a natural group distinguished by having certain small bones (hyoids) embedded in the base of their tongues and separated from the base of their skulls by cartilaginous extensions. This may sound singularly unimportant but it makes roaring possible, as opposed to mere yowling, muttering, and purring. There is some doubt as to the status of the Puma, the Clouded Leopard, and the two Golden Cats in this respect, though the pumas of the Andean region are known to make a pumping roar just like a lion, and the Golden cat of West Africa is reputed to make noises indistinguishable from those of a leopard, notably its so-called "wood-sawing." With the exception of the last two, these animals are so well known and so well illustrated herewith that no description is needed. However, there is a very great deal of misconception about all of them in the popular mind, and some facts may be of interest.

Lions (*Panthera leo*)

There are many racial groups of lions (Plate 98) still living today, but individual members of all of them vary a great deal. Full-maned or maneless males of a variety of color combinations crop up everywhere, while their general appearance may change radically with their physical condition and age. Shorn of their manes, they are not the biggest cats, being exceeded by the north Manchurian Tigers. They may be the King of Beasts in appearance, and they are certainly regal in demeanor, but they are terrified of small children, and flapping laundry on a line, and they won't go within twenty feet of a tiny animal called a Zorille (see page 205). Although hunters, they may even lie down back-to-back with Antelopes during the day, and they appear to make little more than one kill per month on an average. The females do most of the killing, the males just wander around, grunting, to drive the game to her. They never molest men unless suddenly startled, bullied, wounded, or driven to regular man-killing by disease, excessive hunger, or a strange contagious delinquency that sometimes affects whole lion populations and notably the unmated juveniles. Senile individuals usually subsist on small rodents, scorpions, and other little things rather than plump native women and children. Lions do not live in the "jungle" nor in any true forests or even woodlands. They are savannah, grass-plain, scrubland, and semi-desert animals, and they adapt well to cultivated land, finding our domestic stock much easier to kill than wild game. During World War II, when the able-bodied, native, male population of East Africa was conscripted, sport came to a standstill, and stock-raising was pushed, lions became a pest and had to be declared vermin since they even wandered into the main streets of large towns like Nairobi. In early historic times lions inhabited Eastern Europe as far north as Roumania and possibly Italy, the whole of the Near East from Turkey to India, Arabia, and the whole of Africa from Gibraltar to the Cape, outside the Equatorial Forests. Today, they are exceedingly rare or extinct in North Africa, gone from Europe, Turkey, Palestine, and Arabia, and are probably now finally extinct in upper and lower Iraq and in Persia. In India, where a British officer once killed four hundred in a few years'

hunting in one central area, they are now confined to a few pairs in the so-called forest of Gir in the Gujarat. The normal litter seems to be only about three cubs but the youngsters grow very rapidly and start breeding in their second year. There are still an awful lot of lions in Africa, and they are a positive drug on the animal market in America, since they breed well in captivity.

Tigers (*Panthera tigris*)

Although the tiger (Plates 101 and 102) appears externally to be very different from the lion, it is extremely difficult to tell them apart if skinned and laid side by side. Further, they will interbreed, producing "Ligers" or "Tigons," according to which is the father. There are lions without manes and tigers with them, but despite their close similarity in so many respects, they are very different animals. The tiger was originally a northern and probably an Arctic animal and appears to have originated in what is now eastern Siberia, whence it spread south and west in two separate emigrations. The first entered Manchuria, spread to Amuria, Korea, China and then to Burma and India on the one hand, and via Indo-China and Malaya to Sumatra, Java, and Bali on the other. It has never reached either Ceylon or Borneo. The second stream passed north of the great central desert belt of Asia, west of Mongolia and Sinkiang to the Kirghiz, Afghanistan, Persia and the Caucasus, in all of which areas it is still found. The largest tigers, with a thick undercoat and very long, pale overfur, come from the colder areas of Siberia ranging from the Altai right across to the Stanovoi Mountains. Despite the fact that tigers inhabit the true jungles of the south which are comparatively cool, they still suffer miserably from the heat, and whenever possible make a practice of bathing and swimming to cool off. They vary much in ground color from bright reddish-orange to almost white, and the black stripes are always unlike on the two sides. Albinos are fairly common and there are all manner of semi-albinos, a pure black-and-white female having been killed in Orissa. Very large males of just over ten feet have been recorded. There are usually four or five cubs in the womb, but seldom more than two survive and the parents may eat the rest. The young stay with their mother for a full year. The Romans used them in their arenas, but they soon discovered, as did Indian princes who pitted them against bulls and buffalo, that they invariably made every effort to keep out of their antagonist's way. Even recognized man-eating tigers are usually timid, and all of them do everything they can to remain out of sight especially of man. Their favorite food in India is the great Nilghai (see Ungulates) but they will eat almost anything they can kill and, like lions, they can subsist on mice, locusts, fish, and other small creatures. Contrary to belief, they are not good climbers and only the young make much attempt to ascend trees; however, they will scramble aloft in an emergency.

Tigers are great travelers, especially in cold weather when they are much more active, and they usually follow paths for, despite their delight in bathing, they abhor dew wetting their fur. They hunt by ear, have a bad sense of smell, and very poor vision, apparently being unable to differentiate game from the bush as long as the quarry stands still. Unlike the lion, they do not kill with a swipe of the paw, but leap and hug the victim, and then bite its throat. They drag their kills to hiding places and rest up near them till they have finished eating them, even if quite rotten. They will eat 200 pounds of beef at a session and drink enormous amounts of water. When hunting, they make a strange noise like a bell, called "titting." Their greatest enemies are packs of feral dogs and the wild Dholes (see page 199), which worry them to death, but the elephant beats them to a pulp and then kneels on them, and a Water Buffalo was once seen to kill a pair in a single left and right sweep of its great horns.

Leopards (*Panthera pardus*)

Before anything else is said about leopards (Plates 99 and 100), it is essential to dispose of the age-old argument about the names "panther" and "leopard." Fairly important men have been challenged to duels for either affirming or denying that there is a difference—*i.e.,* that there are two different animals. There are not: the two names denote the same animal or animals—for they vary greatly—though they may be used to differentiate between large and small, or between light and dark individuals in any one area. All the Great Cats that can roar are now officially *panthers,* as their technical name implies. The identication of the leopards, however, presents considerable difficulties.

Despite their wide variation, they would seem to constitute but one widely spread species, all members of which could interbreed, while they have been reported to mate also with lions and tigers. Nevertheless, leopards from dry, open, rocky or treeless areas tend to be large and pale-colored, while those from damp, forested lowland or mountainous districts are smaller, with a darker basal color and more profuse and larger black spots and rings.

The Leopard, although just as timid as all the other great cats, is quite a different animal for, besides being a night-hunter, it will not discriminate and may just as readily jump a man as a baboon. Also, they are marauders, and being monkey- and bird-catchers, they find men's domestic birds and mammals a source of great interest. When surprised thus invading human domain, they are apt to become somewhat hysterical and launch an attack. The leopard is probably faster in action than any other mammal, and it can become quite fearless once aroused. Leopards show a tendency to melanism—*i.e.,* black mutations—especially in certain areas, and one of a litter may at any time come out pure black. This mutation, however, does not breed true, and black leopards may show marked sterility. Authenticated cases of albino leopards are not recorded.

Leopards once ranged from the British Isles to Japan and south throughout the whole of Asia except the uplands of Tibet, including Ceylon and the Indonesian islands. They are now extinct in Europe except in the Caucasus, and they are not found in Siberia or Japan. They are still found all over Africa apart from the true deserts of the Sahara and Kalahari, and are far too prevalent even in the fully cultivated and even the industrialized parts of the Union of South Africa. A desiccated leopard corpse was once found at an elevation of 17,000 feet on Mount Kilimanjaro in East Africa, but what it was doing there, far above the snow line, or what it had fed on, remains a mystery.

The Snow Leopard (*Panthera uncia*)

Closely related to the ordinary leopards, this animal is yet quite distinct, having a strangely-shaped skull that can be spotted by any expert, and long, plumelike fur covering a dense woolly undercoat. In color it is whitish grey tinged with yellow above and pure white below. The head and limbs are spotted with darker grey and the body is marked with diffuse dark rosettes or rings. Snow Leopards grow to an overall length of about eight feet, of which three is the tail. This animal replaces the ordinary leopards throughout the mountainous districts of that great area in central Asia north of the Himalayas to the steppes, and east to the Altai Mountains, Amuria, and the island of Sakhalin. It ranges up to the snowline in summer, but never comes much below six thousand feet in winter, except in the extreme north of its range, where it lives at sea-level.

The Jaguar (*Panthera onca*)

The great spotted cat of the New World (Plate 103) is differently constructed, being shorter limbed, broader chested, and heavier in general build. There is much debate as to the size this species attains, but many published accounts to the contrary, it does not appear ever to reach the dimensions of a tiger, though it may weigh as much. Although a quiet and retiring beast, the Jaguar is much more likely to give battle if interfered with, and it can be a rather persistent adversary, making false disappearances, lying in ambush, and stalking even armed men for long periods. The traditional way to tell a jaguar from a leopard depends on no more than a cursory inspection of the pelt, that of the Jaguar having the spots arranged in rings and in more or less horizontally parallel lines on either flank while each ring contains either a single or a small bunch of spots. This holds even when the jaguar is a pure albino, a not uncommon mutation especially in the Guiana Massif, since the dotted rosettes can still be discerned in certain lights. The Jaguar is the best tree-climber of the Great Cats, and in some areas, when the forest floor is flooded for months on end, it is almost wholly arboreal. However, it also lives out on treeless prairies, scrublands, on semi-deserts, as in southern Argentine and the southwest of the United States, and in rocky mountains where there is no water. In the Amazonian area it is almost as aquaceous as the tiger. The species is found from California, Arizona, and New Mexico in the north, to lower Patagonia and almost everywhere throughout the intervening territory. Throughout this area it is quite inaccurately, most misleadingly but almost aggressively called the "Tiger" or "El Tigre" by everybody.

NOT-SO-GREAT CATS (*Profelis*)

If the cats have to be broken down at all, and for the

benefit of the average person if not for the specialist they must be, we immediately run into difficulties, for there are four other comparatively large species that are manifestly greater than all other cats but which either lack the roaring apparatus, or have it far less developed. These four species have much in common, but this is probably the result of nothing more than their common intermediate position in size. They are distributed over four continents.

Pumas (*Profelis concolor*)

Some overenthusiastic college students prosecuting field studies on this animal for their theses, once solemnly described more than a hundred species of Pumas (Plate 105). These cats have also been called Cougars, Mountain Lions, or Catamounts, and by a variety of Spanish names including "Lion," at one time or another and in various countries throughout their vast range from Alaska to Tierra del Fuego and from the west to east coasts of both North and South America. Today, they are restricted to the western third of North America—unless the vague reports that from time to time come in of species killed in Nova Scotia and the Adirondacks are valid. There are, however, a few left in Florida. In Central and South America they are still extremely numerous and are found in almost every type of country from bleak mountaintops to steaming, flooded, equatorial forests, open pine ridges, tangled mangroves, open savannahs, and the pampas. Nonetheless, all these animals are of but a single species.

The Puma is the greatest coward of all the great or not so great cats. Despite voluminous fictional tales and innumerable accounts published as fact, the number of authenticated cases of deliberate attacks upon humans by these animals is so paltry as to be almost non-existent and most of these are open to some doubt. The animal is a retiring beast and a small-mammal- and bird-hunter, though it will run down bigger game and kill stock if available, and especially when young and unprotected. Nonetheless, systematic slaughter of these animals is not warranted, and stock-raisers can fairly easily eliminate confirmed marauders. The Puma is not as great a wanderer as the Jaguar; it normally stays within a limited home range where game is sufficient to support its family, but individuals have turned up in most unexpected places. In South America these cats vary enormously in habit and habitat, but they seem to prefer open country to the closed-canopy forests, and they are commonest in drier, mountainous areas.

Clouded Leopard (*Profelis nebulosa*)

In southeastern Asia, ranging from Bhutan in the Himalayas via Assam, Burma, Indo-China, the island of Formosa, the Malay Peninsula, to the islands of Sumatra, Borneo, and Java, there exists a beautiful cat that reaches the dimensions of a leopard. This animal is arboreal, has an exceptionally long tail, an elongated head, and the longest fangs of any cat.

The color is basically grey or greyish-yellow, with large, sub-angular, black, boxlike markings filled with dark brown. The face is striped, the limbs spotted, and the tail, which is very thickly furred, is irregularly ringed. The Formosan species has a much shorter tail than the others. Nothing definite or particular is known of its habits except that it appears to be a bird-eater and sleeps in trees.

Golden Cats (*Profelis temmincki and P. aurata*)

The last two cats of outstanding proportions are found in Eastern Tibet and the Himalayas south to Burma and Malay, and in West and Central Africa respectively. Both are known popularly as the Golden Cat; the Asiatic species as *Profelis temmincki*, the African as *Profelis aurata*. They are very alike in external appearance, being a smooth, rich, reddish brown, lighter on the flanks, and white below with, in the Asiatic form, a few dull grey spots on the throat and chest, and more profuse black spots all over the white undersides in the African. Their heads are comparatively small. The African animal grows to the dimensions of a medium-sized leopard. They are ground-living, high-forest animals, and appear to maintain regular dens in caves and among rocks or to live in the largest holes excavated by other animals. They are bird and rodent eaters, and regrettably little else is known about them. Their chopped whiskers are used as a man-killing poison by many African tribesmen.

LESSER CATS (*Felis*)

The arrangement of the true cats herein being used is manifestly artificial since mere size is actually of no genealogical significance. The great roaring cats form a distinctly related group; the not-so-great cats probably do not. The Lesser Cats obviously are a composite conglomeration, and this for two reasons. First, they vary very considerably in anatomical as well as general structure, and secondly, there are several among them that appear to be but miniature forms of great cats—*vide*, the Marbled Cat (see below). There is undoubtedly, in fact, some better way of arranging these animals both great and small so that their true relationships would be brought out, but unfortunately we simply do not know enough about the lesser cats, either alive or dead, to do this at present. There is even doubt as to how many different kinds of cats inhabit Asia, Africa, or South America, which of those that are known may be mere forms of others, and which are distinct species. There are at least a dozen groups of species.

Ocelots (*Felis pardalis, etc.*)

The commonest cat of South, Central, and southern North America, is the so-called Ocelot which when fully grown varies from the size of a very large domestic cat to, in the case of some Guianese and Amazonian specimens from the true continental rain-forest belts, almost that of an Indian Leopard. There is, in fact, no decision as to whether there are more than one species, since the colors and pattern of their coats vary even more widely than their size, there being both large-blotched generally dark-colored and small-blotched light-shaded forms in open country, dark and very dark forms in closed forest and an extremely light-spotted rather than blotched form on the isolated savannahs in the rain-forest belt. This last may even be quite a different animal, and not even a true cat, since its legs are long and straight, its fur fluffy, its tail very short, and its ears large and pricked like those of the African Servals. It runs down game in tall grass and makes a whining noise. The Ocelots are simply large, handsomely colored cats, preferably tree-dwellers, and subsist on any animal food they can catch and kill. They make nice but unreliable pets and are said to have mated with domestic breeds. The pelts of those killed in cooler climates are fairly valuable furs. The markings are fundamentally longitudinal bands of rich brown enclosed in jet black lines; the bands, however, are broken up to form a crazy pattern of interlocked marblings and rows of spots. The pupil is highly contractile and forms a vertical slit.

Very similar in appearance to the Ocelot, but not necessarily of close relationhaip are two cats found in northeastern Asia. The first, named "the sad" (*F. tristis*) because its eye-stripes end in blobs on the side of the muzzle like vast tear-drops, is large and blotched; the second (*F. scripta*) is much smaller and marked with irregular interlocking blots and spots. They inhabit the temperate mountain forests of China from Manchuria to Sze-Chwan.

Leopard-Cats (*F. bengalensis, etc.*)

Almost all tropical countries are inhabited by one or more distinct species of small spotted cats. There is no reason to suppose that they sprang from a common ancestor and then spread throughout the world; rather, it would appear that a coat of a light color marked with dark spots is a common protective device that has been developed by numerous different animals quite independently throughout the millenia and in all parts of the world. Nonetheless, the resultant small cats are notably similar both in appearance and habits wherever they now live. They are either forest-dwellers or they roam about under trees in open woodlands or on savannahs. Most of them fish with their paws and some are almost semi-aquatic.

Starting in the Oriental Region, there are at least three and possibly half a dozen distinct spotted cats, known as *Felis bengalensis, rubiginosa,* and *viverrina,* or the Leopard-, Rusty-spotted, and Fishing Cats respectively. The first is arboreal, lives in holes in trees by day, and is found all over India, in Indonesia, the Philippines, Formosa, and coastal China north to Amuria, but not, strangely, in Burma, Ceylon, or Indo-China. It has a striped face, spotted body, and ringed tail. It is

about two feet long. The Rusty-spotted Cat is not a forest but an open-country inhabitant, found only in Ceylon and southern India. It is reddish grey, marked with longitudinal lines of brown blots. The Fishing Cat is a semi-aquatic marsh-dweller of an elongated, civet-like build and covered in almost even lines of dark spots on a dull grey ground-color. It ranges from Ceylon through India to China and south into Burma. It grows to almost three feet in length and there are authenticated cases of its having carried off human babies.

Turning now to Africa, we encounter a rather curious state of cat affairs. The niche filled in India by the Leopard-Cats is here occupied by cats of another group (see below) but there is at least one small, vividly-spotted, arboreal cat found in both the west and north central parts of this continent, and also in the eastern forests of Mozambique. This has been called the Tiger-Cat, but it does not appear to have any scientific name at present (unless it be *Felis servalina*), being exceedingly rare in museum collections and having been dropped out of the catalogue of these animals at some time during the numerous reclassifications to which the family has been subjected by experts in the past century. Nothing is known of its habits.

The remaining spotted cats are South American and there is much confusion and debate as to how many there are and what they should be called. The best known is a neat little animal of almost perfect form, coloration, and disposition known as the Margay (*Felis tigrina*). It is smaller than a domestic cat and has a slightly longer tail. It is a bird-, lizard-, and frog-eater, but also takes insects. It is a bright cream-yellow in color and regularly and immaculately spotted with jet black. Then, in the forests on both sides of the Andes there is at least one small spotted cat, and possibly three others. One, known as Geoffroy's of the Wood-Cat, is found south of the Amazon basin and in Chile; the others are known only from around the north and western fringe of that area.

Tabby-Cats (*Felis lybica, etc.*)

Another group of small cats which may, in this case, actually be closely related and which certainly look much alike, are typified by the common wild cats of Africa—*Felis lybica* and *ocreata*. The former, which has been preserved almost unchanged in the Abyssinian domestic breed, was the ancestor of the house cats of Europe, having apparently first been domesticated by the Ancient Egyptians and spread to Greece, Rome, and western Europe where it was still regarded as a very valuable and protected animal as late as the thirteenth century. The Egyptians called this animal the "Mu" and trained it not only to keep down mice in their granaries, but also to retrieve waterfowl shot with bow and arrows, because it was originally a water-loving animal. Cats do not dislike water: they dislike cold water.

From the study of bones found in caves, some zoologists believe that there was once in Europe a wild cat with a blotched coat pattern which, by cross-breeding, gave us our tortoise-shell and calico types, but which is now totally extinct. Whether this be so or not, it seems

Margay

fairly sure that the true Wild Cat (*Felis sylvestris*), which is still found in most out-of-way parts of Europe (except Scandinavia and Italy) and even in Scotland, played no part in the development of our domestic breeds. The Wild Cat is a beautiful but very shy and savage animal with a rather fluffy, yellowish fur marked with a diffuse tabby pattern, and it has a pink nose and blue eyes. There are authentic records of its having attacked human beings. A form of the European Wild Cat is found in the Altai Mountains of Siberia. Tabby-marked cats with blue eyes also range almost all over Africa. Whether there is a similar cat in the Oriental Region is open to question, and what animal, if any, in that area gave rise to the yellow-eyed Burmese and its semi-albino blue-eyed form we know as the Siamese, is not known. In fact, all domestic cats of India, China, and other parts of Asia may have been carried by man from Egypt, or they may have been developed locally—and possibly from a species named the Waved Cat (*F. torquata*) which is very common throughout northern India. This is shaped exactly like a domestic cat with a tapering tail, and has an ashy grey-brown pelt marked with diffused, wavy bands, and lines of spots like a Tabby. However, these may be domestic cats that have returned to a wild condition. Why one breed in Korea and another on the Isle of Man in the British Isles should have lost their tails is another enigma.

Somewhat like the tabbies but with longer fur and shorter tail, is a curious species isolated on the open pampas of the southern Argentine, known as the Grass-Cat (*F. pajeros*). Its coat is marked with tabby-like, wavy stripes forming diffuse rings on the limbs and tail. It is probably no close relative of the true tabbies, but in appearance it stands somewhere between them and the next group.

Desert Cats (*F. manul*, etc.)

East of the Carpathians and the Caucasus, and extending throughout the desert, treeless, and other drier parts of Asia, including Persia and northern India, there are a number of pale-colored, long-furred, rather flat-headed cats. All over central Asia the species known as Pallas' Cat (*F. manul*) is found. There is some belief that this may be the ancestor of the so-called Persian breeds of domestic cats or have participated in their development, but this is unlikely. It is found even on the highlands of Tibet and is a hunter of rodents. In India it is replaced by the Desert Cat (*F. ornata*) which, with closely related species in the southern Himalayan mountains and in Turkestan, has shorter fur and a slender rather than a bushy tail, and which, most confusingly, has a pale, creamy-grey coat covered with small black spots forming lines. The limbs and upper side of its tail are, however, roughly barred like the Tabbies'. In Arabia, southern Palestine and across north Africa to Algiers; apparently all over the Sahara to the region of Timbuctoo and the eastern Sudan; and, in the other direction, northeast to the Transcaspian area of south Russia, there is to be found a very distinct small feline known as the Sand-Cat (*F. margarita*) (Plate 104). This has a very flat head with ears sticking out on either side horizon-

African Wild Cat (Felis ocreata)

tally, and carries its body low to the ground. Recent observations have shown that it is a truly desert animal living in holes and being entirely nocturnal. It is a pale, sandy-cream in color with just a suggestion of tabby-like markings on the limbs and tail tip.

Plain Cats (*F. planiceps* and *F. badius*)

There are two strange little cats, found only in southeast Asia, that have unmarked coats, ranging in color from dark brown above and white below (the Flat-headed Cat) to a rich reddish-chestnut all over (the Bornean Bay Cat). These animals have small heads, short legs and very short tails, and inhabit the forested areas of the Malay Peninsula, Borneo, and Sumatra, though it appears that the latter is confined to Borneo. Their exact affinities with other cats are not clear, and very little is known about them.

Marbled Cats (*Felis marmorata*)

Throughout the considerable area stretching from the forested valleys of Tibet, the eastern Himalayas, Assam, Burma, Malaya and perhaps Indo-China, to the islands of Borneo and Sumatra, there are a number of cats about two feet long with tails about two thirds the length of the head and body. These are colored and marked exactly like the Clouded Leopard. There are very distinct forms, if not separate species, in Tibet and in Borneo. These animals present zoologists with a real problem and one which clearly demonstrates that we have no real understanding of the true relationships of the numerous forms of cats as a whole. The puzzle is the connection, if any, between these animals and the Clouded Leopard. Are they simply miniature forms of that animal or are they just Lesser Cats that have come to look like it for similar reasons and through identical causes pertaining to the environment that they both inhabit? They even seem to have habits similar to their greater relative and to vary in details of form, color, and pattern from place to place just as it does.

LYNXES (*Lynx*)

Though there is really no anatomical distinction between the Lynxes and the preceding cats, these tufted or plumed-eared, small-headed species with heavy bodies that tend to rise to the base of the short tail, are

[165]

somehow different. Zoologists can define these differences no better than can the ordinary person, yet nobody ever mistakes a lynx, even if it is called a caracal or a bobcat, or just a wildcat. They do not move, sit, or lie like other cats, and they are exceptionally savage. There are four distinct kinds.

Jungle-Cats (*Lynx chaus*)

This is probably an appropriate place to define the word "jungle" since it has an important bearing on the names of many animals and may provide a proper understanding of their distribution and habits. Today, the word has somehow come to mean the tall, virgin equatorial forest with giant trees festooned with creepers. Unfortunately, the word is derived from a Persian word, *djanghal,* which means dry scrub in a desert and was first used in the English language in that sense in India at a time when many tropical animals were first being given popular names. The animal to which the name *Lynx chaus* is given is spread over an immense area from the Caucasus in the west, throughout the Near East and eastern North Africa, and Persia, to India, where it is found from about the 8,000-foot level in the Himalayas to the coastal lowlands of the extreme south. It also occurs in Ceylon, and east into Burma. Although found in high forest, it prefers open woodlands, cultivated land, and scrub of all kinds. It is ashy, grey-brown in color, with a tabby pattern of darker, wavy stripes, and has a slight plume of longer dark hairs on the tips of its ears. Though considerably larger, it interbreeds with domestic cats and is said to have two litters a year. It is, in fact, very close to the Tabbies and may indicate that all the lynxes are a derivative of that group.

Caracals (*Lynx caracal*)

Throughout Africa south of the range of the Jungle-Cat but for the most part in even drier and more wholly desert areas, and in parts of India there is found this altogether different animal. It is slender, with a short sleek reddish-brown coat, and exceptionally long ears ending in long dark plumes that may curve over like tassels. The ears are white inside and black out, and the tail has a black tip. It used to be a common pet in wealthy Indian households and was trained to hunt. It is amazingly

agile, can run down gazelle and catch birds on the wing by leaping into the air and striking them (Plate 108).

Northern Lynxes (*Lynx lynx, etc.*)

If you take a map of the world and draw a line from the Pacific coast of North America east along the southern border of Canada, thence across the Atlantic to the southern boundary of France, Switzerland, Austria, and then continue it on to the Caucasus south of the Caspian Sea to the Himalayas, east to the borders of China proper and finally north to Manchuria and the shores of Amuria on the Pacific once more, you will have defined the southern limits of the Northern Lynxes. Just how many distinct species there are has always been and still is open to debate but, although they vary somewhat from place to place, all are large, long-furred, light colored, and have huge fluffy feet and long-plumed ears. They are almost Great Cats in size, the largest being the Siberian which can reach forty inches in body length with about a seven-inch tail. They are forest animals and are strong enough to pull down large deer. Their range in central Europe has now been greatly reduced and they are very rare south of the Canadian border.

Bobcats (*Lynx rufa, etc.*)

The name Bobcat is American and belongs to the group of short-haired lynxes found throughout the United States and northern Mexico (Plates 106 and 107). We here use the name to cover also the shorter-haired, spotted European representative found in Spain, Italy, Greece, Turkey, and the islands of Sardinia and Sicily, and known as the Isabelline Lynx. This animal is to the Northern Lynxes of Europe what the American Bobcats are to the Canada Lynx, namely small southern representatives. The appearance and habits of the Bobcats are too well known to require any comment except for a word of caution: they vary enormously from place to place, since they may inhabit almost any type of country from bare hills to sub-tropical swamps, and they will eat almost anything alive. Curiously, bobcats will also endeavor to mate with domestic cats especially if housed together in captivity, but the kittens very rarely if ever survive. All types of lynxes have been tamed but if not taken when very young are exceptionally aggressive and dangerous, being very agile and usually going straight for the eyes of their opponent. Unlike other cats they may attack humans on sight and will single out persons who have molested them.

SERVALS (*Leptailurus*)

Normally included in the great mass genus *Felis* but really very distinct, and showing perhaps some likenesses to if not actual affinity with the lynxes, are a number of long-legged, short-tailed, brightly spotted cats of equatorial Africa, known as Servals. These animals are singularly uncatlike in many respects, living among the grasses of the savannahs, standing straight upon their long legs and running down small game. Their large acutely pricked ears and rather large bright eyes equip them for hunting both by night and day, using sound and sight respectively. They eat birds and small mammals but will catch lizards and take insects and other

Canada Lynx

[166]

small food. They come in quite a variety of forms, with rather long and fluffy or very short and sleek fur, colored basically dull yellowish-grey to bright yellowish-orange, and with the black spots numerous and intense or sparse and diffuse. In some places they appear to enter the true forests to hunt and are often found on "lakes" of grass isolated deep in such territory. The young are kept in a "form" usually hollowed out in a dense clump of dry grass.

JAGUARONDIS (*Herpailurus*)

Finally, we come to a very remarkable and puzzling little group of Central and South American catlike animals the classification of which provokes endless debate. Their species may be two, three, or more in number but all vary widely in color and to some extent in form though all are small-headed, sleek, long-bodied felines with short legs and long tails. The least known is called the Colocollo and comes from Bolivia and Chile west of the Andes. It is either grey or brown basically, with some diffuse darker markings between spots and streaks, and lighter below. Some specimens lack these marks either above or below and have the pelt tashed with silver. All have a black line from the eye to the angle of the mouth. The second species, called the Jaguarondi, is found in the Southwest of the United States and thence south throughout Central and South America but for the most part outside the tall forests. It comes in grey, dark brown and red varieties and is plain-colored all over. Whether the so-called Eyra is a different animal or only a shorter-bodied, longer-legged, red forest form of the Jaguarondi has not been decided. The author inclines to the belief that it is, for it is a tree-climber, has an exceedingly small head, rounded face, and brown as opposed to yellow eyes. One of these cats is said to have been domesticated by the Amerinds of Paraguay prior to the arrival of Europeans.

CHEETAHS (*Acinonyx*)

The Cheetahs, of which there are two and possibly three forms, are common to Africa (Plate 109) south of the Sahara, and to India, but they used to be numerous throughout Syria, Mesopotamia, Persia, and the lowlands all around the Caspian. They are plains and desert dwellers, preferring low hilly country. In 1927 a most unexpected creature, which was promptly called the King Cheetah, turned up in Southern Rhodesia. It was beautifully marked with large confluent blotches in contrast to the normal small isolated spots. It was at first thought to be an entirely new species but is now regarded only as a casual variety or phase of the normal coloration that may occur in any litter. Cheetahs are not true cats despite their leopardine appearance. Their claws are like those of dogs and, as in dogs, only partly retractile. They are one of, if not the swiftest of the animals that go on all four legs, and can run down Blackbuck and gazelles, a feat of which no dog or other animal is capable. Cheetahs have been clocked at over 60 m.p.h. and there are claims that they have hit 75 m.p.h. For this reason and because of their beauty and their unexpected tamability, they have been semi-

[167]

Serval

domesticated for centuries in India and in even earlier times in other countries of Asia and in Egypt. The animals are caught wild and when full-grown, and take about six months to train; they then become completely docile although regularly fed fresh blood from the catches they make. They are kept chained, not caged, and are hooded before being taken out to hunt, the hood being removed when game is in sight. The animals then either dash straight at the quarry or stalk it belly to ground until within about a hundred yards when they rush out and run the animal down, knocking it off its feet with a swipe of a forepaw and then seizing it by the throat with their teeth and pulling it down. They then wait till their owners come up and kill the beast. Cheetahs have thus proved to be a most successful means of killing Coyotes in the Southwest of the U.S.A. but enough trained animals to make a dent in the number of those pests would strain even the U.S. Treasury, since cheetahs are worth about $3000 a pair at the time of writing.

Jaguarondi

A NOTE ON THE FELINES

The cats have, in this book, been described somewhat more extensively than any other group because they are better known and more widely appreciated than any other mammals apart from our domestic animals and, perhaps, the Great Apes. Nevertheless, it must be clearly understood that they have in no way been completely or even adequately covered and not only for lack of space. Most species of wild cats are rare in fact, and still rarer in zoological collections. They are also retiring and clever animals that avoid men at all cost and are hard to observe in the wild state. There are many distinct species that we have not even mentioned.

VIVERRINES

We now come to an extremely bewildering array of medium to small-sized mammals which are predominantly but by no means exclusively carnivorous, which are confined to Africa and Asia—except for one species of Genet that is found in Spain and some Mediterranean islands—and which display a greater variety of more completely distinct forms than any other family of mammals. They are an ancient lot that have to be divided into five major and one minor grouping—namely, the Civets, the Palm-Civets, the Hemigales, the Galidines of Madagascar, the Mongooses, and a very odd creature known as the Fossa. In all, there are thirty-six recognized genera alone, and probably more than a hundred species distributed among these, and at least half of them are most distinctive. None of them is popularly known in the Americas and only two—the civet and the mongoose—are commonly known even by name. Worse still, they are not well known scientifically and there is still much discussion as to the identity of many and the classification of the whole. Nonetheless, this is a most important group of animals and forms a not inconsiderable percentage of all mammals. The Germans most succinctly call them "Half-Cats" but this applies more precisely only to the first three sub-families.

CIVETS (*Viverrinae*)

Five very different kinds of animals are included in this grouping but there are actually eight distinct genera. One is a fairly well-known and economically very valuable animal of especial interest to all women unless they are for some reason averse to expensive perfumes. Another is one of the least known and most beautiful of all mammals. As a whole, the Civets are closest of all Viverrines to the cats but they have sharper, more foxy heads and are generally more streamlined. They occur in Africa, extreme southern Europe, and in the Oriental Region.

True Civets (*Viverra and Civettictis*)

The civets of Africa (*Civettictis*) have been separated from those of Asia on technical grounds though they all look much alike, thereby depriving us of what many felt was the most beautiful-sounding of all Latin names for an animal—to wit, *Civettictis tangalunga*, a species from Indonesia. The African Civet grows to almost three feet in length but is small-headed, exceptionally narrow, stands high, and has rather long back legs. The color is a brindled grey with black and white facial markings and a white-edged dark collar. The tail is bushy and grey, the legs and underside black. The animal is of a savage and most unreliable disposition and because of its tremendously powerful neck muscles is almost impossible to handle. Yet these animals are kept in tall stockades in Abyssinia and other parts of Africa and are caught and "milked" regularly. The substance obtained looks like rancid butter and comes from anal glands, and has a very strong, pleasant, aromatic odor. It is packed into cow horns and shipped all over the world at a very high price to perfume manufacturers since it is the best fixative for other animal and plant essences and forms the basis for the most expensive makes. The Oriental species, which is also grey with complex markings of diffuse bands of horizontal and vertical stripes and spots and which has a black crest along the back that can be elevated, is found from India to China and all over Burma and Indo-China. Other smaller species live on the Malabar coast and in the Indonesian Islands. They also give "civet."

The Rasse (*Viverricula*)

This is a smaller animal without a crest but having civet glands and being of somewhat similar color and pattern. It has sharper claws, climbs trees, and eats a variety of fruits and animal food though preferring snails, frogs, insects, and lizards. It is found all over the Oriental Region and in Ceylon, Sumatra, Java, and on Socotra, the Comorro Islands and in Madagascar whence it was carried by the Malayan colonists many centuries ago. It is kept as a pet, for its civet and its fur.

Genets (*Genetta*)

The animals in question are exceptionally lithe and agile, appearing to sail through the air especially when jumping upwards from a standing start. Their form has been quite aptly described as like a "miniature Dinosaur wearing a leopard-skin coat." How many different species inhabit Africa has never been decided for they come in a considerable variety of coat patterns, all having lines of dark spots on a lighter ground, light spots under the eyes, a black median dorsal crest, and light- and dark-ringed tails. There are often two or more species in one area but the lighter coated, sparsely spotted varieties stay in open forest while the darker ones inhabit the tall moist jungles. They are mostly arboreal and eat all manner of small animal flesh. One species is found all around the Mediterranean and in the Near East, where it is still domesticated. This animal was the original "cat" of Ancient Egypt, having been tamed and trained to catch mice and being called "Basta." Later, when the Egyptians found that the Mu (*Felis lybica*) would retrieve fowl from water as well as catch mice, the Genet was relegated to a vaguely holy status and forgotten. It is interesting to note that the "cat" of ancient Greece was neither of these but the Polecat (*Putorius*).

The African Linsang (*Poiana*)

The Linsangs which are described below are an Asiatic genus, but there is an animal in West Africa that displays the essential features of these animals. These are interlocking but irregularly arranged spots, a long lance-shaped naked area of pink skin running up the "heel," and a strange arrangement of alternating wide and narrow dark rings, themselves alternating with white rings of even width on the tail. Such details are perhaps unimportant but serve to point up the almost incomprehensible variety of themes that Nature can ring on one simple basic plan. The *Poiana* has the habits, shape, and much the appearance of the genets but it is an entirely different animal.

Linsangs (*Prionodon, etc.*)

Ranging from the eastern Himalayas, through Burma, to Malaya and the Indonesian islands of Borneo, Sumatra, and Java, the true Linsangs fill a natural niche equivalent to that of the Genets in Africa. They are even more attenuated in form, have tails as long as or longer than the head and body, and even shorter legs. Their coats are various shades of yellowish to greenish ochre, marked with broad transverse bands of square spots or blotches which tend to run together horizontally, so forming a regular checkerboard pattern. Their heads are of a rich russet red with black markings, the underside invariably lighter and in some species white, the tail always ringed with alternating white and black bands of equal width. They eat birds, lizards, and frogs and spend more time on the ground than do the Genets. When stalking their prey they creep along belly to ground and fully extended and, being so long and slender, they are often mistaken for heavy-set poisonous snakes. The large Burmese species in which the dorsal crossbands form solid saddles is now considered to constitute a separate genus (*Pardictis*).

The Water-Civet (*Osbornictis*)

From time to time, even today, new animals are discovered even in the most anciently populated areas of the earth. Sometimes these are by no means tiny obscure creatures but of fair or even large size—for example, the second largest land mammal, Cotton's White Rhinoceros, only discovered in 1910. In 1916, the Lang-Chapin Expedition to the Congo obtained, in the Ituri region, specimens of an entirely new type of civet, an animal of considerable beauty and quite unlike anything else known. Adults measured 18 inches with a huge bushy tail of 14 inches. The body was shaped like a very heavy genet or otter; the head was typically viverrine but the feet were fully webbed. In color the beast was a bright rich reddish brown all over with dark and light facial marking not unlike a raccoon, and with almost black limbs, feet and tail. The animal inhabited stream banks and was semi-aquatic. It has never been collected since that date and nothing is known about it. It was named after the famous American zoologist Dr. Henry Fairfield Osborn.

PALM-CIVETS (*Paradoxurinae*)

The next batch of Viverrines have been named the Palm-Civets because the best known and first recognized —the Musang of Malaya—commonly inhabits the massive heads of these trees. These animals are less catlike, not quite so agile or predaceous, and most of them live on a mixed diet with vegetable matter predominating. There are six distinct forms or genera, five found in the Orient and one, again, in West Africa. There was once continuous land from the Atlantic to the Pacific across the Indian Ocean and this was probably covered with equatorial forests. When it broke up, not only the incoming seas but also the arising of a continuous mountain chain down the east side of Africa separated the fauna into two parts, one part far to the west in Africa, the other in the Orient. Later, the Congo Basin was flooded from the west and so formed still another barrier and drove the western animals still further to the west. There is still some discussion as to the number of genera of Palm-Civets inhabiting the Oriental Region, for specimens of the rarer species are seldom seen in museums.

Musangs (*Paradoxurus*)

Anybody who has lived in Indonesia will recall the misery of sleepless nights spent on a hot hard bed under a "pan" roof while these gay little animals romped back and forth over the ceiling, fought, thumped and banged for hours on end. Musangs are common from the forested southern slopes of the Himalayas, south to Ceylon,

W. T. MILLER

Genet (immature)

where there is a distinct species, and east through Assam, Burma, Malaya and Cochin-China, and south to the great islands of Borneo, Sumatra, and Java as well as a number of the smaller islands. They are rather nondescript little animals as the illustration shows, colored mousy grey-brown, with dark face, limbs and tail, but always with light markings under the eyes. They live in all kinds of country but mostly where there are trees, and they infest human habitations, becoming a great pest, attacking poultry, marauding kitchens and making a most awful racket at night, especially when there is no moon. Besides the Singhalese and Indian species, there are others in Assam and Burma, another in Malaya, and others in the islands. They are all about two feet in length, with tails that match the head and body. One species from the Celebes has distinct rings on its tail and altogether looks rather like the West African *Nandinia*.

Masked Palm-Civets (*Paguma*)

Much larger than the Musangs are a genus of Viverrines that apparently range from eastern Tibet through southern China to Formosa and probably occur in Indo-China and certainly on Borneo. Adults may measure almost two feet and have a tail of equal length. These may compose species of two genera, since some have naked pads under their feet while others have completely furred feet. These animals are very powerful and are completely omnivorous, killing animals twice their size but gorging on fallen fruits of all kinds, digging for insects and earthworms, and catching a lot of fish. In color they are of a smooth dark brown sometimes washed with silver and all have distinctive light and dark facial markings. The coat of mountain varieties is very long and almost fluffy.

Small-toothed Palm-Civets (*Arctogalidia*)

An altogether different animal, much smaller and more weasel-like in form, with very small teeth—except for the canines, which are enormous and recurved—inhabits the same territory as the Musangs from the eastern Himalayas through Burma and Malaya to Sumatra and Borneo. It appears to be more competent on the ground than the Musangs but otherwise has much the same habits. What particular niche it fills is still not known. There is a distinct species in Java.

Musang

Small-toothed Palm-Civet

Celebesean Palm-Civet (*Macrogalidia*)

The Palm-Civet of the Celebes is different. It is clothed in short close fur of a dull neutral grey-brown color but has a rather strongly annulated light and dark tail. The hairs on its neck and on the crown of its head are directed forwards. In details of teeth and other anatomical features it proves to be a very distinct animal. In habits it is like the Musangs and is just as common and bold in playing about native houses and even modern hotels.

The Binturong (*Arctictis*)

One of the most astonishing and paradoxical animals known, the Binturong, is not uncommonly displayed alive in zoos and nobody quite knows what to make of it. It is a large animal—sometimes measuring up to six feet in overall length, is heavily built and looks even larger because it is clothed in long coarse hair. The ears bear large tufts of hair that droop and give the animal a rather woebegone expression. The tail is almost as big around at the base as the animal's body and thence tapers gradually to the tip. Oddest of all, the tail is prehensile, a device not known in any other mammal outside of Australia and South America. Binturongs are

Binturong

fairly common in the high forests over a very wide area from the central Himalayas, through Burma, and Malaya to Sumatra and Java but they are seldom seen, being nocturnal, very retiring, and almost perfectly camouflaged. In life their fur is a strange dark green color resulting possibly from some algal growth such as is found in the Sloths. They have very small teeth for their size and eat fruits, bamboo and other shoots, insects, tree frogs, and any other small animals they can catch. They move rather slowly, trundling along like bears, and walk on the whole foot unlike other civets, which mince along on what is really their toes. They also make bear-like growling and rumbling noises.

West African False Palm-Civet (Nandinia)

This small animal is not unlike a miniature Binturong, being clothed in rather long, hard, very thick fur, having very short legs and being a climber rather than a jumper. It is brown in color, with very diffuse dark spots and always has two vague white spots on the shoulders. It is an exceedingly common animal in many parts of West Africa: the author once inadvertently asked a native hunter to obtain some without specifying how many and the man was back in three hours with forty. They are always reported to be carnivorous or at least omnivorous but captive specimens seem to select a purely fruit and vegetable diet. There may be as many as four young in a litter and the babies sometimes ride on the mother's back. They usually live in giant hollow trees in association with numerous other small animals.

HEMIGALES (Hemigalinae)

For the most part because no two others who have studied and attempted to arrange this group agree, we are here following Simpson in his arrangement of the Viverrines. The half-dozen odd and very distinct kinds of animals herewith brought together actually have little in common, and the Hemigale itself should, we personally believe, be placed near the Linsangs. Combined, the following genera display six different divergences from the civet pattern.

Hemigales (Hemigale, etc.)

In general shape these animals resemble genets but the arrangement of the pattern on their coats seems to be a development from that of the Linsangs. *Hemigale*, which has no English name, inhabits the Malay Peninsula and Borneo and is a very beautiful animal, being greyish-yellow marked with black above and creamy white below, and with the light parts of the face and neck washed with red. The pattern on the back is very singular, being composed of five diamond-shaped saddles that form dependent triangles on the sides, an arrangement otherwise met with only in the Yapok or Water-Opossum of South America. The base of the tail is ringed, the rest very dark brown. Nothing specific is known of the animal's habits except that it is a hunter and is arboreal. A closely related animal named *Diplogale* is of an even dark brown color and has up to the present been found on only one range of mountains in Borneo. The third member of the group is named *Chrotogale* and comes from Tonkin. In external appear-

West-African False Palm-Civet

ance and even color and color-pattern it looks just like *Hemigale* but its muzzle is very elongated, the teeth are small, and the whole skull is compressed almost to the extreme extent found in the Anteater-Civet (see below).

The Otter-Civet (Cynogale)

Throughout Malaya, Borneo, and Sumatra, but nowhere commonly, may be found a fairly large Viverrine —with a body up to thirty-three inches in length but with only an eight-inch tail. It is semi-aquatic and in fact behaves just like an otter, diving for fish and shell-fish, and hunting frogs and other riverine life. Their muzzles are broad and flat and bristling with stiff whiskers, and the hind feet are partly webbed. In color they are grey-brown without pattern and the fur is very sleek and glossy.

The Fanaloka (Fossa)

This animal is most misleadingly named since there is a completely different animal, commonly known as the Fossa (*Cryptoprocta*), described below. It looks like a Genet and is grey in color with vivid white spots above the hind angle of the eyes. The neck and back bear longitudinal black lines that break up into close spots above the tail. The flanks are spotted black on grey but the chin, throat and undersides are dirty white, and the feet yellowish. The tail has narrow half-rings of red on its upper side. This animal is seventeen inches long and has a nine-inch tail. The hairs of the mid-back ridge cannot be elevated as in the Genets. The animal is confined to Madagascar.

Fossa (Cryptoprocta)

The Anteater-Civet (Eupleres)

Oddest of all civets is a small, dark brown animal, also found in Madagascar. It has a short but very thick, well-furred tail and sturdy, rather long back legs but the body dwindles away forwards to a very small head ending in a long sharp snout. The ears and eyes are small, and the front legs are short and very slender and have tiny paws. The animal feeds on insects and earthworms and has minute teeth. It was for long thought to be some kind of very ancient Insectivore but is undoubtedly a Viverrine and is probably closest to the next group—the Galidines. The animal appears always to have been rare.

GALIDINES (Galidiinae)

In some respects bridging the gap between the civets and the mongooses, these obscure little Madagascan animals are quite unlike either and are distinctly primitive. Four genera comprising half a dozen species are now recognized but practically nothing is known of their habits except that which may be inferred from their general build, their teeth, and the type of territory in which they live. This is said to be the dense ground cover under the giant fern forests. However, the various species are distributed all over the great island of Madagascar, whereupon may be found almost every type of environment from tropical swamps to bare rocky mountaintops, temperate woodlands, grasslands and bare sand-dunes. Their general form is that of the mongoose which, with the possible exception of the bushy tail, may be regarded almost as the original plan for all mammals. The commonest form (Galidia) is represented by two or three species, all of which have irregular dark longitudinal stripes covering the olive-grey body from the neck to the tail base. The tail is black at the base and in one species pure white and plumed throughout its terminal two thirds. In another species it

is a vividly ringed black and white. The members of another genus have plain brown tails, and muzzles almost as pointed as that of the Anteater-Civet.

FOSSAS (Cryptoproctinae)

There is in Madagascar a unique carnivore called the Fossa (Cryptoprocta ferox) which is classed with the Viverrines as much on account of the place where it is found as because of its appearance and anatomy. It is generally shaped like a Genet but grows to a length of six feet, of which almost half is tail. It is of a uniform bright brown color and its face, as may be seen from the photograph, is distinctly foxy. Nevertheless, the animal is a close relative of the true Felines in many respects, notably in the structure of its feet, which have gruesome, recurved, fully retractile claws, five on each foot. Its teeth are a strange combination of cat's and civet's. It is entirely nocturnal and a predator and normally has a most unpleasant disposition. However, it is amenable to taming to the point of domestication. Oddest of its features are its eyes, the pupils of which are so highly retractile as to disappear altogether, while the cornea covering the iris is not only transparent but has a high refractive index, giving the animal in daylight the appearance of suffering from advanced cataract. While the Fossa is undoubtedly more a Viverrid than anything else, it is a most primitive carnivore and may have sprung from the mutual ancestors of the Cats, Civets proper, and the Mongooses. It has certain characters in common with the Galidines and Eupleres.

MONGOOSES (Herpestinae)

This is one of the most puzzling and, to the nonspecialist, muddling of all mammalian groups. Almost everybody in the English-speaking world has heard of the mongoose, made famous by Rudyard Kipling as Rikki-tikki-tavi, and everyone knows that "it kills snakes." Unfortunately, there is no such thing as "The Mongoose," there being about a hundred animals that may lay claim to this title. Many of these are extremely colorful, others are obscure to the point of dreariness, and all of them are rather nondescript in the true sense of that word. They are just little furred animals with four legs, a pointed head, small ears, and a fairly bushy, tapering tail; in fact, they resemble a child's drawing of "an animal." Some are more doglike than others; some are more weasel-like. They are distributed throughout Africa, the countries bordering the Mediterranean, thence east to India and the whole Oriental region from Assam and southern China to the greater Indonesian Islands. There are eleven recognized genera of Mongooses.

True Mongooses (Herpestes)

This is the commonest genus and contains more than a dozen recognized species and has an enormous range. The two best known are Mediterranean and Indian respectively; the first is indigenous to all the countries round that sea, having been domesticated over five thousand years ago in Egypt; the second, the commonest of half a dozen species in India, ranges from Persia to Burma, and has been introduced into the West Indies. The latter is only about half the size of the former and is lighter in color. The Egyptian animals are about twenty inches long, including a six-inch tail, and are of a grizzled brown color. The fur in all mongooses is harsh and rather shaggy. In India there is a large grey species (Plate 110) with white speckles all over the fur and reddish legs, another with yellowish stripes on its neck and a black tail-tip, and still others that are distinct though differing in details of appearance only. There are four distinct species in Ceylon, and another, known as the Crab-eating Mongoose, which is semi-aquatic, ranges from Nepal to China and south to Singapore, and may be distinguished by the white spots above its eyes. There are distinct species in the islands of Java and Borneo and a short-tailed species common to Malaya and several of the islands. In Africa south of the Sahara, there are a similar number of species which come in an almost exactly equivalent range of forms, though there are valid grounds for separating all of them as a distinct genus. One is very large, and there is even a semi-aquatic species.

Mongooses do not have scent-glands like other Viverrines; they have non-retractile, doglike claws and five digits on each foot. They can erect their hair when excited. Enormously agile animals, they do, it appears, actually make a practice of killing snakes. They are not, however, immune to the bites of poisonous snakes, nor do they use any herb as a natural antidote if struck; their almost complete immunity stems from their speed and inborn knowledge of serpentine behavior. When confronting a snake, they deliberately induce it to strike and when it does so step smartly aside and then pounce on its head from above, cracking the skull with one bite. A large King Cobra can often withstand the first bite and throw the little animal off its head by suddenly recoiling but the Mongoose lands on all fours and calmly awaits its next lunge. Non-poisonous, constricting snakes have sometimes defeated a mongoose by throwing a coil or two round the lower belly and banging them about so fast that they do not have time to bite through the snake's long muscles. Mongooses make extremely intelligent and reliable pets if raised from early age and will follow to heel like a small dog. They are death on rats and although they have become a great pest in the West Indies, killing most of the indigenous small fauna, they did accomplish the job for which they were originally introduced, namely eliminate the rats that were destroying the plantation crops. They are absolutely prohibited from being introduced into the United States, even by the Government itself.

Banded Mongoose (Mungos)

Closely related to the above animals but of singular color are small species found in the drier parts of East Africa and known as the Banded Mongooses. The bands run across the body. They are very neat little animals that customarily select the holes under termites' nests as dwellings. They are somewhat communal, two or more families with young living together in one complex of burrows. They are partly insectivorous but kill any small animals they can overcome. They have bright pink noses.

Dwarf Mongooses (Helogale)

This is a very small edition of the True Mongooses but is anatomically distinct. They are found in certain limited areas of Africa and reputed to be active during the day, unlike other mongooses. They have very short tails and naked soles on their feet. The two species come from Natal and Mozambique in S.E. Africa.

Marsh Mongooses (Atilax)

A water-loving species from West Africa proves to be a quite separate kind of animal. It is very short-bodied and the back bears dim bands of light and dark. Little is known of its habits but it has been observed diving for food, while the author shot one in a tree with a tall perpendicular trunk. In life the fur is of a distinctly greenish hue but turns almost black after death. Stomachs of recently killed animals often contain green leaves and other vegetable matter. These mongooses probably have very special habits.

Cusimanses (Crossarchus)

In West Africa there is still another type of mongoose represented by a number of forms. These animals for some unknown reason are called Cusimanses and are semi-fossorial, which is to say that in addition to digging their own burrows like all the foregoing types, they spend the greater part of their time underground and apparently obtain some of their food below the surface. This is not an invariable practice, however, and in some areas Cusimanses live under dry grass. They are unmarked, but the pelt is a strange grizzled mixture of browns, greys, yellows and shades of red caused by the hairs being individually ringed and often ending in yellow tips. They are low-chassied and have almost anteater-like snouts. They are the only Mongooses that relish carrion but will eat almost anything dead or alive and some vegetable matter. They have rapier-like tusks and are extremely irascible. Their very long claws are developed for digging, which they do in the hardest ground and at great speed.

There is a very rare East African form known as Meller's Mongoose or *Rhynchogale melleri* that looks almost exactly like the common Egyptian True Mongoose, but which is a sort of Cusimanse—having no median vertical groove down the muzzle, among other gruesomely technical points of difference. Nothing notable whatever is known about the beast.

White-tailed Mongooses (*Ichneumia*)

Having five toes on both hind and fore feet, these animals belong near the foregoing group. However, they represent the first step towards the doglike animals that follow. They stand straight on long legs and have much the appearance of foxes which they replace in many areas. They live in holes by day and forage by night and actually may run down their prey rather than stalk it and pounce. In color they are basically jet black but the whole head and body may be washed with silver and the tail is pure white. The limbs appear to wear black stockings. However, they vary a great deal in color from place to place and within any one area, so that almost wholly black specimens are found. They are found all over Africa south of the Sahara and in southern Arabia.

Bushy-tailed Mongooses (*Cynictis*)

In the Union of South Africa there is a type of Mongoose-like animal having five toes on the front and four on the back feet. They are rather long-legged and stand high off the ground. Their tails are bushy like those of foxes and have a white tip. The general color is a greyish-yellow. These animals do not live in forests but belong with the White-tailed, the Dog-Mongooses, and the Xenogales which do. Misleadingly these animals are sometimes referred to as the Meerkat by the Afrikaaners, a name that should be restricted to another animal to be described below.

Dog-Mongooses (*Bdeogale*)

These are also known as the Black-footed Mongooses but the name is misleading in view of the coloration of the White-tailed genus. They are found in West Africa and the Congo, and others in the southeastern forest belt about Mozambique. They have only four toes on all feet and stand even higher off the ground. They also vary much in color but in the typical form are silvery white all over but for the black feet. They do not appear to dig their own holes and usually live in the bases of huge hollow trees. They even make a doglike sound that is like a human trying to imitate a bark. Despite their typical adaptations for a ground-living and -running existence they can climb trees rather well.

Xenogales (*Xenogale*)

In the Congo may be found some large mongooses shaped like small African Civets, being sharp-faced, narrow-bodied, small-headed, and bushy-tailed. They are black in color all over and they too stand high upon rather long legs. Their habits appear to be similar to those of the White-tailed species. Although very unusual animals in shape and appearance, nothing particular is known about them apart from their anatomy.

The Meerkat (*Suricata*)

A most odd, almost idiotic-looking animal from South Africa was named Meerkat—i.e., "Marsh or Lake Cat"—by the Hollanders when they first colonized

Black-footed Dog-Mongoose

Meerkat

the country. Actually, it lives in dry places and often among bare rocks. It is sleek-furred and, as the photograph shows, looks like a cross between a Lemur, a Dog, and a starved Raccoon. A more ridiculous sight can hardly be seen than that of a mother of this species, followed by a train of young, walking over a flat surface. The animal's back legs seem to be a size too big, the head two sizes too small, and the tail designed for another animal altogether. They also have only four toes. Meerkats are communal animals and make extremely good pets with most of the attributes of both dogs and to a certain extent monkeys since they are very lively and use their hands to manipulate small objects. Although inhabiting the dry Karoo plateaus, they are inordinately fond of water and, like the Ring-tailed Lemur, may sit for hours under a dripping faucet.

HYAENINES

Of altogether different form and looking more like large ungainly dogs, the Hyaenines have nonetheless a common ancestry with the Viverrines. An extinct creature, known from fossil skeletons and named *Ictitherium,* and a living animal known as the Aard-Wolf, bridge the gap between the two groups, while Hyaenas have some anatomical features in common with Mongooses. The Aard-Wolf has actually evolved backwards as it were, but to a point where early zoologists thought it might be a kind of Civet. In the past, Hyaenas of various kinds lived all over Europe south of Scotland and Scandinavia and throughout central Asia to China. They are still found throughout Africa, the Near East, and India.

AARD-WOLF (*Protelinae*)

This rather astonishing looking animal is found all down the eastern side of Africa, south around the Kalahari Desert and up the west coast to Angola. It is nocturnal and sleeps by day in Aard-Vark holes, under giant termite nests or in holes excavated by itself. The tall crest is permanently erected. These animals have five toes on the forefeet, and four on the hind feet whereas the Hyaenas only have four on each. Their teeth are reduced both in number and size to minuteness, and are set wide apart, and the animal is virtually an insect-eater though it can chew very rotten meat and newly born animals. It is nowhere common.

HYAENAS (*Hyaeninae*)

There are two types of hyaena still living, one the Striped (*Hyaena*) found in India, throughout the Near East and in Africa north of the Sahara, down through the Sudan to the drier parts of East Africa. The other, spread all over Africa south of the Sahara, is a larger animal and is known as the Spotted Hyaena (*Crocuta*) (Plate 111). There is much debate about a third type called the Brown Hyaena, some believing it to be a separate genus, others that it is a species of *Hyaena,* or that it is only a race of the Striped Hyaena, and still others that it does not exist at all, being merely old individuals of the Striped species. There is also the everlasting question as to whether the so-far-uncaught, large animals known to the natives over a wide area as Nandi

W. T. MILLER

Aard-Wolf

Bears are simply very large examples of this elusive Brown Hyaena, myths, or rare survivors of a very strange carnivorous Ungulate, known as the Chalcicothere, which only became extinct about the dawn of the historical era. There is some valid reason for believing that there is some large animal still to be caught around the Nandi forests of East Africa and subfossilized bones of *Chalcicotherium* have been found in most recent strata in the same area.

Striped Hyaena (*Hyaena*)

This animal, in the typical form, is still fairly common all over the range described above but notably so in parts of northern India. It is distinguished by the tall crest along its back and half a dozen black upright stripes on its flanks, black cross-stripes on its legs, and a black throat and ears. Otherwise it is a dirty yellowish-grey. The tail is short and bushy and the front legs are much longer than the back. It is a cowardly carrion-feeder and loathed everywhere because it digs up human burials. When it knows it is safe it will attack small animals and has been known to carry off children. It lives in holes that it digs for itself but uses caves, tombs, or ruins whenever available. If taken young, however, it makes a very amiable and loving pet.

It is the opinion of the author that the so-called Brown Hyaena is a separate species of the genus *Hyaena,* found around the coastal areas of south Africa from southern Kenya on the east to Angola on the west. It is now very rare and probably always was so, and is extinct in South Africa proper. It is covered in brown shaggy fur, has a long lighter-colored mane, a non-erect crest, and white spots on its limbs. The ears are very long and pointed.

[175]

Spotted Hyaenas (*Crocuta*)

There is a widely held belief that Hyaenas either change sex, or are all hermaphrodites. This has come about through the virtual impossibility of defining the sex without dissection, something that is by no means unique among mammals and is often a puzzle with quite common creatures. It is of course untrue. Similarly strange is the appearance of the twin pups of the Spotted Hyaena, which are jet black and take an excessively long time to wean. These animals are much larger than the foregoing, much more common, and have front and back limbs more nearly equal in length. Relatively, they are terrifyingly powerful brutes, with jaws, teeth, and muscles to work these with, that put the largest cats to shame. They can crack an ox leg-bone as we do a match, and they completely demolish the largest bones of hippo and even elephant. They are singularly unpleasant beasts of cowardly disposition but dangerously bold and sneaky. They will attack lone men, anything injured, carry off children or any smaller animal, domestic or otherwise, if left unprotected, yet a dying man only has to say "boo" to a gang of them and they will flee. Unlike the Striped Hyaena they travel in large packs and they eat on the spot rather than drag provisions to their lair. They stink and are genuinely dirty, a most unusual thing in Nature, yet the author can attest to the really cleanly and attractive qualities of pups raised from shortly after weaning. Despite their enormous strength, they make docile and apparently trustworthy pets. They are noisy creatures, however, and give out with the most bloodcurdling howls, growls, and barks often ending in an insane laugh. Like all Hyaenas they appear to be lumbering brutes but they can cover rough ground at an astonishing speed, and lope along on flat terrain far faster than a horse. They have a curious habit of spreading their dung around their food at certain times and for no apparent reason, for at other times they will bury it like a cat. They are doing rather too well under modern conditions and are rather intelligent animals that quickly learn man's ways and adapt to them.

PROCYONINES

One main stem of the vast Order of Carnivore contains the Raccoons and their allies (herewith called the Procyonines), the Canines, and the Ursines or Bears (which are nothing more than vast tail-less dogs). The earliest and most primitive Procyonines seem to merge, as do the Viverrines, with the Mustelines or Weasels. Thus we get a family tree starting at the bottom with weasellike types and giving off two side-branches: one the Viverrine, with Felines and Hyaenines as offshoots; the other, the Procyonine, with Ursines and Canines as offshoots. The Procyonines are primitive, and they have developed along two quite separate lines themselves and on opposite sides of the world. The majority are South American with two kinds in North America. The others, the Pandas, are isolated in Southeast Asia.

RACCOONS (*Procyoninae*)

There are six genera of animals that may be classed as raccoons; two are dog-shaped and have short tails, two are cat-shaped and have long tails, and two have bodies like big squirrels with long bushy tails. The pairs are called Raccoons, Coatimundis, and Cacomixtles respectively. They are all from the western hemisphere.

North American Raccoons (*Procyon*)

The Common North American Raccoon (Plate 123) is almost too well known and is certainly well enough illustrated herein to obviate the necessity of a description. It is a remarkable form of wildlife that has managed to survive the impact of modern man and is rapidly showing signs of becoming a truly domesticated animal in the United States. Despite gross and quite unwarranted persecution it is thriving in the most highly industrialized areas of our country and has even become a nuisance on the island of Manhattan which it invades via the few bridges leading from the mainland, intent upon the contents of garbage-cans. If taken young "coons" may be readily tamed but about half of them remain unpredictable. They do molest small farm stock and poultry and transmit sundry diseases but they are otherwise innocuous. Though slow and deliberate, they display many qualities of dogs, and they are vastly more competent than those animals at looking after themselves, for they climb, swim very well, and use their hands with almost as much dexterity as monkeys.

Just how many different kinds of these Raccoons there really are is open to debate but they range from southern Canada to Panama and may extend into South America as far as Trinidad on the one hand and Guayaquil on the other. They may also once have been indigenous to the Bahamas and perhaps some other West Indian Islands and they were certainly introduced to and kept as pets on some of the Greater Antilles by the Amerinds in pre-Columbian times. There is a very strange, dark-furred pigmy race on some islands in that region today, and they may still actually be found wild on some of the Lesser Antilles. Further, they are extending their range rapidly northward and positively vast individuals are common in the northern Middle West, where they have long thick coats with a dense underfur. In the tropics they are sleek and short-furred and the farther south one goes the slenderer and shorter-furred they become.

These animals are in many respects equivalent to that competent living fossil, the Common North American Opossum, in that they can subsist on almost anything, hide from anything, and defend themselves from most predators, while they are themselves predaceous, aggressive, and tough. They also breed very regularly and strongly. A very large and interesting book could be written about the Raccoon and, with its industrious energy and resourcefulness, it deserves to be elevated to the status of the National Emblem in place of the parasitical, carrion-feeding Bald Eagle. Nevertheless, these animals are ingenious and persistent rather than truly intelligent and survive more by giving ground than by aggression. It comes as a surprise to many to know that in the South and in the tropics they are one of the most vociferous of all night animals and in the mating sea-

[*continued on page 193*]

98. *Lion*

99. *Leopard*
VAN NOSTRAND
FROM NATIONAL AUDUBON

100. *Leopard*
VAN NOSTRAND
FROM NATIONAL AUDUBON

101.
Amurian Tiger
LA TOUR

102. *Indian Tiger*
VAN NOSTRAND FROM SAN DIEGO ZOO

103. *Six-month-old Jaguar Cub*

VAN NOSTRAND FROM NATIONAL AUDUBON

104. *Arabian Sand Cat*

MARKHAM

105. *Puma*

LA TOUR

107. *Bobcat Kitten*

106. *Bobcat*

108. *Caracal*

109. *African Cheetah*

COMMANDER GATTI AFRICAN EXPEDITIONS

110. *Indian Grey Mongoose*

KING: CARROLL
EXPEDITIONS

111. *Spotted Hyaena*

YLLA FROM RAPHO-GUILLUMETTE

112. *European Wolf*

MARKHAM

113. *Coyote*

114. *Colombian Bushdogs*
PINNEY

115. *Dingo*
AUSTRALIAN
INFORMATION BUREAU

116. *Grey Fox*

PORTER

117. *North American Red Fox*

CHACE FROM OPL

118. *Fennec*

MARKHAM

119. *American Black Bears*

120. *Himalayan Bear*

121. *Polar Bear*

LA TOUR

123. *Common Raccoon*

SUKERT FROM PHOTO LIBRARY INC.

→

122. *Syrian Brown Bear*

MARKHAM

124. *North American Cacomixtle*

125. *Lesser Panda*

E. P. WALKER

[continued from page 176

sons they give out with screams, as well as mewing, growling, and whistling.

Crab-eating Raccoons (*Euprocyon*)

An altogether different animal is the red-coated, long-legged Crab-eating Raccoon of South American seashores and river banks. It is semi-aquatic, goes about in large family parties, and is crepuscular rather than purely nocturnal in habits. In color it is a rich brick red, with rounded white-fringed ears, a black mask, white muzzle, and a vividly ringed black and white tail. Its headquarters is the Amazon Basin but it extends north to Venezuela and Colombia and south to the southeast coast of Brazil and Paraguay and it is found on the eastern slopes of the Andes. It is a rather savage and aggressive animal and gives the oddest of all animal noises when alarmed—a long-drawn-out whistle very near the upper limit of human hearing and not unlike that of the Pika. These animals have been caught swimming across arms of the open sea as much as four miles from land.

Coatimundis (*Nasua*)

These are elongated raccoons with a snout that might almost be called a trunk since it is so long it can be twitched about to an angle of forty-five degrees. The tail is ringed and long, the legs short, and the ears small. They have very long front claws. In habits, these animals are of great interest because they show a true community spirit and at least the rudiments of what we call a logistical approach to the problems of life. They are found all over the Americas from southern Oklahoma to the southern tree limit in the Argentine. They travel about in small or large bands of all ages. When they invade a territory, they comb it completely from tree-tops to the bottoms of holes, turning over everything and devouring anything edible. They are rather aggressive also, and a large male, using its claws and rapier-like canine teeth, is a match for almost any dog.

There appear to be two principal types of coatis, each, however, being an aggregate species. The first, found from North America to the desert limits on the West Coast of South America, is a brindled color with a raccoon-like dark mask, complex light facial markings, a lighter underside, and a tail with vague light and dark rings. The second is South American and a forest animal, ranging from Panama to southern Brazil. It is basically red in color, with an orange belly, light rings around each eye, and a strongly contrasted black-and-white-ringed tail. However, innumerable species have been described from both areas and the animals appear to vary clinally to a marked degree from place to place. Even a small river will divide one color variety from another and in some localities there may be two or more quite distinct species. The forest types have softer and sometimes even silky fur.

These animals are readily tamed and often kept as pets. They are more reliable but even more mischievous and destructive than raccoons, particularly of cigarettes, which they almost invariably open with their claws. They were often reported as escapees from zoos in the southwest, but although long known to be indigenous, were not popularly recognized as such until a forest officer reported 400 on a road in New Mexico.

Mountain Coati (*Nasuella*)

This is another odd, extreme type of an otherwise common and widespread animal, found only in the mountainous area of the northwest Pacific forests of South America. It is a diminutive Coati with a very short tail, colored olive greenish-brown, and has an exaggeratedly long and slender, trunklike muzzle. Its teeth are small and weak and seem to indicate a more wholly insectivorous diet. Nothing notable about its habits has been described.

Cacomixtles (*Bassariscus*)

Going down the scale, we come next to some strange alert little animals that are in many respects raccoon-like but are apparently very primitive and ancient types. They have squirrel-shaped bodies and immense bushy tails, vividly ringed with black and white. The so-called "Ring-tailed Cat" (Plate 124) of the southwestern half of the United States (from Oregon to Alabama and south to Mexico) is really a very common animal but seldom seen on account of its nocturnal habits and extraordinary nimbleness. Further, it can squeeze into slits between rocks or cracks in trees, sometimes hardly an inch wide, that one would suppose a good-sized beetle could not enter. The bodies of these animals are clothed in very soft, dense fur of a gingery brown color. To the south, in the forests of Central America, there are other larger species, silvery grey in color. All have foxy faces and pronounced black and white facial markings. They are arboreal animals but in the Rockies they often live in open desert country among boulders and in Mexico City they infest houses and parks. They are the all-time escape artists, can open any device but padlocks, and can squirm through the minutest cracks and crannies. They are predators and very agile at catching sleeping birds but they also eat fruits, insects, and earthworms. They have been seen fishing with their paws and burrowing in loose soil after Pocket-gophers. They are extremely savage little animals and survive well in captivity especially if kept in pairs.

SUSCHITZKY

[193]

White-nosed Coatimundi

The Cuataquil (Bassaricyon)

Of all idiotic scientific names for an animal this takes the prize: it means literally the "Fox-dog" or "Dog-dog" as *bassara* is an ancient Thracian word for dogs and foxes, and *kyon* meant a dog in classical Greek. The animal is question has caused a great deal of confusion in scientific records, completely bamboozles the non-specialist, is usually overlooked, is seldom represented in museum collections, and yet appears to be fairly common. The trouble is that it looks very much like a Kinkajou, being of an overall gingery, yellowish brown, but having a squirrel-shaped body like a Cacomixtle and an exceedingly long, fairly bushy tail. It is an entirely arboreal and nocturnal animal found so far only in certain limited areas of Central America. It may, however, have a much wider range. The author has owned two live specimens bought as Kinkajous in the United States, though the animal is reputed scientifically to be exceedingly rare. When compared with a Kinkajou it is seen to be altogether different, being shaped like a Cacomixtle and having a tail that is not at all prehensile. It appears to be a fruit-eater, is comparatively slow-

Kinkajou

SUSCHITZKY

moving, and makes nests of dry leaves in holes in trees. The face is foxy and the muzzle rather long. It is the most primitive Procyonine known.

KINKAJOUS (*Cercoleptinae*)

Of altogether different appearance, mien, and habits from the raccoons, are a group of animals known very popularly but quite inaccurately in America as "Honey-Bears" or correctly as Kinkajous. They are vicious little tree-climbers, called variously by the inhabitants of the countries where they are found "Night-Walkers," "Mico-de-Noches," or by a host of other names. Zoologists have not done much better, since they have given them a variety of names (ending today with *Potos*, which is most muddling in view of the Potto, a lemur from Africa), and have bandied them about between various families and even orders of mammals. They are, indeed, very nondescript in that they seem to combine the qualities of many other animals and, although most distinctive, they do not fall into any particular category. The appearance of the head and face may be seen in the magnificent accompanying photograph: the rest of the animal is hard to define. It is clothed in very dense, soft, short, gingery-yellow fur, which may be bright yellow below and adorned with a jet black stripe down the mid-chest. Some are almost grey and washed with silver, others bright yellow, and a kind from central Brazil is rich chocolate brown with a cream belly.

These animals are predators and will eat any other animal they can catch, including insects, but they also subsist on fruits, green nuts, some leaves, and a lot of fungi. Above all, however, they relish honey, upon which they gorge, using their narrow six-inch tongues to extract it from bees' nests. They are really very common though most dangerous pets because their honey-eating apparently leads to an insatiable appetite for alcoholic liquids of all kinds. When they are inebriated, they go quite mad, and will attack their owners, latch on with their sharp claws, prehensile tail, and vicious teeth, and continue biting like no other mammal.

PANDAS (*Ailurinae*)

The affinities of these Asiatic animals is most puzzling to all except Americans, who, even if not specifically interested in natural history, almost invariably spot Lesser Pandas as kind of 'coons. Even zoologists for a time considered the Giant Panda to be a strange kind of bear. The latter is wholly and the former almost wholly vegetarian and both have specialized in bamboo-eating.

The Lesser Panda (*Ailurus*)

This delightful little animal, looking rather like an animated toy especially when it trundles along on its somewhat pigeon-toed, bearlike feet, is nonetheless not recommended as a house pet. It may be tamed to take food from the hand and even climb to one's lap but its claws have cutting edges like razors, and it cannot bear to be touched. It can move with lightning speed and give a terrible bite. It comes from the Himalayas west of Nepal to Yunnan and Sze-Chwan in China and south at least to Laos, but always in the mountains ranging

ZOOLOGICAL SOCIETY OF LONDON

Giant Panda

from the 6,000- to 12,000-foot levels. It is an arboreal animal and usually lives in holes in trees but spends much time on the ground foraging. The parents live together at all times and are usually followed by a line of young ones often of various sizes (Plate 125).

The Giant Panda (*Ailuropoda*)

Though one of the rarest large mammals, this animal is well-known to almost everybody as a result of the flood of publicity on live specimens that have been exhibited in European and American zoos during the past three decades. They are bear-shaped and bear-sized animals that live exclusively on bamboo-shoots in forests of those plants growing in limited areas on the mountains in Sze-Chwan and Kansu in China. The Giant Panda has been known to the Chinese since time immemorial but its existence was considered a myth by Europeans until a late date. A few skins were obtained by missionaries and thought to be pelts of some strange bear. A live specimen is said to have been taken to Paris in the eighteenth century but the creature was not officially described until 1870, and its anatomy was not investigated until this century. The first specimen obtained in its native haunts was apparently shot by Kermit and Theodore Roosevelt simultaneously on an expedition in 1929.

CANINES

We herewith come to "Man's best friend" and a large number of associated animals that are well-known to almost everybody. Despite the universality of Domestic Dogs, there is a great deal of misunderstanding about this group. A very strange question about them that nobody appears ever to consider is why one of the purest predators, with rather unpleasant and treacherous habits, should have become the most reliable of man's mammalian allies—not just a domestic pet, but a veritable associate, slave, and true friend? The dog's status is very odd if you really think about it but there may be an explanation such as the one that follows.

Long before the last or current Ice-Age, the ancestors of Man came down out of the forest trees and took to running about on open plains. They were slow, small-toothed, omnivores, grubbing for roots, collecting fruits, and above all eating any little animals they could catch. Their best meals came from the carrion left by the great cats and it was there that they met the dog—also a carrion feeder. Man took to following the dogs to the cats' kills. Then they took to living together in caves and other lairs. In fact, they literally grew up together.

Dogs, it comes as a surprise to most people to learn, are most primitive and in some respects almost basic mammals, being in their most typical form the kind of animal that lived sixty million years ago in the Eocene period. Their unspecialized structure and catholic habits are probably the secret of their worldwide success and survival. And they are found almost everywhere—even in Australia, in the form of the Dingo—and on distant islands like the Falklands. Apart from four very distinct animals—the Dholes, the Hunting-Dog, the little Bushdog, and the Bat-eared Fox—all canines are very much of a oneness. Nonetheless they may be split up for convenience into nine lots.

TRUE CANINES (Caninae)

There are five groups within this sub-family—first the wolves, dogs and jackals; second, the foxes; third, a group of jackal-like animals from South America; fourth, a creature called the Maned Wolf and fifth, a strange beast called the Raccoon-Dog. Together, these animals are worldwide in range.

Wolves (Canis)

If you skin a wolf, a coyote and a domestic dog, you would be hard put to it to identify any one of them even if you were an anatomist. It appears that, before man started tampering with Nature and produced an array of different-sized, -shaped, and -colored domestic breeds of what we call dogs, almost the whole land surface of the earth was parceled out among different kinds of doglike animals, with the range of each seldom coinciding with or overlapping that of any other. Each stayed in its own ecological niche. Confusion arises from the fact that dogs in the wide sense of that term have gone ahead and developed what we may call "nations" without changing in appearance very much. Thus, wolves are the indigenous dogs of the sub-arctic and temperate and desert areas of the whole Northern Hemisphere; and Jackals, the tropical regions south of the deserts in Africa and Asia. In South America, their place is filled by a distinct grouping of jackal-like animals (genus *Dusicyon*).

It would be purposeless to attempt a description of all kinds of Wolves and Dogs since this would entail end-less technical minutiae. Besides, almost everybody knows what a wolf, as well as a dog, looks like. There is, however, a very great deal of misconception about the habits of all these animals. These, of course, vary greatly from place to place but also, it now appears, from time to time. Wolves are far less conservative than most other mammals and they are highly adaptive. Thus, they sometimes live alone, sometimes in family parties, and at other times in packs. They are also great travelers, since their main source of food supply may itself fluctuate (lemmings, for instance) or migrate regularly (deer of various kinds). Therefore, individuals or whole races of dogs may emigrate from one area into another and, being very rugged, quickly adapt themselves to a change in climate. They are actually and fundamentally nomads, except at the breeding season, when they almost all use permanent lairs in which to raise their young.

Wolves were once found all over Europe, Asia, and North America but man has driven them north to the boreal forests and the Arctic—except in Asia, where the wolves of India have not been so affected. There are really only three, and possibly no more than two, "species" (in the wider sense) of Wolves—namely, the Wolf and the Coyote. There is no real "Wild Dog" known, the animals so called in eastern Asia being entirely different creatures (of the genus *Cuon*), as may be seen below. Thus, we come to a very strange impasse; to wit, are all domestic dogs merely altered forms of the wolf, or are they the descendants of some animal that is now totally extinct? If the latter, where are the fossilized bones of that animal? A third alternative is that our domestic dogs may have been developed from different strains of wolves, jackals, and other extinct animals at different times and in different places. Whatever the truth, we are nonetheless confronted with the bizarre fact that there probably is no such thing as a "dog" (in the strict sense) and never has been!

Many scientists have tried to separate the American, Japanese, European (Plate 112), and other Asiatic wolves but it seems that the same variations and mutations occur in all of them. The Indian Wolf (*Canis pallipes*) certainly appears very different but there are intermediate races in Afghanistan and other areas where the ranges of the two meet or once must have met. The Coyote (*Canis latrans*) really is different (Plate 113) and, be it noted, has a range overlapping that of the wolf. There are innumerable recognizable races of the wolf, and white, black, grey, brown, red, and yellow individuals are often mixed together. Apart from the Coyote, which is an animal of the open prairies, wolves live in almost every type of country from the bleak tundras to stony mountain tops, baking deserts, forests of all kinds and even equatorial jungles. They are predators and scavengers and are not nearly so dangerous to man or domestic beasts as is popularly supposed. Actually they are very shy, retiring animals.

The only True Canine that is not a wolf is the Dingo of Australia (Plate 115). This is probably just an ancient breed of Domestic Dog, taken to that country in boats by either the original or later human immigrants. However, this sleek, yellow animal is undoubt·

edly an ancient breed and might be somewhat close to the basic stock from which our other tame stocks were originated. Today it is a truly wild animal though the Australian aborigines catch the pups and tame them for hunting. The animal is very odd in appearance and habits and not only in that it lives in Australia and is the only large non-marsupial mammal in that country.

Jackals (Thos)

Throughout the drier parts of Africa, sub-tropical and tropical Asia but seldom, if ever, in the really virgin, moist equatorial forests, the wolves are replaced by smaller doglike animals with much the appearance of foxes. Nobody can state categorically how many valid species of Jackals there are for they too vary considerably and particularly geographically. There are four distinct kinds and innumerable races, sub-species, and other doubtful types. The common species (*Thos aureus, lupaster, anthus,* etc.) ranges from southeastern Europe, through Asia Minor and the whole Middle East to India, Assam, Burma and Malaya and Ceylon. It also is found all over Africa north of the Sahara and as far south as Kenya on the East coast. A second very large kind (*T. simensis*) called a wolf is found in Abyssinia. The Black-backed Jackal (*T. mesomelas*) ranges from the Sudan to the Cape on the east side of Africa, and the Side-striped Jackal (*T. adustus*) from the Cape north to the Gabun on the west and to Kenya on the east, and also occurs throughout the Congo area.

Jackals are scavengers and more or less omnivorous. They may assemble in packs and hunt down quite large animals. They are nocturnal and sleep by day in holes unless it is very hot, whereupon they will lie soaking in water all day. Their name has come to be applied to all forms of unpleasant hangers-on—a result of their habit of following the large cats, making a special noise when doing so, and then eating up most of the feast as soon as the cat's back is turned. They are otherwise most notable for their appalling call which, in the case of the Indian race is said to be *"Dead Hindoo . . . where, where, where?"* They will all interbreed with domestic dogs and the progeny of this union in turn with domestic dogs, other jackals, or even wolves so that in India there is a colossal mix-up of doggy creatures.

Foxes (Vulpes)

Foxes are extremely numerous and widespread animals. There are about a dozen species, one of which, the Common or Red Fox (Plate 117), is found all over North America south to Mexico, in Iceland, in Eurasia north of and including the Himalayas, and throughout Africa north of the Sahara. This animal comes in an extraordinary variety of sizes, some races being twice as large as the smallest; they also vary from bright brick-red to orange, yellow, and all shades of brown, and the Cross, Silver, and Black Foxes are only mutations of this species. The back of their ears and lower limbs are usually darker, and the tail tip may be white or black. The fox is a burrowing animal and not nearly as smart as tradition would lead us to believe. It kills game of all kinds but does a great deal more good than harm by controlling rodent pests. It inhabits all kinds of country but is not a true forest animal. More has probably been written about the fox throughout the ages than about any other wild animal but a very large percentage of all this verbiage is purely fabulous. On the other hand these animals do many of the most extraordinary things attributed to them, such as de-fleaing themselves by backing under water.

From eastern Europe to India and Siberia there is also found the Hoary Fox (*V. canus*) which is grey above and white below, and south of this animal's range, from Arabia, through Iraq and Iran, and in the western part of India the Desert Fox (*V. leucopus*) takes its place and a form of this species, the Corsac Fox, ranges across Central Asia to Siberia. In India there is a small species (*V. bengalensis*), that is very common. Then, between the equatorial forests of Africa and the Sahara there is another desert form, the Pale Fox (*V. pallidus*), which is light buff in color. America too has an extreme form in the Kit Fox (*V. velox*) of the northern prairie belt, which is a small short-legged animal with an immense bushy tail. Finally, there are a group of close-furred, rather slender-tailed foxes in Africa that have enormous ears and are often misleadingly referred to as fennecs, which are much smaller and quite different animals. One of these is called the Kama (*V. cama*) and inhabits the Kalahari Desert and adjacent dry areas; another is found along the north fringe of the Sahara; and still a third (*V. familicus*) or Ruppell's Fennec is spread all over the Near and Middle East. These big-eared foxes live on jerboas and gerbils, as well as lizards, and insects.

Fennecs (Fennecus)

These are probably the "cutest" animals to be found, looking like tiny, fluffy, toy foxes and being unique in that they always look like baby animals (Plate 118). They are nocturnal and range in color from pure white, but for the black nose, and huge dark eyes, to various pale cream-yellows washed with rusty brown on the back. Fennecs are true desert animals and spend much of their time underground to avoid the heat of day and the cold of night. They are communal animals and their holes are often linked together. They sometimes congregate in large numbers at the scarce waterholes. They are found all over North Africa and the Saharan region to the eastern Sudan and in Palestine and Arabia. To add to their charm, these little creatures make pathetic whimpering noises when alarmed.

The Arctic Fox (Alopex)

This is the "Blue Fox" of the fur trade and is altogether different from other foxes, having a shorter muzzle, small rounded ears, a ruff on the throat, and fully furred soles to its feet. Anatomically also it is unlike other foxes. Further, it changes color completely each spring and fall. In winter it is pure white, in summer either a beautiful mauvish grey-brown all over, when it is truly a "blue fox," or parti-colored with a most astonishing pattern. When the animal is sitting down with its bushy tail wrapped around its chest, it appears to be dark grey-brown washed with mauve, but when it stands up, all those parts previously concealed, even the underside of the tail, back of the legs, lower legs, groin, armpits, and neck, are disclosed as pure white. This arrangement serves two purposes, camouflage by blending with the ground when sitting and by breaking up the animal's outline when it is standing, both very useful devices on the barren wastes of the Arctic where the creature lives. It is found all over the Arctic Circle and south to about 55 degrees north latitude. It feeds on lemmings and fish and lays away stores of the former for the winter in the natural deep-freeze created by the first snowfalls, and by this means can stay active during the perpetual night of midwinter even on the actual Arctic Ocean ice-raft. The explorer A. R. Glenn, who wintered under the ice in Spitzbergen, has described the terrifying noise, like the footsteps of a giant approaching, when one of these animals came to sniff at the ventilating shaft above their heads in midwinter. In Iceland they may stay blue all year round.

The Grey Fox (Urocyon)

This is a very common animal (Plate 116) throughout the United States and Central America, ranging to the northern part of South America, but is seldom seen. It stands more erect than other foxes, has a slightly elliptical pupil, and is a facile tree-climber. It inhabits a niche that is subtly different from that occupied by the Red Fox, and before the clearing of the continent for the European type of farming, it probably lived—as it does in Central America today—only on what is called the "Pine Barrens" or "Pine Ridges" and on other savannah-like tracts. Wherever the southern pines occur, as far south as Venezuela, there the Grey Fox will be. In

many areas they run in packs and chase squirrels up trees. They have now taken to infesting agricultural land and they even make their lairs under buildings.

South American Jackals (Dusicyon)

There was for long much heated debate among zoologists as to how different animals of this group should be recognized, and to which of these which names should be applied. The matter was finally settled and appeared simple: now, at the time of writing, the whole business is about to break out again because collectors are penetrating many hitherto unvisited parts of the continent and quite a number of entirely new and unexpected forms have been brought out alive. Unfortunately these have gone into the retail pet trade, and so have not been scientifically examined. There are at least half a dozen very distinct species groups, in addition to the recently extinct form from the Falkland Islands. The latter presents zoology with one of its greatest mysteries. How the animal in question ever got to those remote inhospitable islands is quite beyond comprehension, for even if a pregnant female got adrift on something which could remain afloat in the intervening windswept sea, the currents all run in the wrong direction. This animal looked and behaved like a small wolf, and was exterminated by the sheep-farming settlers for that reason. An apparently closely related form, if not a race of this animal, known as the Colpeo, still inhabits the stunted forests, open hills, and cold deserts of lower Patagonia, Tierra del Fuego, and southern Chile. This animal was partially domesticated by the indigenous natives of the area—one of the most primitive peoples on earth—but there is no evidence that they ever reached the Falklands. This animal has the longest, sharpest muzzle and the longest, bushiest tail of all canines. It is of a reddish hue especially in the deserts.

Throughout the rest of South America there are sundry little greyish jackals with long, rather shaggy black-tipped hair, bushy tails and sharp faces, and with very neat little feet upon which they can only be described as "trickling along." The savannah and open country species used to be called Azara's Dog, and the coastal, swamp and forest forms, Crab-eating Dogs, but there is a third form from Guiana that is exactly like a fox externally, and a fourth, a long-bodied, close-haired species from the upland savannahs of Amazonian Colombia. Finally, there is the little-known Small-eared Dog (D. microtis) from the Upper Amazon. More aggravating still, there are reputed to be other species in the dry Caatinga, and in the wet Tupai forests of eastern Brazil. All these animals seem to behave more like foxes than jackals, but they will run in packs and pull down larger game. The Crab-eating Dog really does eat crabs and has been seen diving for shellfish in shallow water.

The Maned Wolf (Chrysocyon)

As the photograph shows, this is the most outrageous-looking of all canines. It is colored rather like the typical Red Fox. When seen alive its proportions are hardly believable and must surely have been developed for getting through or between things, though just what

Maned Wolf

these may be nobody has thought to observe or record. It inhabits the broken forests on the southern edge of the Amazonian Basin, Paraguay, and northern Argentine but, being nocturnal, is seldom seen. It is a rodent-eater but takes any small live things it can find, including insects, and it is almost as swift as the Cheetahs. It apparently does not dig, and it swims very well. Altogether this is one of the oddest of mammals and we would like to know more about it.

The Raccoon-Dog (Nyctereutes)

Quite the most remarkable of all canines is an animal found in Amuria and the islands of Japan. It looks very like a raccoon, with small round and fringed ears, a heavy body, short legs and tail. It is brindled grey and yellow and black on the throat, belly and legs. Instead of a mask, the eyes are ringed by black spectacles, and the muzzle is dark. The tail is not ringed. It lives on mountains during the summer but descends to valleys to hibernate in winter—a unique procedure among Canines. It can dig its own burrows, but usually takes up residence and hibernates in the holes of foxes and other animals. These animals are omnivorous, catching small rodents, some birds, and eating quantities of insects and fish for which they dive or which they scoop up with their paws. Individuals may be active all winter or during mild periods, when they forage in deep snow. They appear to be distantly related to the bush-dogs of South America.

FALSE CANINES (Cymocyoninae)

We herewith come to an assemblage of animals that really have nothing in common except that they are all not quite "dogs" in the sense that the foregoing animals are. They may be distantly related or they may each be relics of now otherwise extinct branches of the Canine stock. They are three in number and found in Asia, Africa, and South America respectively.

Dhoies (Cuon)

There are three species of Dholes; one covering Siberia south to the Sinkiang-Mongolia desert belt; a second, in East Tibet, the Himalaya complex and India but not Ceylon; and a third, from Burma to Singapore and on the islands of Borneo, Sumatra, and Java. They are more like the average person's conception of a dog than are any of the True Canines. However, they have two less teeth, twelve to fourteen nipples instead of ten, long fur between their pads, and concave rather than straight or convex bridges to their noses. They are clothed in coarse hair which is long and fortified with a thick underwool in colder climates. In color they vary much, but the Siberian types range from white to red-brown and change seasonally; the Indian types are variegated, and the Malayan bright red with white undersides. They do not dig or live in holes but among rocks or in thick brush, and they associate in gangs for hunting. They are slower than wolves and jackals, but although they average only about three feet in length and have rather short legs, they can pull down even Water Buffalo and the huge fleet Nilghai. They are bold, apparently rather proud animals, behaving like the better

Indian Dhole

Cape Hunting-Dog

breeds of domestic hounds and are quite untamable. Though they are often called "Wild-Dogs," they have no relation to our domestic animals and are not "dogs."

The Hunting-Dog (Lycaon)

This strange African animal has one of the most peculiar coats of any mammal, being like that of tortoise-shell or calico domestic cats, a mixture of black, white, and orange. Further, it has only four toes on its front as well as back feet. It is about five feet long, of which just over a foot is tail, which has a white, terminal, dumbbell-shaped tuft. Their hunting tactics are a byword. Associating in large packs they run down the largest game and even old lions by tireless and continuous runs, the lead dogs retiring when spent and those at the rear, which have saved their strength by cutting

[199]

Bat-eared Fox

corners, taking their place in relays. There appear to be special killers who finally go for the throat of the quarry, and no amount of wounds or deaths among their number deters the pack. They do not like human beings and have the utmost contempt for domesticated dogs, which they hunt down on sight. They make lairs in holes for the females to raise their young. They are found all the way up the eastern third of Africa.

Bush-Dogs (*Speothus*)

Around the fringes of the open savannahs that occur throughout the great Amazonian forest area of South America, there dwell the funniest-looking little doggy animals shaped like fat Dachshunds but with slightly bushy tails (Plate 114). They have small, rounded, furry ears, are clothed in coarse sparse hair and make the oddest chirruping, clicking, and whistling sounds. They dig and live in burrows and forage at dawn and dusk, mumbling to themselves all the time. They are carnivorous and appear to be rather intelligent little animals. Most of them are reddish-brown but some have shoulder capes of light yellow and others are black below and all over the flanks.

BAT-EARED FOXES (*Otocyoninae*)

The last of the canines is a beautiful little fox-shaped animal from the eastern side of Africa. It has enormous ears and a very bushy tail, is colored dark grey and black with a silvery wash, and has a pale yellow underfur. It has more teeth than any other Canine and is structurally very different from the rest of the tribe. It lives in holes, is nocturnal and its food consists of all small animal life, including insects. It has in all from forty-six to forty-eight teeth, a characteristic found in no other mammal except Marsupials and some Tenrecs (and, of course, the whales—see below). It may, indeed, be a very primitive Carnivore that retains certain basic features of all mammals.

URSINES

Although nobody can mistake a bear for any other animal, it is actually almost impossible to distinguish bears technically from the Canines, more especially when a

number of extinct animals are taken into account. Some of the latter are absolutely intermediate, and between these and the dogs on the one hand and bears on the other, are further creatures showing an almost complete intergradation between the two groups. Bears are nothing but gigantic, tail-less Canines, and are, comparatively, a very recent development. Bears were, until historical times, spread all over North America from the Arctic to Mexico and in the Andes in South America, all over Eurasia including the islands of Nippon, Formosa, Borneo, Sumatra, and Java, and along North Africa. Today, their range is greatly restricted so that they occur only in limited and isolated areas throughout their original range. For the most part these are mountainous areas such as the Pyrenees, Alps, Carpathians, Caucasus, and Scandinavia in Europe. In India, Malaya, Siberia, Canada, and throughout two north-south belts on the west and east coasts of North America respectively they are still quite universally distributed.

Bears form one of the smallest, most homogeneous, and yet most variable of all groups of mammals. There are only seven clearly separable forms and these are all very much of a oneness except for two—the Sloth-Bear (*Melursus*) of India, and the famous Polar Bear (*Thalarctos*) of the Arctic. Even these are insufficiently distinct to warrant their being given family or even subfamily status, so that the breakdown herein employed is almost arbitrary.

TYPICAL BEARS (*Ursus, etc.*)

There are five groups of species that may be brought together under this heading—the Brown, Spectacled, Himalayan, Sun, and American Black. Though each is assigned a separate genus, all are closely akin. Yet, they present certain conundrums to even the most interested, principally because of their popular names, as will be seen. Much further confusion arises about bears because everybody who knows or thinks he knows something about the species inhabiting his country is wont to assume that the bears of other countries behave in a like manner. Unfortunately, bears are remarkable improvisors, highly adaptable, and most ingenious creatures so that their habits vary to fit the circumstances. Thus, the same kind of bear may truly hibernate in one country, do a lot of sleeping, or remain active all winter in another. It is best, therefore, to attempt to clear up the worst confusion first.

Brown Bears (*Ursus*)

These would much better be called "Dish-faced Bears" because they range in color from almost pure white to jet black, and may show almost any intermediate color in the browns, be grizzled, or even particolored. Just to make matters more confusing, the American Black Bear may be brown. They are the bears of Europe and Asia, north of the Himalayas—though there are species *in* that mountain complex—and the northwest part of North America. There are numerous races, and some of these may perhaps be called true species, but they all vary in appearance and habits in the most confusing manner. Just about the only way to define them is to say that they comprise all the bears not included under any of the six titles that follow.

The outstanding race is found in Alaska, and is quite popularly known, since it is the largest terrestrial carnivore on earth. Large males measure up to ten feet long and stand four feet tall at the shoulder. Another type of "Brown Bear" is the so-called "Grizzly" but this is not a species or even a race: it is a "type," or rather, just a name given to grizzled-coated and grizzly-looking bears found, or once found in certain areas of western North America. Most of these grizzled populations are now extinct, having, apparently, been dependent for their food upon the great herds of Bison that once roamed America. There are still populations of bears that may be called Grizzlies, but these populations need not necessarily be related. There are also "grizzlies" in the Tian Shan area of eastern Asia. A very large form of Brown Bear from Tibet may actually constitute a separate species (*U. pruinosus*); and there is a very outstanding type in the Himalayas that seems to be a true race and to breed true, known as the Isabelline Bear; it is a beautiful grey-white with a creamy under-fur. There are huge races in Eastern Siberia and in Kamchatka, and there is another readily recognizable race from Syria (Plate 122). There is reputed to be, or until recently has been a small form inhabiting the Atlas Mountains in North Africa, but no authentic specimens are known in museums. In Europe, where they were once widespread, bears are today found only in the Pyrenees, Scandinavian, the Carpathian and Caucasus Mountains and in some limited areas in the Alps. They are still fairly common in Russia and adjacent countries right down to the foothills of the Himalayas.

The habits of bears are almost impossible to describe because they are so enormously variable. On the whole, bears do not normally and deliberately molest humans, but they are altogether more dangerous animals than the great cats and may, if encountered unexpectedly, make most determined attacks. Moreover, despite their lumbering appearance and awkward gait, they are extremely swift and tremendously powerful. They climb trees, can swim very well, and get over any kind of ground at a great rate. They eat enormous quantities of almost any animal or vegetable material, but relish fish and have a great partiality for honey and other sweet things. They appear to be as immune to stings of bees as they are to almost everything else except fire, rock avalanches and, presumably, lightning. They lay up huge stores of fat in their bodies in fall, and most of them become sluggish in winter, retiring to sleep for long periods in holes under the snow which become lined with their frozen breath. The young are born before their final emergence from these winter quarters and are comparatively minute, those of the largest being not larger than a small rabbit. They often have twins and sometimes three young and these follow the mother about for many months.

The Spectacled Bear (*Tremarctos*)

The first odd bear we encounter is found, most unexpectedly, in the Andes Mountains of South America.

Malayan Sun-Bear

It is a fairly small, shaggy, black animal with yellow rings round the eyes and often a cream-colored muzzle, throat, and chest. If it is more closely related to any one other bear, its anatomy would point to the little, sleek-furred Sun-Bear of Malaya. There are two distinct races of Spectacled Bear: one in Bolivia, the other in Chile.

The Sun-Bear (*Helarctos*)

From the eastern Himalayas to Sze-Chwan in China, and thence south throughout Burma, parts of Indo-China, and Malaya, and in Borneo, there is found a small bear, black in color but for a light muzzle and a strange, yellowish, ring-shaped mark on the chest. These animals grow to approximately four and a half feet in length and have a tiny, two-inch tail. Although the fur is coarse, it appears rather sleek, and the young have soft, shiny coats. The head is very short, wide and flat, giving the animals an almost benign expression— at least for a bear. They are tree-climbing forest animals, and although quite commonly offered for sale as pets, practically nothing definite is recorded about their habits.

Moon Bears (*Selenarctos*)

South of the range of the Brown Bears in Asia but north of that of the Sun-Bear, and spread along a belt of territory from the borders of Persia to Assam and on to China, Japan, Formosa and Hainan, there are to be

found a number of races of large black bears (Plate 120) with a white mark on their chests shaped like a widely opened "V." Their upper lip is often white and their claws, which are rather small, are black in color. Although seldom growing much more than five and a half feet long, they are of very great comparative weight and become grossly fat in the autumn. They have rather large ears low down on the sides of their heads, and they are forest-dwellers, except about the desert area stretching between Persia and Pakistan, where they live among rocks and scrub. They are more essentially vegetarian than other bears, but will gobble up vast quantities of insects and honey, and occasionally take to killing domestic animals. They are singularly indifferent to humans and may be readily approached, while, in turn, they boldly maraud orchards when fruit is ripe. These bears go about in pairs or family parties, sometimes with cubs of two succeeding litters. The father marches along in front, followed by the mother and then the cubs in descending order of magnitude, presenting, especially when following a mountain ridge, a most comical sight.

American Black Bear (Euarctos)

The small, common bear of North America (Plate 119) can only be distinguished from the Brown Bears by the shape of the head seen in profile: the upper line of the muzzle arches slightly upwards; that of the Brown Bears downwards. As stated, it may be various shades of brown, and black and brown individuals may be twins. There are areas where its range overlaps that of Brown Bears, and there are races of *black* Brown Bears. These animals are forest-dwellers but they have adapted themselves rather well to the invasion of their territory by civilized man and, after a serious decline in their overall numbers, they made a remarkable recovery in certain localities and now appear to be more than holding their own all down the eastern mountains from Maine to Florida. They are medium-sized animals with comparatively short claws—the Brown Bears are distinguished by having, among other characteristics, enormous claws, sometimes over half the length of the sole of the foot. Large adult Black Bears have been recorded as weighing as much as five hundred pounds. They apparently do not go into true hibernation but do indulge in prolonged sleeps in the north in the winter. They are completely omnivorous and can consume such extraordinary things as wasp's nests, piles of skunk-cabbage, briars, river-clam shells, and even pine cones. It should be stressed that these animals are at all times very dangerous and unpredictable. They should never be offered food in the wild, however friendly they appear to be, since they adopt the attitude that, when no more food is forthcoming, the giver is holding out on them; whereupon they may proceed to make a search for more, and often about the donor's person. During such an operation the human party is likely to be seriously hurt and very possibly killed. The strength of Black Bears is almost unbelievable, and it takes quite a good horse to run them down. Up to three cubs, each about the size of a rat, are born in late winter.

SLOTH-BEARS (Melursus)

Not sufficiently different from other bears to warrant being treated as a separate family, but nonetheless distinct enough to call for very special mention, is a long-snouted, tropical species from India known as the Sloth-Bear, or Bhalu. The snout is almost anteater-like and the animal is, in truth, highly insectivorous, and, in accordance with this habit, the teeth are small and there are only two of them in the front of the upper jaw. The body is most ungainly, with huge feet and positively enormous claws. Its fur is black, shaggy and long, especially over the shoulders but the face is almost naked and pale grey. There is a white chevron on the chest and the claws are white. The body is about five feet long and has a four-inch tail. It is still a common animal in parts of India south of the desert belt and is found in Ceylon. It is nocturnal and sleeps in caves, especially in river banks, by day. Although it eats leaves, flowers, and fruits, it is a very clumsy climber of trees, but it will struggle anywhere to get at white ants or bees' nests. It is, in fact, well on the road to becoming an ursine anteater. When attacking termites' nests it has a curious habit of blowing away the earth and dust after ripping the nest open with its long claws. The brood galleries of the insects are then licked up to the accompaniment of cataclysmic puffings and belchings.

POLAR BEARS (Thalarctos)

This is one of the best known of animals (Plate 121) and is really, in many respects, different from all other bears. Its bodily proportions, with small head and narrow forequarters, and its habits are ursinely unique. It is found all about the Arctic Ocean and upon its ice-raft, and it is an extremely dangerous animal. Its principal food is seals though it is just as omnivorous as it can be and eats much seaweed and terrestrial vegetable matter in the summer, when it may roam quite far inland. It appears not only to regard anything alive that invades its icy domain as appropriate game but also to resent interference on the part of men in particular. Polar Bears will stalk people among the broken surfaces of ice-floes with remarkable persistence. In the water, they are, of course, deadly, and they have attacked small boats. There is still much debate as to whether these animals hibernate below the ice or migrate in winter. Strangely, they are known to do both and it is possible that the pregnant females retire for the winter while the males move south. They are not all white by any means; some are yellow, and others may be a dirty grey. It is constantly stated that the last are the dangerous animals but that the big white ones seldom, if ever, attack men. They can weigh over seven hundred pounds and yet can sometimes overtake a fleeing man. They kill by biting, not swiping.

MUSTELINES

This is the last great group of Carnivores; and, indeed, great it is. It is also the most primitive and apparently the oldest, and that from which all the other Carnivores were derived. *Mustela* meant a weasel in Latin, and this is the weasel family, but it usually comes as a great surprise to most people to learn that such animals as skunks, badgers, and otters belong herein. Mustelines are worldwide in range outside of Australia, New Zealand, and true Oceanic islands. They vary in size from the miniscule Least Weasel of North America, which can dart through a hole no more than an inch in diameter, to the Saro of South America which is an almost wholly aquatic otter that may measure over six feet in length. It also contains some nondescript animals, the relationships of which baffle all but the profoundest specialists. There are four principal kinds of Mustelines—the Weasels, the Badgers, the Skunks, and the Otters.

WEASELS (Mustelinae)

There are a seemingly endless number of weasels, most of them belonging to a great genus named *Mustela*, which is often made to embrace the Polecats and the Mink. However, both these are every bit as distinct from the little, long-bodied True Weasels as are the Martens, Tayras, Grisons, Striped Weasels, and Zorilles. With the exception of the Zorilles, all these animals are elongated and short-legged. They are predators par excellence, usually entirely carnivorous but may, in some cases, take carrion.

True Weasels (Mustela, etc.)

One simply cannot give any adequate description of this great host of little, seldom seen animals within the present compass. Apart from the Stoat of Europe, and some other northern forms which turn pure white in winter, and are then known as Ermines, they are all very much alike in form and even in coloration (Plate 127), being sleek-furred, and usually some reddish-brown color above, and white below. They range in size from less than six inches to almost two feet, and many of the larger ones tend to have a dark mask and other light and dark facial markings. Weasels are found from the barren ground north of the Greenland Icecap throughout the Americas to Patagonia where there is a very odd type (*Lyncodon*). They appear to be absent from the Amazon Basin, though this is not a certainty. In the Old World they range from the Arctic seashores throughout Europe and Asia south to and including the Himalayas, Assam, and the mountains of inner Burma. They are not known in the lowlands of India, Indo-China, or China, They are enormously strong and quick little beasts that can sometimes kill animals over a hundred times their size. One species, only a foot long, is trained by the tribesmen of northern Burma to kill large geese and even small goats, which the weasels do by biting through the main arteries in the neck of the unfortunate quarry. They do ghastly damage to domestic animals and especially poultry wherever they appear, and one weasel can just about clean out a henhouse unless it is promptly caught. They will bite and fight anything however large; yet many species, and perhaps all, can be tamed and trained, and often even if wild-caught and adult. Were it not for weasels, rodents might well have defeated man's early attempts at agriculture, and despite their deliberate destruction of game bird's eggs and young, they should be most rigorously protected.

Many kinds hunt in small packs, darting about like vast insects, never still, examining every nook and cranny, and ganging up for the kill. There are persistent tales of weasel funerals, when, it is said, dead members of a pack are ceremoniously dragged away by the others. Despite ridicule, several most careful students in Europe have recently stated that they believe this to be true. Most weasels have litters of about five young and sometimes two or three times a year. The babies look like tiny cocktail sausages.

Polecats (Putorius)

Of very similar form to the True Weasels are a number of larger, heavier, slower-moving animals found on the more open plains areas of North America, Europe, and Asia. They are parti-colored, have bodies lighter in color than their tails and limbs, and pronounced dark masks. Their fur also is much longer, having a dense underwool and a rather sparse, long, shiny overcoat. The typical species of Europe is the original and true Polecat (Plate 126), from which the domesticated hunting Ferret was produced, and the fur of which was once well known as "fitch." Ferrets are usually of pure albino stock that breeds true, but they may be colored like the wild animals, while there are also intermediate forms that may or may not breed true. Ferrets are fine ratters and in some places are trained to work with a Falcon to get the pestiferous European Rabbit out of its warrens. From southeast Europe through the Near and Middle East to India and China their place is taken by the Mottled Polecats (*Vormela*), and in North America by the Black-footed Ferret (Plate 128), which ranges from Montana to Texas and lives principally on Prairie-Dogs. Polecats do not hunt in packs and are far less wanton in their killings than weasels. They make the most delightful pets but the females become irascible when pregnant, which is their normal condition throughout the year, unless actually nursing a litter.

Minks (Lutreola)

Of slightly shorter and more compact form, and with even thicker and more lustrous fur are the semi-aquatic minks (Plate 129) of Eurasia and North America. They are now apparently extinct in Europe west of the Baltic States, east Poland, and south of the Carpathians. They are still found throughout Finland and in Russia south to the Black Sea and east to the Ural Mountains. Beyond that point their place is taken by another species (*L. sibirica*). The European form is almost identical to the American, though usually having a white throat, chin and upper lip. Various races are recognized on both sides of the Atlantic according to size and shades of color. In America the largest are in the northwest, and the smallest in the northeast, while all those in the north are dark. The farther south one goes the lighter and redder they become, as a general rule. Strangely, the fur of the mink was regarded as hardly worth a trapper's effort to collect until this century; now it is more prized than anything but the true Sable, and millions are raised in captivity, while a great variety of new colors have been produced in this caged stock

Tayra

through taking advantage of rare mutations. Minks subsist on fish, frogs, and other water animals and live in holes near streams and ponds. They have a litter once a year and the babies follow the mother about for weeks.

Martens (*Martes, etc.*)

These are the arboreal representatives of the weasels though the most famous, the Siberian Sable, spends most of its time on the ground and is not a very good tree-climber. There is much popular confusion about these animals because of their silly English nomenclature. There are six distinct species—the Pine or Baum Marten of the forests of Europe, exclusive of Spain, Greece, and central Asia from the Caucasus to the Altai Mountains; second, its close relative, the American Marten, often called "sable," of the Canadian and Alaskan coniferous forests; third, the Beech or Stone Marten, which ranges from the Atlantic south of Scandinavia to the Urals in Russia and southward to Sikhim in the Himalayas; fourth, the true Sable of Siberia, once ranging from the Urals to Kamchatka but now hunted back into a few of the northern mountain ranges; fifth, the beautiful Indian Marten, which ranges from Kashmir to south China and thence south throughout India, Burma, Indo-China and Malaya to Sumatra; and, sixth, the Fisher or Pennant's Marten of the temperate North American woodlands. They are all beautifully furred animals with long bushy tails, and only in the case of the Indian Marten—and then only in the southern parts of its range—do they lack a thick, woolly undercoat, this coat usually being of a lighter hue even to pure white in the Stone Marten. The Sable is usually almost black, sometimes with silver flecks, but it may be pure white below and often has a white chin and throat patch. The other species display variously shaped chest patches ranging in color from white to deep orange. These may envelop the whole chin, throat, chest, and even the inner side of the forelegs. The pattern culminates in the Indian Marten, which is really very odd, having a dark face and body, except for a light-brown saddle over the fore-back and back-neck, a white chin and throat, an orange chest, and an underside like the pale part of the back. The Pine and American Martens are very retiring, elusive creatures; the Beech Marten,

however, often resides among buildings. The Fisher was almost exterminated when the forests of North America were first cleared but is now making a comeback. It was called the Fisher not because it fished, but because it stole fish out of traps.

Tayras (*Tayra*)

These are short-haired, rather long-legged, flat-headed, small-eared representatives of the Martens that have taken up residence in the tropics, ranging from Mexico to the Argentine, and being very common animals. They are as at home in trees as on the ground and, despite their big, blunt, doglike claws, can literally gallop through the forest tree-tops. A plethora of different kinds has been described, but it is possible that they are all races of one species, their coats varying in a most remarkable serial—or, in what is called "clinal"—manner according to the temperature, humidity, and altitude where they are found. Thus, on low coastal plains and in swamps they may be pure black or have only a small reddish-yellow star on their throats but, as one moves inland and up on to the drier hinterlands, progressive lightening of the pelt occurs. This starts with the head becoming brown and the throat patch turning yellow and spreading to the armpits and up behind the ears, eventually forming a complete collar. Meantime, the body begins to lighten also, first by a sprinkling of silver and then by going either pale brown or grey. In some Tayras from the dry upland savannahs, the head is pale cream, the neck-ring, chest, and undersides are white, and the body is a pale fawn color. The Tayras of the Pacific Sierras of Mexico are longer-furred, and this makes them look bigger. Their tails are not so bushy, or plumed, and they are dark brown in color with a stippling of grey on the back. They are very like the Fisher Marten at the extreme south of its range in the Rockies.

Grisons (*Grison, etc.*)

These animals live on the ground and although they have elongated bodies, short legs and tail, and are generally weasel-like, show the digging habits, and the strange coloration that is so typical of the badgers—namely, light above and dark below. It has been suggested that this arrangement is to camouflage the animal in moonlight, making it less visible to birds of prey from above and yet not making it stand out from the point of view of its food, which is on the ground below it. Grisons are absurd to watch in the forests, where they wriggle in and out of holes in tree roots, chasing rats and other small fry. They look like animated sausages glistening in the moonlight. They are very ferocious and the most appalling gluttons, more than one owned by the author having literally burst from overeating. They are also found all over Central and South America and on all types of land from swamp forest to bare upland plateaus.

Striped Weasels (*Poecilogale and Poecilictis*)

These are most incredible-looking little animals well justifying their other popular name of Snake-Weasels, being enormously elongated and having very short legs. The first is only about nine inches long and is vividly striped black and white with a white top to its head and a rather short, tapering, white tail. It occurs here and there all over central and south Africa, and up the eastern third of the continent to Kenya. Nothing specific appears to be known about its habits other than that it behaves like the True Weasels and is predaceous and extremely agile. Its head is minute and the jaws are reduced so that there are only very few teeth. The second type (*Poeilictis*) comprises several species from North Africa and isolated areas throughout the Sahara south to the Sudan. These are somewhat sturdier animals with more weasel-shaped jaws, and have from three to five stripes along the back. Some are very vividly black and white and closely approach the Zorilles in pattern.

Zorilles (*Zorilla*)

There is a mechanism in Nature that produces similar results by evolution of similar animals from two or more quite different stocks. This is one of the best illustrative cases, for the little animals, as the photo shows, look exactly like American Skunks, and, what is more, they have similar habits, carry their tails in the same way, and above all stink just as badly if not worse; in fact they perhaps surpass all other animals in this respect, the mink and Grey Fox not excluded. There are a number of species found all over the drier parts of Africa outside the equatorial forests and even in isolated areas within them. They are found in North Africa, Egypt, northern Arabia, Palestine, Syria, Turkey and even in Europe around Constantinople. Zorilles can dig their own holes but usually occupy those of other animals or hide under thick brush. They are rather slow-moving, are absolutely fearless, and appear to have a great dignity. A famous game warden has described the piteous sight of nine full-grown lions waiting patiently to take a meal from a fresh-killed zebra, while one of these little animals sniffed about for some hours, nibbling the vast meal and even taking a nap with his back to the corpse. Every time a lion approached within ten feet the Zorille simply raised its tail and the great cat retired. Their aim is precise and the volatile oil they squirt can blind other animals. In the Sudan, the natives call them "The Father of Stinks."

BADGERS (*Melinae, etc.*)

We are taking some liberties with established practice in grouping herewith all the badger-formed animals under one head. It must be stressed that the Ratels or Honey-Badgers of Asia and Africa are given a separate sub-family status, and the Wolverine is still left as an appendage to the Weasels. Were it not for the habits of Skunks they might also be included without stretching the laws of classification. Badgers are heavy-bodied, long-legged, short-tailed Mustelines that dig and, except in one case, are terrestrial.

Wolverines (*Gulo*)

In historical times these powerful and aggressive creatures were found all over the northern two-thirds of North America, Europe, and Asia. Today, their range is greatly restricted, having been pushed back towards the Arctic in all three continents. They are the

Zorille

largest of the Mustelines, big males measuring almost three feet and standing over twenty-four inches at the hips; they weigh as much as forty pounds. They are rather bearlike in stance and movements and have very large, semi-retractile claws. The short tail is bushy. They are left alone even by the large cats and bears, and they in turn avoid only those animals. They can pull down deer and even, it has been reported, young Moose. All sorts of lurid tales are current about the physical powers, cunning, and appetite of the Wolverine, and it is even called the Glutton, while its Latin name, translated directly, happens to mean "the swallowing thing who has gone blind" or as we might say, "The Blind Glutton." Nevertheless, the animal is just a large weasel in the widest sense, and bold in proportion to its size. It is normally classed with the Weasels but has much in common with the long-legged Badgers.

Ratels (*Mellivora*)

Often known also as Honey-Badgers—a distinctly redundant name, since all badgers go for honey—the Ratels are inhabitants of the tropical belt of Asia and Africa. They are of particularly gruesome appearance and rather fearsome habits, for they will stand and fight on occasion, a thing that is very rare in the wild state. The Indian species is light above and black below, a type of coloration noted in the little Grison of South America. The South African type is similar but has a pure white band between the grey of the back and the black undersides. In the Congo there is a pure black form and in West Africa another with a white head. These animals have tremendous, incurving front claws with

Wolverine

Ratel

which they dig their own holes and rip open rotten trees and banks in which bees are housed. They hold the honeycombs in the forepaws and lie down to eat. In Africa they follow little birds called Honey-Guides, which do actually, as their name implies, lead them to their favorite food. They also eat insects and small ground-living animals. The West African species appears to be essentially a fish-eater.

Eurasian Badgers (*Meles*)

These animals which are so very well known to all Europeans and which have figured so largely in their folklore and in children's books starting with Beatrice Potter's famous Mr. Brock, are very different from the one we call the Badger in America. There are several distinct species spread all over Eurasia, except northern Scandinavia and the island of Sardinia, and Asia north of the Himalayas. One species, pale in color, covers the Near East and Persia, there are two species in China, and another in Japan, the latter also figuring largely in the folklore of that country. Nocturnal animals of most shy and retiring habits, they sleep all day in large and comfortably appointed burrows, known as setts, that they dig themselves. In northern latitudes they go into profound hibernation, and in the hottest parts of Persia they aestivate. Badgers are almost as omnivorous as bears and have perhaps the most formidable set of teeth for their size of any mammal. In Denmark it was the custom when hunting these animals to wear two pairs of boots separated around the ankles by a layer of charcoal, since it was said that a badger would never let go until he heard cracking bones. The fur was used for shaving brushes, but it is not so soft as that of the American animal.

Sand-Badgers (*Arctonyx*)

In the eastern Himalayas, Assam, Burma, and southern China there occur two species, and in southeastern Tibet a third species of Badger-like animals with long, naked snouts ending in a flat, piglike front. They are rather long-legged and their tails are longer than those of the preceding genus. They live in rocky places and preferably among low hills, and even in virtually desert areas. They are also nocturnal and great excavators, enlarging natural holes among tumbled rocks and boulders. They are more insectivorous and eat earthworms as well as some fruits and roots. They do not see well and their sense of hearing is not good, so that they use their long nose like an antenna, to detect danger.

The Teledu (*Mydaus*)

In the mountainous parts of Borneo, Sumatra, and Java there is to be found a most evil-smelling creature

Sand-Badger

about fifteen inches long with a tiny, fluffy bobble for a tail. It is dark brown in color except for a white stripe from the top of the head to the tail. Its fur is very hard and dense, and forms a crest on the back of the neck. It also has a long, naked, piglike muzzle, the tip of which can be twitched about. It has large stink-glands by the anus and can eject a foul liquid to considerable distances. Like the Zorilles and Skunks it appears to be fully aware of its virtual invulnerability, so it minces about fearlessly and is readily tamed, seldom resorting to extreme measures unless profoundly molested. It is a nocturnal grubber for insects and other small fry, and lives in holes.

American Badgers (*Taxidea*)

On the great west-central plains of North America from Michigan and Ohio to Saskatchewan and Washington, and thence south to Texas, California and Mexico there still live very considerable numbers of these large, low-chassied animals (Plate 130). Males may measure up to three feet, of which only about six inches is tail. They are very flat, broad animals and are fearfully strong. They have been known to raise a horse and rider standing on a platform. Their principal food used to be the Prairie-Dogs and Pocket-Gophers of the plains, which they dug out of their holes. Today they have taken to living on farmed land and eat not only all animals they can catch but also a considerable amount of vegetable matter if fresh meat is lacking. They have ridiculously long front claws for digging and can accomplish extraordinary feats of excavation at incredible speed. Their resting and nursing quarters may be thirty feet below the surface. In midsummer they give birth to about five young at a time. Their skins are exceedingly thick and tough and are about a size and a half too big for their bodies, which aids them in getting through tight corners. And they are really tough animals that can literally take almost any beating. They have a strange habit of going into some form of profound sleep or coma, so that they appear dead and are somewhat rigid. If mistakenly handled when in this state, however, they may suddenly snap out of it fighting and inflict ghastly wounds. They are fearless if cornered and will then stand and fight, pressed to the ground and hissing.

Tree-Badgers (*Helictis*)

These animals, also known as Ferret-Badgers, are herewith retained in the general badger group more through convention than because of anatomical affinities. They should be placed at the other end of that group and in a class by themselves because in some ways they bridge the gap between them and the marten-like animals. They have nondescript bodies, long, pointed faces with badger-like markings, short limbs, and rather long, tapering tails. There are four distinct species, all found in the Oriental Region. One is brown in color and ranges from Nepal to Singapore and the island of Sumatra; a second is grey and comes from lower Burma; and the other two inhabit southern China, Formosa, and Hainan. Unlike badgers, their coloring is darker above than below, and one Chinese species is colored brilliant orange on the chest. Their teeth are quite unlike those of badgers. All are tree-climbing forest-dwellers and appear to take an omnivorous diet.

SKUNKS (*Mephitinae*)

The skunks form an exclusively New World and very distinguished little group of animals. Most misleadingly they were dubbed "polecats" by the first English-speaking settlers and this has caused endless confusion. They produce large quantities of highly volatile oil in a pair of pigeon-egg-sized glands under the skin below the tail and on either side of the anus. This liquid, which is pale yellow, may be forcibly squirted at will by the animals to a distance of several feet and forms a fine mist. It has a concentrated aromatic odor that is almost universally held to be highly offensive but which tests have shown is almost as universally declared to be rather pleasant when not too concentrated and its true origin is not disclosed. Although most of the horror of these animals is thus demonstrably psychological, the fluid can cause serious or permanent harm to the eyes if received at close range. All skunks appear to know the power of these defensive weapons and they are on the whole loath to use them unless pushed too far. They are also highly self-confident and very indifferent to other animals and people, and they cannot seem to get used to automobiles although they are intelligent and quick to learn. They make very fine pets, but as with all animals, each individual is of a different temperament so that, like men, they range the whole gamut of personality from blithering idiots, to thieves, bruisers, geniuses, and sloths. They may be deodorized when young, but those that have not been operated on make the best pets and seldom "let go" except at bothersome dogs or incredibly stupid people who persist in teasing them. Even then they give fair warning by elevating their bushy tails and sometimes by tipping forwards onto their front feet and balancing upside-down like acrobats. There are three genera of skunks, one typically South, one Central, and the third North American.

Hog-nosed Skunks (*Conepatus*)

These are the most badger-like of the three, being essentially burrowing animals, having low, rather flattened and corpulent bodies, and small, rather pointed heads with naked, piglike muzzles. The coloring is basically white above from the top of the head to and including the whole tail, and black on the flanks and below. These animals range all over South and Central America from Patagonia to the extreme southwestern part of the United States. They are not forest animals but inhabit forest clearings, open woodlands, mixed growth, scrub, grasslands, and deserts. It is not generally realized that they occur in suitable localities in the Amazonian region as well as all down the west coast. Like all skunks they are omnivorous, live in holes or under things by day, and trundle about at night in search of food.

Striped Skunks (*Mephitis*)

This is a somewhat misleading name since the arrangement and extent of the black and white areas of their pelts is subject to the wildest variation even within a single litter. They may range from pure black to pure white, have black or white tails, single or double white stripes, a white cap, or an endless number of other designs. Further, the species inhabiting the Southwest and Mexico, the Hooded Skunks (Plate 132), may have the same color arrangement as the Hog-nosed Skunks. The range of this genus is from Hudson's Bay in the north, to southern Guatemala in Central America. They are exceedingly common in most parts of their range and they are little affected by modern civilization, being at home in suburban areas of the largest cities. They like to nest and spend the winter under buildings. Their oil is a good substitute for "musk" and "ambergris" (see pages 249 and 215) as a fixative for perfumes. Up to ten, minute, blind young are born in early May and are nursed for many weeks before being led out in a solemn little procession by the mother on their first foraging expedition. There is only one prettier sight and this is the little communal dance that these animals sometimes put on in the mating season.

Spotted Skunks (Spilogale)

Overlapping the range of both the Hog-nosed and Striped species and holding an intermediate geographical position, is a group of smaller and otherwise quite different skunks (Plate 133). They have beautiful silky, as opposed to rather coarse, fur ornamented with a complex arrangement of broken white belts forming a maze-like pattern, and they almost always have pure white tails. They are much more sprightly than their larger cousins, and are more prone to perform acrobatics on their front feet, running along with bodies perfectly balanced upside-down. Spotted skunks range from a line drawn across North America from Puget Sound in the west, to Maryland in the east, and thence south to Panama. They too dwell in all kinds of territory, but in the tropics are more often found in real forests than are either of the others. They can also climb, and the author was once greatly surprised to see a mother and three young emerge from a hole in a tree thirty feet from the ground and go off through the branches just like arboreal animals. They are often abroad by day.

OTTERS (Lutrinae)

The Otters form a very compact group of animals spread all over the world except for Australia. There are endless described species, but all are much alike so that only five may be separated as distinct genera, and these on slim grounds in three cases. Two are very strange indeed, and one of these, the Sea-Otter, may well form a distinct sub-family of its own.

Common Otters (Lutra)

The most puzzling aspect of the otters is perhaps that we do not really know how many different kinds there are nor the limits of the ranges of those that we do know. The Common Otter of Europe (Plate 131) with variations appears to range all over that continent up to the Arctic Circle, and thence right across Eurasia north of the Himalayan massif to Amuria and also through Mesopotamia and Persia to India south thereof. The common otter of India is smaller and greyer in color

and ranges on into Burma. In India, however, is another species called the Smooth Otter—a really silly title—which extends east to Malaya, where it is domesticated for fishing. Then there is a very large species with a hairy nose found in Malaya and on Sumatra. Common Otters, per se, are not found in Africa, but a species or group of closely related species ranges from sub-arctic Canada to the Argentine in the New World. Just how many kinds there are in South America is not known at all, but there are at least two in the Amazon Basin. There is a distinct small form in Central America and west of the Andes south to Ecuador; and there is a very special small, pale-colored species in southern Chile which is as at home in the sea as in fresh water. In Asia, Africa, and South America there are other otters of other genera, which will be described later.

The habits of otters are, as it were, obvious and have been widely publicized in many movies, yet there are curious aspects of their behavior that are not so widely known. Otters prefer to be in water, and although sometimes making long treks across waterless country, always dash for water when alarmed. They are not entirely nocturnal and some races fish by day and night. They can stay underwater for long periods and when they come up for air just protrude the nostrils and then dive again, letting out a little line of bubbles. The power of otters is not appreciated. If one of these animals can get its nose between any two movable objects, up to positively vast sizes and weights, they can pry them apart. Even welded wire can be opened up if the spaces are large enough for the insertion of their flattened, sharklike snouts. Otters are not exclusively fish-eaters; on the contrary, they eat more crayfish, snails, frogs, young water birds, water rats and suchlike than fish. They are almost as impervious to cold as are seals, and provided there is water available, they frolic about in snow and ice at the lowest temperatures. They are, it seems, truly playful creatures so that their mud- and snow-slides are used by them much as similar devices in playgrounds are by human children, though there may be deeper reasons for this singular behavior. Otters make remarkable pets and a whole volume could be devoted to their habits under artificial as well as natural conditions. Nonetheless, they are dangerous animals, and for all their playful antics, can give terrible bites.

The Simung (Lutrogale)

This is also known as the Smooth Otter—a term that is not redundant in spite of the sleek fur of the Common Otter. This is a large Otter that is found throughout India, even in the Central Provinces, from the extreme west to Burma, Assam, the whole Indo-Chinese-Malayan area, Sumatra, and Borneo. Although a truly aquatic animal it makes long treks overland and has turned up in the most unexpected places. Its distinguishing feature is the end of its tail which is flattened from top to bottom—namely, has a flange along either side.

Clawless Otters (Amblonyx)

Besides the Common Otter of India, there are in that

country also two other completely different animals of the Otter family. The first is somewhat exaggeratedly called the Clawless Otter of which there are three distinct races or species. The first occurs in Southeast Asia and Sumatra and Java, the second lives in north India from Sikkim in the west to Assam in the east, and the third is found in the hill country of southern India. To distinguish it from the Common Otters would involve technical matters outside our present scope but it may be said that these animals, although sometimes inhabiting the same streams have quite different habits and fill different natural niches.

Small-clawed Otters (*Aonyx and Paraonyx*)

As was mentioned above, the Common Otters are not found in Africa, their place there being taken by some species of different genera. These are of two kinds, the first type (*Aonyx*) is of very large size and is also most misleadingly known as the Clawless Otter—just as is the Simung of the Orient. Large males may measure well over four feet, of which only about sixteen inches is tail. It has claws but they are rudimentary. These animals are spread all over Africa in forested areas south of the Sahara, and along almost all rivers even in the driest areas. However, they are seldom seen except in South Africa, and even natives have never seen or heard of them in many areas. The same may be said of the Spot-necked Otter (*Paraonyx*) which apparently has just as wide a range. That these two quite different otters should be found in the same place is possible only because they have quite different habits and food preferences. The first is essentially a crab-eater—freshwater and land crabs are very common in Africa—and is also a diver. It spends most of its time on land in marshes. The second is almost purely aquatic, can stay below water for a very long time, and is a fisher. Its specific name derives from its having a number of irregular, reddish-orange spots on its throat.

The Saro (*Pteroneura*)

In the sluggish sidestreams, creeks and in the tributaries of the great rivers of the Amazonian Basin, the Guianas, and the upper Plata River river system, there dwell the largest of all otters—incredible animals that may grow to six feet in length. They are undeniably otters, but their bodies are so elongated, their legs so short, and their heads so widened and flattened that they have more the appearance of bottom-feeding fish. What is more, their mouths have actually retreated under their snouts just like those of sharks. Strangest of all are their tails which have lateral flanges so that the whole is shaped like a spear. They are as truly aquatic as any seal and they have many traits in common with those animals—standing upright in the water with head above the surface to look around, barking, and humping over mud-banks on their bellies rather than using their legs. Saros fish by day, when they are very conspicuous, and by night, when they are silent and extremely stealthy. Like all otters they are very curious and have been known to swim up to canoes and launches in which people are sleeping and snake their way aboard. When alarmed they slither into the nearest water on their bellies with a serpentine motion. Their principal food is river mussels, but crabs and some fish are also taken.

The Sea-Otter (*Enhydra*)

The last of the otters, of the Mustelines, and of the Carnivora is a preposterous animal, called very rightly the Sea-Otter, that once inhabited the whole Pacific coast of North America south to Oregon and east to the Asiatic coast and south thereon to the Kurile Islands. It is an otter, but its appearance and habits are very strange. The back feet are enormous and fully webbed; its body is seal-shaped, the front limbs very short, and the tail much reduced. It is clothed in a very fine dense underwool and a thick overcoat. It dwells in the vast kelp or seaweed beds that are strung out all along these coasts, and feeds on sea-urchins, starfish, some shellfish, crabs, and other marine life for which it dives to considerable depths. It has the most engaging habits, not least of which is floating on the surface on its back and using its broad flat chest as a lunch counter on which to lay out its food, manipulating it with its forepaws and cracking shells and other hard items with its powerful jaws and huge teeth. The animal is about four feet long. Mothers sleep on their backs on the water, clasping their babies to their chests with their forepaws. These animals, in fact, spend much of their time on their backs, in which position they will play with unopened shells or bits of seaweed for hours, batting them back and forth between the hands. Only a single young is born at a time and it is dependent upon the mother for a long period.

Whales

(*Cetacea*)

WHALES are the most exotic and one of the most widely known, by name, of all groups of mammals but are probably the least understood. A great deal that has been published about them even in scientific literature is pure make-believe, while popular literature anent these animals is often completely inaccurate. Only in this century have new and often most tedious methods of investigation begun to bring to light some true facts about these mammals and their habits. The commonly accepted story of the whaling industry is in an even more deplorable condition but more through incorrect emphasis than for lack of accurate records. Man has been hunting whales on the high seas for at least ten thousand years and among mammalian products those derived from whales stand second in importance only to those derived from domestic mammals.

There are about a hundred known species of whales divided very clearly into two groups known to science as the *Mysticeti* and the *Odontoceti* and in popular parlance as the Baleen or Whalebone Whales, and the Toothed Whales respectively. There was previously a third group known as the *Archaeoceti* or Ancient Whales but this is now, we presume, extinct. However, entirely new kinds of whales are still from time to time being thrown up on beaches in various parts of the world, and we cannot state categorically what is or is not still living in the seas and oceans. These Ancient Whales had several features, such as sets of teeth differentiated like ours into various kinds—front, eyeteeth or canines, premolars and molars or cheek-teeth—that point the way to land-living ancestors. This is important to anyone interested in animals and especially to serious students of the mammals because, not only is there a great mystery about the origin of the whales, there is also involved a much wider puzzle which requires investigation. This—the time factor in evolution—cannot, however, be discussed here.

It is still not absolutely sure that the two living groups of whales had a common origin. They might represent a vivid case of parallel evolution from different sources but the most primitive Baleen Whales known from fossilized skeletons display some features that would seem to have been derived from Toothed Whale ancestry. On the other hand, there are some remarkable facts which, when considered together, point to an extraordinary possibility for the origin of the Toothed Whales. First, the tails of the Saro or Giant Otter, the Beaver, the Manatee, the Dugong (see page 290), and finally those of the little Platanistids (see page 219) indicate various steps by which the whales may have evolved their tails. Secondly, one of the Platanistids, known as the Boutu, has, almost like those of seals, front flippers with considerable flexibility in the fingers, and by using these can actually hump over mud-banks in the shallow freshwater swamps or flooded forest floor where it lives for part of the year. Thirdly, and most curious of all, some of the porpoises have a few scattered or regular lines of small bony plates embedded in the skin of their backs, while certain extinct dolphins were partially covered with interlocking bands of similar bony discs almost like armadillos. Putting these salient facts together with many other more technical details we find ourselves confronted with the rather strong probability that the Toothed Whales, at least, originated in tropical rivers from semi-aquatic land mammals having bony protective coverings rather than mere fur. What these animals may have been, however, is not known and nothing like them has been found in the fossil state. Whatever the ancestors of the whales may have been, they must have lived and presumably become extinct more than sixty million years ago.

Whales today are quite unlike any other mammals, though they still have warm blood, breathe air, and suckle their young with milk produced in the mother's body. This milk, incidentally, is not unlike very rich cow's milk and is squirted into the mouth of the young under considerable pressure as the mother lies on her back on the surface of the sea. The babies seize their mother's teats, which lie in a closed pouch, with the front of the mouth held above the surface of the water. All whales are born without any external trace of back limbs but there are tiny bones buried in the bodies of some that represent the remains of the hip-girdle and the larger leg-bones. The tail is not just the terminal portion that bears the two horizontal tail flukes but comprises the whole hinder third or so of the animal, starting above the anus. This is a solid, streamlined cone of immense muscles and straplike sinews that are tougher than steel and more resilient than nylon. These move the whole back end of the animal up and down at high speed and with tremendous energy, all of which is transferred into the flukes. These, however, bend in a most complex manner that results in a sculling motion that is, in the aggregate, semi-rotary, so that the animal achieves what is virtually the same action as the screw of a ship. The result is a forward drive that is almost incomprehensible even to engineers. Whales, and notably the Rorquals, are perfectly streamlined and can be of gigantic size. A female Blue whale once pulled a twin-screw, steel-hulled, ninety-foot whale-chaser with her engines going fullspeed astern, forwards at an average speed of five knots for eight and a half hours, despite the fact that there was a quarter-ton harpoon embedded in the animal's back and half a mile of four-inch rope between this and the ship. Right Whales have been known to smash their jaws on rocks by diving too fast in water that was too shallow, and an enraged Sperm charged, stove in and sank a steel whaler in three minutes. Killer Whales can break through four-foot icefloes by charging up from underneath although a large bull of this species is only about thirty feet long.

The size of whales is a topic for endless debate which often waxes furious because there is a widespread belief that there is only one animal called "The Whale," instead of over one hundred kinds. The smallest species of whale is about four and a half feet long when fully grown; the largest whale ever recorded was a female Blue Whale measuring one hundred and thirteen and a half feet in length. It used to be believed that whales averaged about one ton in weight per foot of their length, but three whales have now been weighed in detail (that is to say, each part separately—flesh, bones, blubber, blood etc., etc.) and it turns out that the Rorquals and most other whales weigh one and a half tons per foot of length and the Right Whales about two tons. Thus the giant female Blue Whale probably weighed over 170 tons. This is equivalent to 35 elephants, or 2380 human beings, or 136 million Pigmy Shrews, which are the smallest known living mammals. Yet this enormous thing can leap clear of the water when breaching.

Another source of debate is the speed at which whales can swim. Recent work done with underwater radar has provided some definite figures showing speeds of up to twenty knots but it is known that some whales can far exceed this speed in short bursts since they have often been seen to precede ships doing over that speed and then to dart ahead and vanish in a second without any apparent effort. How fast they may travel when coming straight up from deep dives has not yet been ascertained. And this leads to another rankling question, namely, to what depth do whales dive. It is obvious that when cruising along, feeding, the Baleen Whales have no cause to go deep since their minute food is on the surface layers of the sea. However, the Toothed Whales and notably the octopus- and squid-eaters like the Sperm and the Bottlenose apparently descend to terrific depths and can stay below for over an hour. A sperm was once dredged up from a depth of over 5000 feet entangled in loops of a heavy marine cable which it had broken but which had held it down till it drowned. Most whales have a regular respiratory rhythm when feeding or undisturbed, surfacing to exhale the hot moisture-laden air from their large lungs in the form of the famous spout, rolling under for a brief period, and then appearing again to breathe once or twice. After they have done this a few times they will go below for a period of five minutes to half an hour to feed or just swim.

The food of whales is sometimes astonishing. The largest appear to feed exclusively on small crustaceans, called krill, that are somewhat shrimplike in appearance, half an inch to an inch long, and that swarm in cold seas in uncountable millions. Some whales, however, may feed on much tinier creatures known as *Calanus,* less than an eighth of an inch in length, that swim through the water by using their antennae like oars. Then there are fish-eaters, and the Killers which will take almost anything from huge bites out of the larger whales to shellfish. One was caught while gagging on the fifteenth full-grown seal it was swallowing. Some of the largest whales, notably the Sperm, appear to feed exclusively on octopuses and squids, including the giant *Architeuthis* which may itself have a body twenty feet in length, have thirty-foot tentacles and weigh over ten tons. Incidentally, circular scars made in the paper-thin skin of the Sperm by the suckers of these giants measure about three inches across but whales with similar scars over a foot across have been taken, so that we are led to wonder to what size octopuses or squids may grow in the depths. There is one small whale—a species known as *Sotalia teuzii*—found in the Cross River in West Africa that apparently feeds on water plants and other vegetable matter derived from the land.

The anatomy of whales is as odd and confusing as are their habits. First, if you stand on top of a whale's skeleton and look down at the head you will find that in many species the left side is quite unlike the right. The single "S"-shaped blow-hole of the Sperm, for instance, is always on the left side of the tip of the snout. This asymmetry may be carried out even to the surface coloring of the skin so that the left-hand side of the jaw may be white, the right black. The Beaked Whales or Ziphioids display the greatest asymmetry. Secondly, the nostrils of whales have, in all but the Sperms, moved back on to the top of the head and the nasal passage goes directly to the lungs without joining the throat. Thirdly, the skin has been reduced to about the thickness, and much the appearance, of a sheet of carbon paper in the larger whales but is backed with a thick layer of fibrous tissue impregnated with oil, a substance called blubber. In the Porpoises, most Dolphins, and the White Whales the skin is thickened to form a very tough leather. The eyes of whales are compara-

tively small and the ear-openings of the largest will just admit a pencil but they have extremely acute hearing both above and below water. Most whales are completely hairless but some have a few thick bristles on their chins—the Common Porpoise just six—while the curious Grey Whale has lines of hairs all over the upper part of its body.

Whales are found in all the seas and oceans of the world except completely landlocked bodies of water. There are freshwater species in most of the larger rivers of the tropics, and there is one lake-dweller in China. Most of the larger species are cold water animals but the Sperm prefers tropical waters and almost all of them have been found wandering across the Equator. There are species exclusive to the Arctic or the Antarctic regions and there may be separate populations of other species in these two areas which never meet or mingle, but most species are worldwide in distribution.

BALEEN WHALES

Baleen is the proper and the best name for a very remarkable structure unique to these whales and made of a remarkable substance that is very like the bill of the Platypus and also our own fingernails. It is often called "whalebone," or even "whales' bones" or "fins"—all of which is exceedingly misleading, since the substance comes from the roof of the animal's mouth. Baleen grows in triangular tapering sheets with the longest side fringed. These sheets hang down from the upper jaw on either side of the roof of the mouth in a continuous series from front to back like the plates in a battery. The fringed edge is on the inner side facing down into the mouth. The mouth of Baleen Whales is immense, the lower jaws forming a huge hoop from which a vast sac is hung like a pelican's pouch. The lips curve upwards on either side and the rows of baleen plates hang down into the lower jaw sac when the mouth is closed. Between them lies a tremendous tongue that may weigh upward of two tons. The whole device is arranged so that the animal can swim through the water with its mouth open in places where there are shoals of krill and other small food, and literally sweep tons of them up against the matted surfaces presented by the fringed edges of the baleen plates. The animal then closes its mouth and brings the tongue up. This forces the water out between the baleen plates and the floppy lips, and leaves the food filtered out on the baleen fringes to be licked backwards by the tongue and swallowed. Baleen plates are really greatly enlarged and multiplied editions of the ridges you can feel on the roof of your mouth.

There are three families of Baleen Whales of about a dozen species in all. Economically they are far more important than the much more numerous Toothed Whales. They are the most "modern" whales, yet one— the Grey Whale—shows some very primitive characters.

RIGHT WHALES (Balaenidae)

These whales are so called simply because they used to be the right ones to hunt and this because they floated after death and could therefore be towed ashore or cut up at sea. There are two genera, the first containing two very distinct, and also possibly two other, species; the second, only one small, rare, and obscure animal. Those in the first have no fin on the back and are of great girth in comparison to their length. The larger species were nearly exterminated by the end of the last century through constant whaling operations starting with the Basques in about 1000 A.D.

True Right Whales (Balaena)

There used to be large populations of Black Right Whales inhabiting the North Atlantic, North Pacific, and the South Pacific from New Zealand to Western Australia. These separate populations did not apparently meet anywhere but all behaved alike, migrating annually towards the Equator in the winter, breeding just after they turned back to the Poles, and spending the summer in the far north or south. Those in the North Atlantic and Pacific divided into two groups; each fall one lot went down the west and the other down the east coast of those oceans but joined up again the following spring. Those in the southwest Pacific came from the Antarctic in the southern autumn over a single route—actually a deeply sunken ridge that connects the Antarctic continent to the immense sunken plateau upon which Australia, New Zealand and the west Pacific islands lie. Reaching this, they fanned out over that plateau for the cool season and then, in the southern spring, returned whence they came. Whether the three populations represent distinct species is open to some debate but they are probably no more than races. Black Rights are usually jet black all over but they may have white areas or even be pure white below. Most noticeable is the so-called "bonnet," a strange circular area on the tip of the snout, or beak, composed of hardened degenerate skin into which marine worms bore and to which all manner of other parasites cling. Its purpose is unknown but there is a belief that it is either caused by hitting the bottom or to cushion them if they should do so. This is most unlikely as they are not known to dive on their backs. This animal was the mainstay of the whaling industry of Europe and colonial America from the Dark Ages till the end of the seventeenth century.

The other Right Whale is known as the Bowhead, Greenland, or Arctic Right Whale and was discovered by the Basques only about the turn of the fifteenth and sixteenth centuries. It was then grossly exploited by the Hollanders and later the British in the North Atlantic until almost wiped out in the mid-19th century. At that time, a new stock of these whales was discovered in the North Pacific and was similarly eliminated by the Yankee whaling fleets working out of San Francisco. This animal is larger than the Black Right, yields more oil for its bulk, and has the largest baleen plates—measuring in some cases over thirteen feet in length, and being present to the number of over three hundred in one animal. In the last century the value of baleen, mostly for use in womens' stays and hooped skirts, rose to such an exaggerated figure that the plates from one Bowhead would pay for the outfitting of a whaler for a season. As a result, the whales were slaughtered for their baleen alone, the whalers not even bothering to take the blubber oil.

Bowheads are Arctic animals and always stay near the ice-front, following it north in summer. They are even vaster in girth than the Black Rights and their heads, which have a pronounced bump on top, comprise more than a third of the total animal. There is a distinct "neck," unlike the Black Right, and the tail is much slimmer. In color they are usually black, with a white chin and often a white band around the tail just in front of the flukes. They used to breed in late spring in the Davis Straits and the Sea of Okhotsk but they are now very rare animals.

The Pigmy Right Whale (Neobalaena)

This little, twenty-foot relative of the True Rights is known principally but not exclusively from the New Zealand–Australian region. It has a small head and a sickle-shaped back fin. Its ribs are remarkable, being greatly flattened, so that the animal's vital organs are almost completely encased in solid bone, yet they are loosely attached to the backbone so that the whole barrel-like structure can expand enormously. It is believed that this animal is a deep diver and that this mechanism has something to do with rapid descents and ascents. The baleen is ivory-white in color. This whale is very rare but turns up in the southern fall along with the Southern Black Rights, and then vanishes again.

RORQUALS (Balaenopteridae)

These are the mainstay of the modern whaling industry and are the animals which people today call simply "whales." They were the "wrong" whales of bygone times since they sink when killed and it was not till Svend Foyn of Norway invented the harpoon-gun and the modern steam whale-chaser that they could be hunted, killed, inflated with air, and towed to shorestations or floating-factories for processing. They are the antelopes of the sea as opposed to the Rights, which are more like rhinoceroses. They are also more varied in species, having the mighty Blue Whale and also the little Piked Whale which seldom exceeds thirty feet. They are worldwide in habitat but are oceanic animals and seldom enter bays and inlets, though the smaller species will chase cod and other fish into water only a few feet deep. They have curious fluted throats, structures which are hard to explain. Imagine the animal's undersides from chin to about mid-belly being slashed with a sharp knife in parallel lines from front to back, then keep the slashes open while skin grows over each side of the cut right to its bottom, and you will have some concept of the structure. What these are for has been much debated but is not known. The most interesting suggestion is that these animals use their mouths like a parachute in reverse for making sudden stops since they have no brakes and cannot swim backwards. The pleats allow the whole front end of the animal to open out like an umbrella.

Finwhales (Balaenoptera)

There are five species in this genus—the Blue, the

Finner, Bryde's, the Sei, and the Piked Whales. They all have several other popular names, the best known of which are, in the same order, as follows: Sibbald's, the Common, the South Atlantic, Rudolphi's, and the Lesser Rorquals. Their Latin names were once in an even greater muddle, both the Blue and the Finner having been called *B. musculus* at one time or another, but this is now straightened out to the satisfaction of about fifty percent of zoologists. The Blue (*B. musculus*) is the largest of the species and is not blue at all but a steely dark grey above and pale grey to white below, with irregular, slightly darker grey mottlings. It has a small dorsal fin far back. The Blue is basically a southern animal but used to occur in the Arctic in some numbers and apparently roams all over the temperate seas during the winter and even crosses over from one hemisphere to the other. It lives on krill (*Euphausia*) but the young take small fish. They are shy creatures and have an acute sense of hearing. The young are about twenty-three feet long when born, after a ten-month gestation period. They are weaned in eight months by which time they have more than doubled their length to about fifty feet, and they reach maturity two years later, by which time they average over seventy feet in length. One calf is born every other year, and the gestation period is believed to be only about ten months.

The Finner or Common Rorqual (*B. physalus*) is a smaller, more streamlined whale with a dorsal fin placed far aft and a sharp ridge running from this to the top of the tail. It has one hundred pleats on the throat, as opposed to only about sixty in the Blue. It grows to an average length of sixty-five feet and is dark slate-grey above and white below. It has 370 blue-black baleen plates about three feet long; these, like the skin of the lower jaw, may be white on one side. It eats fish, notably a kind of herring known as *Osmerus,* of which over a thousand have been taken from one whale. Finners are very fast and are apt to show fight. In 1894 the steel-hulled, steam-chaser *Garcia* was charged and sunk by one in Varanger Fjord, Norway; in 1896 the *Jarfjord* was sunk; in 1910 the wooden bark *Sorensen* was charged by one at twenty-five knots and was ripped apart and sunk. They are found all over the world but notably in polar and sub-polar regions though they are washed ashore all over the temperate regions.

Bryde's Whale is a comparatively rare species apparently confined to the South Atlantic region and commonest on the coasts of South Africa. It was long confused with the Sei and even the Finner, since it varies in color from a smooth, over-all medium grey to black above and white below. Nothing much is known about it except that it appears to migrate, disappearing for the summer and staying near coasts in winter.

The Sei or Sardine Whale is the third commonest Rorqual and was much exploited by the Japanese in early times and by the Norwegians when they re-entered whaling in the last century. They are named after the Seje or Coalfish because they make their annual appearance off the Norwegian coast coincident with these fish. They grow to about fifty-five feet and are blue-black above and glistening white below from the chin to the base of the tail. They have about fifty pleats on the throat, and some 330 black, twelve-inch baleen plates with white fringes. They spend the summer by the ice and the winter in the tropics and are generally great travellers. In the Atlantic they feed almost exclusively on the tiny Aate (*Calanus*) mentioned above (p. 211), but in the north Pacific their stomachs are usually full of a small shrimp or sardines. In the Antarctic their backs are often spotted with oblong white patches caused by a kind of parasite.

The last of the Finwhales is the little Piked Whale, so-called by the Scots because of its sharp-pointed snout. This neat, quick little whale grows to about thirty feet in length and is blue-black above and white below and may always be identified by a clearcut, vivid white band across the jet black flippers. It has fifty to sixty throat pleats and about 325 eight-inch, yellowish-white baleen plates. It associates with larger whales but is a fish-eater and apparently a bottom-feeder, judging from the dogfish and pebbles that are often found in its stomach; like birds it takes the latter, it is believed, to aid digestion. It will chase fish into a few feet of water among rocks, and it darts among ice floes. It has a habit of standing upright in the water with its head out. It is found all over the world and there may be different species in the North Atlantic, the Pacific, the Antarctic, and the Indian Ocean.

The Humpback (*Megaptera*)

One of the oddest-looking whales, this is a species that used to be extremely common off the coasts of Europe, New Zealand, Australia, and California, and was numerous almost everywhere else in cool to mildly warm seas. It is primarily but by no means exclusively a coastal animal. In shape it is very short and "deep," with a large head almost flat on top but with a huge dependent throat sac like a Right Whale. The tail tapers rapidly and is very thin in front of the flukes but these are enormous and have a straight but irregularly slotted back edge. The dorsal fin is small but the front flippers are enormous, sometimes being almost a third the length of the whole animal and being used, like oars, for slow forward movement. The top and sides of the head and the front edge of the flippers are covered with rows of large bosses. Humpbacks are black above and have a variable amount of white below but the underside of the flippers and usually of the tail flukes are white. The baleen is grey. They grow to about fifty feet in length.

Humpbacks pair in the winter and calve the following winter. They eat krill and small fishes, notably capelin. They are at all times heavily parasitized with barnacles, seaweed, and all manner of worms and crustaceans. In order to relieve themselves of the irritations of these, they perform all manner of antics and sometimes scratch themselves on rocks. But they are in any case, it appears, of a most sportive nature and frolic like flirtatious behemoths when mating, smacking each other with their flippers with a noise that can be heard from over a mile away. They are also great jumpers and whole schools will, on occasion, go into a riotous dance. Like Finners they go about in large gangs and sometimes used to assemble to the number of some hundreds. Like all other Baleen Whales they migrate to and from the polar regions in the fall and spring but in reverse directions in the northern and southern hemispheres. Since they spend the cold season in the tropics, the two populations mingle and may cross over from one hemisphere to the other. When migrating they travel purposefully along in a straight line off some coast. Their numbers have been so reduced by whaling that they have completely vanished from many areas.

GREY WHALES (*Rachianectidae*)

There is only one species in this family, known also as the Californian Whale or Devilfish, and to the Japanese as the Koku-Kujira. It is a primitive beast, growing to about forty-five feet in length, and colored various greys dappled with dirty white. The white spots are left when barnacles are scraped off the skin since this animal is also a sort of swimming menagerie of parasites. The Grey Whale has hairs in regular rows all over the top and sides of its head and has only two to four throat pleats. The baleen is small. They have no back fin but in place of it a series of lumps. Their habits are in every way most odd. They spend the summer in the Bering Sea and along the shores of the Arctic Ocean, where they feed enormously on small crustaceans known as Amphipods—the little things that hop when one turns over seaweed on a beach. In the fall they all start south but half go down the west coast of North America and the other half down the east coast of Asia, reaching respectively the coast of Mexico and Korea by the end of December. Calves are born there in late January, then the animals start north again about six weeks later. During all this time and until they get back to the Bering Sea they do not appear to eat anything at all but live off their blubber. During this whole migration their stomachs are always empty and they are in wretched condition when they get back to the Arctic. They swim along close by the shore and delight in the vast kelp seaweed beds just off shore. They come in to water only just deep enough to float them and then will lie in the surf.

In 1846 white men started to slaughter them for their oil and by about 1890 no more appeared off the American coast and they were thought to be totally extinct. However, in 1911 Roy Chapman Andrews rediscovered them in Korea, where they formed a not inconsiderable part of the Japanese whaling industry. Now they have made a comeback to California. They are extremely dangerous creatures when wounded, cornered, or even molested. The males are very much attached to their females and unlike any other whale they will attack unprovoked, smashing with their tails, charging with their heads, and even biting. They will keep up an attack on a small boat and go for people in the water.

TOOTHED WHALES

There are six most distinct families of Toothed Whales and it is not at all certain that they all belong to the same branch of the family tree since they are all so unalike. Their habits are equally various, and they range in size from the mighty Sperm to the tiniest living whale,

which is only just over four feet in length when fully adult. One Toothed Whale has no teeth at all, several have only two, others have them either in their upper or lower jaws alone, while many have more teeth than any other mammal. This multiplication of teeth from the basic set for mammals of forty-four appears to have been evolved through the division of the actual teeth into three parts or more. The first stage of such a process is seen in the cheek teeth of the Southern Seals (see page 158), which have three long pointed crowns like a trident and sometimes three roots to go with them embedded in the jaws. Finally all the teeth came to look alike—just simple pegs—and the jaws developed into long tapering beaks like those of primitive seabirds.

SPERM WHALES (*Physteridae*)

There are only two known living kinds of Sperm Whales but in the past there were all manner of incredible forms, some, found fossilized in the rocks of Argentina, being of immense size and even greater bulk and with terrible teeth in both jaws. They are totally unlike all other whales except perhaps the Bottlenose, and they have the oddest habits.

The Cachalot (*Physeter*)

Perhaps better called simply the Sperm Whale this is probably the best known of all whales and particularly so in America on account of the vast industry founded upon its capture by the Nantucketers in the eighteenth century. This is also the animal on which that brilliant literary fiction "*Moby Dick*" was composed by Melville; a work of magnificent imagination and almost total zoological inaccuracy, but rightfully now become a classic. Americans owe more to this animal than even to the worthy cow, for upon it was founded the wealth and expansion of the New England colonies and of the Republic in its earliest phases. Sperm oil lighted the new cities of America, and spermaceti made the candles that lit its homes. Almost alone, sperming saved the nation after the War of Independence, and the Union after the Civil War, and this latter despite the discovery of petroleum in 1859 and the financial "crash" of 1857. At one time there were nearly eight hundred whalers out of twenty ports from Maine to Virginia in one year and they sailed to every last island, cove and inlet in the world except in the Antarctic.

The Sperm Whale is essentially a tropical animal though it, or rather each regional group, circulates around a wide lozenge-shaped circuit every year. Mysteriously, the males follow one route and the females and young another for part of the year. In the Pacific they apparently all meet up twice a year but then split up for a few weeks into what are called "gams" of both sexes and all ages. At certain times these gams come together on special "grounds" to form vast loosely associated gatherings. The reason for all this movement is the habits of squids which are their food. These creatures in turn depend upon the movements of their food—Pelecypods and other smaller items—and these again have to go where the microscopic diatoms, foraminifera, and suchlike may be *flushing*, i.e., breeding and multiplying in vast masses. Finally, this takes place in accordance with seasonal temperatures, winds, current shifts, ice at the Poles and so forth. The old Yankee whale captains found these *grounds* and plied them for two hundred years. One ship voyaged continually for ninety-nine years and another showed a profit of $652,-000 in twelve years, which at the present day would represent about $1,900,000.

The Sperm grows to about sixty feet today but males measuring over seventy feet have been reported. The animal is about one-third head and this is about half composed of a vast rectangular tank that is filled with spongy tissue and a very light kind of wax called "spermaceti." This was originally used as medicine by the Amerindians, then for candles by the colonists, and it is still used as the best "oil" for fine scientific, military, and other instruments. The bones of the Sperm's skull are all off kilter and the blowhole is always on the left front end of the tanklike head. They have a continuous row of huge peglike teeth on each side of the very narrow, pointed lower jaw but none in the upper. There, instead, they have hard horny sheaths with holes in them into which the teeth fit. They are black all over in color, have very wide, rounded flippers and large tail-flukes, but no dorsal fin. Instead, there is a series of diminishing bumps down the hinder third of the back.

Sperms feed exclusively on squids and their allies, the octopuses and cuttlefish. These have hard, horny beaks like parrots', cartilaginous skeletons, and bony rings in their suckers. These the Sperm Whales apparently have never been able to digest properly so that they set up violent stomachic irritations. The results may be heard for more than a mile when the animals are on the surface and in addition they vomit up great wads of a most mysterious substance called *ambergris*. This is grey, of the texture of very hard cheese, looks like marble when cut, and has a delightful aromatic odor. It is created by a special bacterium, called *Spirillum physeteri,* that lives in the Sperm's stomach and is the finest fixative known for the floral essences used in the manufacture of the most expensive perfumes. As of this writing it still commands a price of about $8 per ounce.

The Pigmy Sperm (Cogia)

The only relative of the Cachalot is a tiny ten-foot creature, with a sharklike underslung jaw, that has never yet been seen alive at sea. It is black above and light grey below and the mouth is startlingly pink. The head is rounded but contains a spermaceti organ. It has only about fourteen sharp recurved teeth on either side of the lower jaw and apparently also feeds on cuttlefish. Despite its rarity, no less than six distinct species have been described but all except one are probably nothing more than variations. This one, called *Cogia simus,* from the Indian Ocean, has a pronouncedly upturned snout, only nine teeth in the lower and two teeth in the upper jaw. The name *Cogia,* incidentally, is believed to be derived from an extremely pompous old Turkish gentleman named Kogia Effendi who was much interested in whales and whose name and personality also gave rise to the expression "an old codger."

BEAKED WHALES (Ziphiidae)

These are the least-known and apparently the rarest of whales though the family comprises five genera and at least two dozen species. They are too small and too rare to be of any economic consequence except in one case, the Bottlenose (*Hyperoodon*), which carries a spermaceti tank in its bulging forehead and upon which a highly specialized industry was once founded. These animals display several very primitive characters but the most interesting (a species of Cowfish, *Mesoplodon*) are only known from a few skulls or incomplete skeletons. Ziphioids are apparently distributed throughout the oceans of the world but individuals are only washed ashore occasionally and may crop up anywhere. Several of them have, however, up till now come ashore only in the southwest Pacific.

Cowfish (Mesoplodon)

There are at least ten and possibly fifteen species of Cowfish, none of which is adequately known. The commonest is Sowerby's Whale (*M. bidens*), which is apparently a North Atlantic species, measures about eighteen feet and is of a brownish color above and lighter below. The first described was washed ashore in Scotland and another stranded at Le Havre in 1828 stayed alive for two days and lowed pathetically like a cow. These animals have strangely curved lower jaws and usually bear only two teeth, but each species has a differently formed jaw and the teeth are variously arranged from the front to the middle of the lower jaw. The jaws of the various species may be arranged in a series wherein the bones become increasingly sturdy and the teeth move progressively backwards. All those that are known in the flesh have small dorsal fins placed far back, small flippers and small pointed heads ending in a pronounced beak. One, Layard's Whale, has straplike teeth that grow out from the lower jaw to a foot in length and may curve over the upper jaw like a pig's tusks and lock so that the animal can hardly open its mouth at all. They seem to be open-ocean animals and eaters of squid and cuttlefish. Nothing is known of their habits.

Cuvier's Whale (Ziphius)

This animal was first discovered in the form of a fresh skull found on the Mediterranean coast of France in 1804. It was thought to be a fossil till another turned up in the same place in 1850; finally a whole animal that had the same skull turned up in New Zealand and was described as being white *above* and black *below*. Very odd indeed! It had two throat pleats forming a "V." Numerous skulls of this or a closely related animal were found fossilized when fortifications were being dug near Amsterdam in Holland. Their skulls and those of all Ziphioids become flinty and almost stony in old age and are easily fossilized, being practically indestructible.

Baird's and Arnuxi's Whales (Berardius)

Most obscure are two fair-sized whales, growing to about thirty feet in length. The first, called Baird's Whale is apparently North Pacific and turns up fairly regularly at certain times of the year in Tokyo Bay in Japan. The other, Arnuxi's, was only known from three specimens before this century and is apparently a southern oceanic form having been stranded in New Zealand, on the Falkland Islands, South Georgia and the South Shetlands. It is of a velvety black color and bellows like an infuriated bull. It has a tiny head but two and sometimes four teeth that are unique among mammals in that they are embedded in cartilaginous sacs and can apparently be elevated at will to form grappling hooks.

Shephard's Whale (Tasmacetus)

An entirely new kind of whale was washed up on the coast of New Zealand in 1938. About the same time three others came ashore in the same country. These specimens varied from twenty-three to thirty feet in length, were shiny black above and white below with, in one case, yellow streaks on the flanks. The head was elongated into a beak with the lower jaw protruding a little and both the jaws carrying about twenty teeth on either side. The dorsal fin was falcate and rather far astern and the flippers were very small, six feet back from the head, and appeared to act merely as immovable stabilizers. The tail flukes formed a half-moon and spread five feet across. Under the throat there were two fore-and-aft pleats about fifteen inches long. The animals were manifestly Ziphioids.

Bottlenosed Whales (Hyperöodon)

This is a common whale previously encountered in colossal schools in the North Atlantic and now making a remarkable comeback after being greatly reduced in numbers by Scots-Norsk whalers in the last century. The Bottlenose is a gregarious animal, normally traveling about in family parties of about ten individuals. When migrating, hundreds come together and travel along side by side. Large males are about thirty feet long and have huge bulging foreheads in which is located a hydrostatic organ like the sperm-tank of the Cachalot but containing another type of oily wax. The young are blue-black, but with increasing age the males often develop cream-white heads, and the old females may be light brown or yellow. They have never been fished anywhere except in the North Atlantic but they have been seen in the Pacific and in the southern oceans. They are the most courageous and aggressive of all whales apart from the male Grey Whale and the terrible Killers. They feed exclusively on one kind of cuttlefish and are very deep divers. Their bodies are highly streamlined and their speed is tremendous. They are also very curious and will gather around ships. This led to the novel practice of harpooning them directly from schooners. However, they will fight and use a wide variety of tactics—breaching, smashing at boats from below, diving to snap lines, slashing with their tails. The herd will never leave a wounded member and they try to support it in the water. They mate in April to May and give birth to comparatively enormous young a year later: a ten-foot baby was taken from a twenty-eight-foot mother. The head oil is of a very high quality as a lubricant. There may be several species, and one known mostly from skulls picked up in the extreme south seems to be very distinct.

WHITE WHALES (Monodontidae)

This small family comprises only two species of small whales both of which are very singular and both of which are exclusively Arctic in range. Although they have much in common anatomically, they are very different in appearance. One, the Narwhal is seldom white and the other, the Beluga, need not be so and seldom is when young. They have no dorsal fin but instead a long inch-high ridge. Their teeth are reduced to a few irregular molars or are absent. They are both bottom-feeders, taking fish, crustaceans, and cuttle and simply crushing the food, not masticating it, before swallowing. Both animals stay in the coldest water and are found all along the Eurasian and Canadian Arctic coasts.

The Beluga (Delphinapterus)

Also known as the White Whale, this animal was once the principal source of "porpoise-hide" and so-called porpoise oil, which is of very superior quality. Large quantities of both have been shipped to Europe since the days of the early Norse, and the animals are still hunted in the St. Lawrence and in Alaska. The result is their numbers are greatly reduced. The young are born black, then become mottled and progressively more yellowish, and finally turn pure glistening white except for the outside edge of the tail flukes, which are greyish-brown. They grow to about eighteen feet in length. There are from eight to ten teeth in the front part of the mouth. Belugas have been caught up the Yukon River six hundred miles away from saltwater. They are gregarious animals and are the rowdiest of all whales, having an extraordinary variety of calls, lowing like cattle, whistling, and making loud phuts. They are fast swimmers and dive to great depths for their food.

The Narwhal (Monodon)

So-called by the Norse (from Ná-r, meaning a corpse), because of their strange mottled coloring that looks like a bloated human corpse long in sea water. These are one of the oddest forms of mammalian life known. They are light grey below and mottled blue-grey and dirty white above, and the old males turn white. They have a few irregular teeth embedded in their gums when young but then lose them all except in the case of the males, which retain the left upper canine only, or, on rare occasions, both upper tusks. These tusks grow straight out of the front of the head and sometimes reach a prodigious size, specimens ten feet long and over seven inches in girth at the base being known. These tusks are quite straight but always have a left-hand or sinistral thread on them and even if two are present in one individual both have this twist. These tusks were once thought to come from a land animal and had much to do with the tale of the Unicorn. They were considered priceless as ornaments, as medicine, and as material for making sword handles, and were exported from both the Atlantic and Pacific Arctic through Europe and China to all the ancient and medieval civilizations of Asia.

Narwhals migrate in vast hordes and are also rather rowdy. They make a sharp whistle when blowing, the bulls can roar, and mothers call their young with a sort of low-pitched blast. They are fast swimmers and deep divers.

DOLPHINS (*Delphinidae*)

This is by far the largest family of whales and contains animals of very widely differing appearance, size, and habits. It can be divided into three parts, the first containing four most singular large animals; the second, the main body of closely related true Dolphins; and the third group consisting of odd, rather primitive forms that inhabit tropical coasts and rivers. The animals of this family have no common over-all characteristics by which the non-specialist may recognize them.

The Killer Whale (*Orcinus*)

This is by many times the greatest carnivorous or predaceous animal on this planet today as far as we know, being surpassed in bulk only by the two largest sharks, which, however, eat comparatively tiny food items. Killers may reach thirty feet in length and are perfectly streamlined but of a barrel-like spindle shape and are thus extremely bulky. The front end of the animal opens right across in a wide gape and the mouth is filled with a continuous row top and bottom of immense, sharp, recurved teeth. With these, Killers bite great chunks out of other larger whales, dive into their mouths and rip out their tongues, or tear off their lower jaws. These morsels they then raise out of the water and chew on like a dog. They are the fastest swimmers and most maneuverable of all whales, darting about aided by their large tail flukes and making sharp turns by using their very tall dorsal fins and foreflippers as stabilizers. They hunt in packs and will even smash up under ice floes to get at seals or men. They eat prodigious quantities; thirteen porpoises and fourteen whole seals were taken from the stomach of one. They are black and white in color and there may be more than one species since those in the Pacific are usually differently marked from Atlantic specimens. They are cosmopolitan in range.

The False Killer (*Pseudorca*)

This is an equally rotund and equally gruesome-looking medium-sized whale, but it appears to be a harmless and inoffensive gregarious creature that meanders about the oceans in small companies in pursuit of fish. It is colored a smooth dark grey-brown all over and has a splendid set of teeth, but it is rarely seen, never hunted, and has only occasionally become stranded. In fact, they were hardly known at all till this century. They grow to about fifteen feet in length and, unlike the Killer, the males are not greatly larger than the females. The fin is small and the flippers are pointed.

Blackfish (*Globiocephalus*)

These animals grow to thirty feet in length and are of a different shape from all other whales. The head is very short and blunt and where the neck should be is the thickest part of the animal; from there back, it tapers gradually to the small tail flukes. The flippers are long and slender and spring from the body just behind the angle of the jaw. The back fin is very far forward. These animals are colored jet black but some have a white chin and throat, and there appear to be several species. They are found all over the Atlantic and the Pacific, and are usually referred to as Pilot Whales because they go about in large herds following a leader. If he runs aground, which is not an uncommon mistake, the whole lot follow him, and once in shallow water they are incapable of getting back to the depths. There has been a regular industry founded on this habit in the Faroe Islands for centuries, the poor brutes being driven into narrow shoaling channels and then dispatched by hand. They are fish- and squid-eaters.

Risso's Dolphin (*Grampus*)

Of somewhat intermediate position between the foregoing and the dolphins of the second grouping, all of which have beaks, is an animal growing to about fifteen feet in length, that has a small rounded head, long tapering front flippers and remarkable coloring. It is actually a smooth chocolate brown with almost black tail-flukes, fin and flippers, and has a pale-cream head and undersides. However, the young are dark grey or brown while the adults are almost invariably marked all over the body with white sickle-shaped patches running at all angles. It is not known whether these patches are an intrinsic feature of the animal's color or the result of the healing of scars invariably received by this species from some unknown cause. It eats squid and octopus, and the beaks of these may make the scars. There is much confusion about the name of this animal—*Grampus*—as this was often applied to the Killer, and is really applicable to almost any whale since its origin is simply "fat fish."

Beaked Dolphins (*Delphinus, etc.*)

There are five genera of beaked dolphins recognized, three of which will be dealt with here. The best known are the Common Dolphins (*Delphinus*) (Plates 134 and 135), of which there are many species distributed all over the tropical and temperate seas of the world. They have been known since time immemorial and have played a large part in the folklore of many lands, notably classical Greece and Rome. Much nonsense has been written about them, including the solemn statement that they like music, especially that made by a "water-organ." Nevertheless, several of the most outlandish of these claims now appear to be perfectly true but some of the currently believed "facts" about these animals are pure bunkum. For instance, dolphins will come to a boat on which loud music is being played on a still sea and they can be attracted by knocking on the hull with a hammer. For some extraordinary reason they can and have been tamed and notably by young boys: there are photographic records of this. They do appear to try to aid wounded members of the school by pushing them up for air. On the other hand, the new and now very widespread practice of calling them "porpoises" is not only wholly inaccurate but ridiculous. The name "dolphin," for these animals, is thousands of years old, whereas porpoises are quite other animals of another family. The animals that are often depicted jumping through hoops for fish in Floridian aquaria are Dolphins, *not* Porpoises.

Dolphins appear to be intelligent animals and since they have been observed at close range in semi-artificial conditions, they appear to be much more like the rest of the mammals than one would ever expect of a whale. Their whole behavior and notably the care of their young is astonishingly homely and is accompanied by audible expressions of all kinds that may be heard under water. Normally they travel in schools and have a tendency to string out single file, leaping out of the waves at regular intervals just as constantly depicted. This is probably to rid themselves of lice rather than for sport. When leaping they cannot bend their backs, being rigid, but depress the head and tail-tip.

Closely related are the Spotted Dolphin (*Prodelphinus*) and the Long-beaked Dolphin (*Steno*). The first is a small, seven-foot, Atlantic species dark above but with its white undersides marked by more or less even lines of dark grey spots. The second, which grows to about ten feet in length and is found all over the world except in polar seas, is black above and white below but has large white spots all over the back and a fine stippling of black on the belly. Both of these animals are open-ocean fish-hunters and very agile swimmers.

Bottle-nosed Dolphins (*Tursiops*)

This is the second commonest dolphin and often occurs in enormous multitudes near coasts. In the past, they used to form the basis of a special industry at Cape Hatteras where they were taken in droves in large seine-nets. They and the Piebald Dolphins to be described below have shorter beaks than the Common Dolphin and these are more upturned. They have a very short upper jaw. These animals feed on shallow-water- and bottom-fish rather than on the swift pelagic species. They are black above and white below.

Piebald Dolphins (*Cephalorhynchus and Lagenorhynchus*)

These two formidable Latin names mean simply the beak-headed, and the flask-beaked ones, which is not at all helpful in identifying the animals. They are closely related, highly streamlined dolphins found in southern and northern seas respectively and there are a number of species of both genera. They are predominantly shiny jet black above and glistening white below but their heads and flanks are marked with a bewildering variety of black and white patterns and in some cases with a slash of brilliant yellow on the flanks. There are two distinct species of *Lagenorhynchus* known as the White-sided and the White-beaked Dolphins, which are very common in the North Atlantic and often appear in the Davis Straits and off the coast of Norway in vast schools. They are swift pursuers of open-water fish but when following shoals in the summer they may enter fjords or even ice floes.

The Rightwhale Dolphin (*Lissodelphis*)

Oddest of the Beaked Dolphins, as of all dolphins, and in some respects unique among whales, is a small eight-foot, jet-black animal with a pronounced down-curved beak but no dorsal fin whatsoever. It is rare but apparently worldwide in range and nothing is known about it. It does have somewhat of the aspect of a tiny model of the Black Right Whale.

Bay Dolphins (*Sotalia*)

There are more species of this genus than of any other among the whales—over two dozen described and probably at least half this number being validly distinct. They are small nondescript dolphins, usually of light color and in at least one case about the color of a Caucasic human child. They are found along the coasts of West Africa, all about the Indian Ocean, among the Indonesian Islands, around New Guinea and Northern Australia, and on the coast of China. In South America they ascend the larger rivers to the headwaters of the Amazon where there are three distinct species; and one species in the Cross River of southern Nigeria in West Africa appears to be an exclusively freshwater animal. Moreover, its stomach has never been known to contain anything but vegetable matter, making it quite unique among whales.

The Irrawaddy Dolphin (*Orcaella*)

The last of the Dolphins and a very curious species is found in the great Irrawaddy River system of the Oriental Region. It is a very small creature, black in color, with small flippers placed far forward, large tail-flukes and a tiny dorsal fin. Its head is bulbous and almost spherical, the eyes are minute, and the small mouth points downwards and has pronouncedly everted lips that come to a point in front. The numerous teeth are absolutely minute. This animal feeds on crayfish, shrimps, small freshwater snails, beetle larvae and other small animal food for which it grubs about on the bottom or pursues through water so muddy as to look like soup. It ascends the rivers to the first cataracts and this carries it over a thousand miles from the sea. There is another marine species in the Bay of Bengal.

PORPOISES (*Phocaenidae*)

There are two genera of Porpoises—the finned, of the North Atlantic and the North Pacific, and the finless, of the Indian Ocean and the West Pacific. There is also a species of Porpoise in the Caribbean but no specimens have been scientifically examined. Another species with a slight beak, a ridge along the back bearing lines of plates and a distinct keel at the tail is found off the southeast coast of South America. They are primitive cetaceans distinguished by having more complex teeth than any others, those of the Common Porpoise being spade-shaped and sometimes having double roots. Porpoises also sometimes display a number of facial bristles and, as has been mentioned, there may be remnants of bony plates or little horny bosses in the skin, notably along the lead edge of the dorsal fin and along the ridge of the back. They are coastal-dwelling fish-eaters, but they have complex stomachs with a crop like a cow.

Finned Porpoises (*Phocaena*)

This was probably the first whale hunted by man since it has always been extremely common in the bays and inlets of Europe and North America. Seldom more than six feet long, it is not too strong to be taken with a light boneheaded harpoon from a canoe. Porpoise teeth have been found in association with very early human burials and dwellings. There are at least two species in the North Atlantic, and two in the Northwest Pacific, one ranging from the Arctic to California, the other south to Japan on the Asiatic coast. They ascend rivers to considerable distances and have on occasion become landlocked in freshwater and apparently bred there. They yield a little oil but were formerly fished for their meat, which although tough was in favor in Catholic countries, having been decreed to be "fish." Polpess-pudding was a favorite of King Henry VIII of England and may at least in part account for his figure.

Finless Porpoises (Neomeris)

In the Indian Ocean and thence east along the Oriental coasts and north to Japan, the place of the Finned Porpoises is taken by another genus of whales, also black above and white below, but lacking any trace of a dorsal fin. In habits they are not unlike the Finned Porpoises and they also have been hunted since time immemorial by many coastal folk. Their backs are studded with small tubercles, and their teeth are flattened from side to side. Sometimes they also have a few conical teeth embedded in the front of the jaws. Even old males seldom are longer than five feet. They are lead-grey in color with bright red eyes with reddish-purple upper lips and patches on their throats. They are found all the way from the Cape of Good Hope to Japan and often ascend rivers to great distances, being caught a thousand miles up the Yangtze in the Ichang Gorge. They feed mostly on prawns and other crustaceans.

PLATANISTIDS (Platanistidae)

Although obscure in every way, the little whalelike creatures grouped in this family are of profound interest to naturalists and zoologists. They all display certain rather extreme characteristics that may be an indication of primitiveness or simply the result of specialization and a retrogression to the form and behavior once pertaining to their ancestors but which they lost long ago. In many respects, the Porpoises show more primitive features, such as their teeth and the bony bosses in their hides, as well as their complex stomachs. It is in their habits, rather, that the little Platanistids seem to point the way to the distant ancestors of all whales. They live in, or are at least based in freshwaters though both the La Plata Dolphin and the Boutu enter the sea, and the former wanders up and down the coasts. They have long beaks almost like birds or the extinct reptilians, the Ichthyosaurs, and tremendous sets of teeth. Above all, they move about in water so shallow that on occasions it is hard to see how they can float; and the Boutu can actually flop or hump over mudbanks, using its flippers for partial support. Finally, the distribution of the Platanistids is very odd and would seem to indicate that they really are leftovers from a once much more numerous and universally distributed ancient group of animals. There are four species, each of a separate genus.

The La Plata Dolphin (Pontoporia)

These little cetaceans measure only about six feet and are rather slender. The record male was six feet two inches and weighed seventy pounds, the record female six feet eight inches and weighed eighty-eight pounds. The babies are tiny, being only about fifteen inches long when born and weighing about twelve pounds. They are colored dirty white but for a black band down the mid-back. There is a thin falcate dorsal fin and the tail-flukes form a sickle. These animals go about in small family parties and feed on fish that, like the Silvery Mullet, swim in shoals. These they dart after and catch with their long bills, which are armed with over two hundred sharp-pointed teeth that have been described as looking like toothpicks. In the southern winter they vanish from the La Plata and apparently go to sea, though schools are found far up the tributaries.

The Boutu (Inia)

In the Amazon Basin, as also in the larger rivers of the Guianas and in the Orinoco there is another Platanistid that has a long, almost cylindrical beak. This is armed with continuous rows of teeth and turns slightly downwards towards its tip. There is no dorsal fin but a long ridge rising to a slight angle about the mid-back. The front flippers are very large and fan-shaped so that the contained digits are visible and the whole may be partly closed. The tail-flukes form a little more than half a circle when viewed from above. The teeth are twenty-six to thirty-two in number on either side of both jaws and not all of these teeth are simple pegs like those of most other whales; some at the back of the jaws have a distinct additional "tubercle." Finally, the muzzle is bespeckled with small stiff hairs. Boutus grow to about eight feet in length and come in a great variety of colors—often within a single school. Those in lakes tend to be dark to black, but in the big rivers they may be various shades of brown or grey, yellow, cream, white or bright pink. They are usually lighter below and most of them are pink underneath. They have to take in air much more often than other whales and they get about in the flooded forest even when there are no streams as such within miles. The author once encountered them in Guiana in knee-deep water among giant tree-boles ten miles from the nearest river.

The Susu (Platanista)

Very strangely, this was probably the first Cetacean to be "scientifically" described and by none other than the Roman columnist—Pliny the Elder. And, for once, his description is excellent, that master romancer having stuck very close to the facts as reported to him by some keen observers. The little animal inhabits the Ganges and Indus River systems of India. It grows to a length of about nine feet, has no back fin, and has twenty-nine teeth in each half of each jaw and a long snout somewhat compressed from side to side. Further, it has a distinct neck and can move its head about to a considerable extent. Its blowhole is a single slit and its eyes are degenerate and without any lenses. They are dark grey to black in color. Susus are found all the way from the tidal mouths of the rivers to the foothills of the mountains and are in places very common. They muck about in the murky waters, probing for bottom fish, river clams, crayfish, and other morsels with their long slightly upturned snouts, and rise to the surface every two minutes or so to breathe, when they make a noise that gave rise to their name—*soo-sooh*. The beak is very sensitive. They go about in small parties and apparently migrate annually up and down the rivers. At times they become somewhat skittish and may leap clear of the water—doubtless to rid themselves of parasites.

The Chinese Lake Dolphin (*Lipotes*)

In the Tung Ting Lake in Hunan Province of China, which is six hundred miles up the Yangtze River, there lives a small cetacean, colored bluish-grey above and glistening white below. It is only about seven feet long and seldom weighs over three hundred pounds. In form it is of rather "normal" aspect except for a foot-long, slightly upcurved beak. Although the lake is connected to the main river, these animals never go downstream: on the other hand, at certain seasons they leave the lake and go up small tributary streams and even into ditches, so that it is hard to see how they ever turn round and get back again. The Chinese fishermen never molest the animal on the grounds that it embodies the spirit of a Princess who committed suicide in the lake.

The Aard-Vark

(*Tubulidentata*)

THE term classification has become associated with the idea of regimentation and thus acquired an unpleasant connotation. Rather, it should imply orderliness and, in science, at the same time be a device for demonstrating relationships. This is to say, classification should bring related objects together, rather than simply be used to divide up a mass of material. The seemingly endless subdivision of the mammals encountered in this book is

essential in order to bring together those which are truly related and to keep apart those that only happen to *look alike*. These remarks are most necessary when we come to an animal so totally different from all others as the Aard-Vark.

This creature—or perhaps better, these creatures, for there are at least two kinds—stands quite alone in the mammalian tree of life, like a single green leaf caught adventitiously on a spider's thread. Being a highly specialized termite-eater, a digging animal, and nocturnal, it incorporates many features of other animals with similar habits belonging to various orders of mammals. Its basic anatomical structure, however, is unlike all of these and gives no clue to its affinities. Further, no extinct links to any other order are known from fossils. All we can say is that it belongs to the same great division of mammals as the Carnivores and Hoofed Mammals. And anent this point, a brief explanation is necessary.

As we are not specifically interested in classification *per se*, we have grouped the mammals under nineteen major heads, or *Orders*. These orders can and should, however, be themselves grouped in a particular way, best explained as follows:

  I. Pre-Mammals—*the Duckbill and Spiny-Anteaters*
 II. General Mammals
    (a) Half-Mammals—*the Marsupials*
    (b) True Mammals
      (1) *Insectivores, Dermaptera, Bats, Primates, Edentates, and Scaly-Anteaters*
      (2) *Rodents, and Hares and Rabbits*
      (3) *Whales*
      (4) *Carnivores, the Aard-Vark, Seals, Hoofed Mammals, Hyraxes, Sea-Cows, and Elephants*

To a certain extent, this arrangement progresses from the primitive to the advanced and, at least in some measure, indicates the "age" of the groups, the oldest being placed first and the most recent to evolve, last. Unfortunately, the Aard-Varks must nonetheless be of very ancient lineage for the most recent kind of extinct animals with which the most conservative expert will link them existed only in the beginning of the age of mammals at least sixty million years ago. If, moreover, the fossilized remains of a certain animal found in Wyoming finally proves to be that of a relative of this creature, we must expect to have to look very much farther back in time for their origin. The origin of the Aard-Varks is, in fact, the most interesting thing about these animals.

Today, they are found only in Africa but in immediately pre-glacial times they apparently dwelt in Europe and Asia as well. There are various races or perhaps species distributed all over Africa south of the Sahara in all kinds of terrain outside of the true closed-canopy forest. In Abyssinia they live at considerable altitudes on bare, rocky mountains and in southeast and West Africa they inhabit isolated humps of solid ground in swamps. In fact, they will live almost anywhere except in the forests as long as there is a sufficiency of their

E. P. WALKER

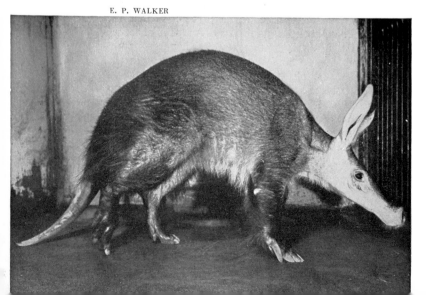

Aard-Vark

termite and other insect food. Even in the forests they may be found on open islands of grass or savannahs that occur like lakes surrounded by tall virgin jungle.

Aard-Varks have to be seen alive to be fully comprehended because they look so different when lying down, squatting, standing up, or walking that no photograph can be wholly adequate. They are immensely sturdy and solid, and the body appears to flow into or be continuous with the tail, which drops away rapidly to a fine tip. The ears are not unlike those of a donkey, the muzzle looks like a trombone, and the rather large eyes always have a somewhat soulful expression and fine long lashes. The limbs are immensely powerful and the claws, especially on the forefeet, are second only to those of the Giant Armadillos and Pangolins in solidity and strength of attachment to the fingers. In skin-covering Aard-Varks range all the way from virtual nakedness to a piglike sprinkling of stiff bristles or a complete coat of glossy black fur. These variations appear to be regional rather than climatic for naked and furred types are found both at high altitudes and on lowlands. Those from West Africa are the most furry.

Although these bulky beasts grow to an over-all length of six feet their food is predominantly termites. To those who have been brought up in temperate climates and have not visited the tropics and especially the equatorial parts of Africa, the word termites implies only minute, white, antlike insects that bore into timber and cause houses to collapse. In Africa these insects, which are not ants but more closely related to such creatures as caddisflies, come in a bewildering variety of shapes and sizes, and display quite other habits. There are those that bore into wood and live underground but there are also others that build what are virtually concrete structures above ground which sometimes reach monumental proportions. Structures 20 feet high have been recorded. These are honeycombed with passages great and small and contain, at the base, half sunk into the ground, a large gallery from the domed roof of which the "combs" or brood-chambers of the animals are suspended. The whole structure is made of tiny pellets of earthy material cemented together with the insects' saliva and is so strong that a pickaxe is needed to demolish it and it can be used for road metalling when mixed with a type of tropical soil known as laterite. This explains the immense claws and tremendous arm muscles of the Aard-Varks, since they have to break open these towers to get at the insects.

It is also the reason for their very thick hides since the "soldier" termites which come rushing out to defend the nests as soon as the outer covering of their shelter is breached, are in some species almost an inch long and have a pair of razor-sharp cutting pincers. To these the Aard-Varks are immune even if hundreds become attached to their naked piglike snouts. What is more, they often put the holes under the termites' nests to their own use and may rest there after a feed although thousands of the insects are still attached to their hides. They are, with the possible exception of the Giant Armadillo, the greatest diggers among mammals, being able to bore their way into the hardest soil at an aston-

ishing speed. Boulders and loose rocks are attacked without flinching, the loose earth being removed from all around and the stone then scooped out with the forepaws. Aard-Varks prefer to feed and work at night and are one of the few animals of the tropics that are abroad in bright moonlight. If caught by the sunrise far from their hole they will go to ground with great industry, digging with positive fury, the earth and stones pouring out of the excavation in a steady stream about two feet in diameter and arcing some twelve feet into the air.

The sense of hearing of these animals is extremely acute and it is believed that they use it to detect the workings of termites below ground and so save themselves the trouble of opening abandoned towers. In captivity they thrive, like other "anteaters," on a diet of chopped meat, boiled rice, raw eggs, and milk.

Odd-toed Hoofed Mammals
(Perissodactyla)

EVEN those of us born in cities come to understand, as soon as we can name any animals, that dogs and cats belong together; that rats, mice, rabbits and guinea-pigs are somehow associated; and that monkeys are classed with apes and ourselves. For some remarkable reason, the average youngster will also instinctively group horses, donkeys, pigs, deer, and even elephants and rhinoceroses, together, though nobody appears to be able to state exactly just why this should be. Early classifiers of the mammals did likewise and both now appear to have some solid grounds for doing so. Moreover, there is an actual relationship between these animals, as we may see from the breakdown in the preceding discussion of the Aard-Vark. Nevertheless, this enormous group of mammals—comprising, as a whole, almost a fifth of all living, and more than half of all known mammals, if fossil forms be included—is clearly divisible into no less than fifteen orders, of which seven still have living representatives. The two outstanding of these are the Carnivora and the Even-toed Hoofed Mammals. The rest are remnants of a former much more numerous glory— the Aard-Vark; the Elephants, the little Hyraxes, the Sea-Cows to be mentioned later; and the Odd-toed Hoofed Mammals.

The last, known to zoologists as the Perissodactyla, —which literally means the same thing in Latin—are today few in number but include some well-known animals. In bygone times they were much more numerous and were represented by all manner of strange beasts, some of gigantic size—one, a form of rhinoceros known as Baluchitherium, was the largest land mammal that ever lived, standing well over eighteen feet in height.

Today, there are three kinds of hoofed mammals with an odd number of toes, the Horses (including the donkeys and zebras), the Tapirs, and the Rhinoceroses. The Horses stand considerably apart from the other two and, despite their great variation in external coloration, are all very much of a oneness. The Tapirs are split into two parts, for one is found in the Oriental Region and the other in tropical America. Of the so-called Rhinoceroses, there are two quite distinct types: one containing three and possibly four species, the other being really an entirely different kind of animal (the White Rhinoceros of Africa) which should properly be called a Ceratothere. For purposes of simplification we separate the *Perissodactyla* into three groups.

EQUINES

With one exception, all the horses and asses, including the domestic breeds, and the zebras are placed in one genus: *Equus*. The single exception is the animal popularly known as Grevy's Zebra, and that is separated on anatomical grounds as *Dolichohippus*. There are only six species of *Equus* but five of these may be split into very distinct sub-species. Unless a very singular, heavy-bodied but extremely light-boned breed of wild horses occurring only in the mountain pine forests of the island of Hispaniola be remnants of the prehistoric wild horses of North America, the Equines are today confined to Asia and Africa though domestic breeds of horses and donkeys are, of course, worldwide in distribution today, south of the Arctic. These animals have but a single toe left on both hind and front feet; short tails bearing long plumes; crests on their heads and necks which may be so long as to form manes; and teeth in the front of both the upper and lower jaws.

The origin of the horse was until recently thought to be better known than that of any other mammal; this was based on a series of complete fossil skeletons of dozens of different extinct creatures, starting with simple animals of the size and shape of small dogs with five fingers and four toes, and ending with the modern Thoroughbred. However, this pleasantly neat evolutionary picture of orderly progression in tooth structure, loss of toes, increase in size, and wrist and ankle elongation has now unfortunately come under grave suspicion. So many side-branches have been brought to light and so many intermediary forms are completely lacking that we can now only say that the classic description is no more than a guide to the probable steps by which the modern horse evolved. For purposes of simplicity we describe the Equines under four heads, as follows:

Horses (*Equus caballus and przewalskii*)

Domestic horses have rightly been made the subject of what now almost amounts to a science in itself. From the tiny Shetland Pony to the gargantuan Shire they all appear to have been derived from one, two, or at most three separate wild species. One of these, Przewalski's Wild Horse (Plate 136) is still alive; the others are extinct as wild animals, if they ever existed. There were undoubtedly species of wild horse other that Przewalski's

in Europe until the end of the Old Stone Age of human culture but to what extent these entered into the formation of the domestic breeds still extant is not known and therefore open to debate. There were true wild horses in North and Central America at the same time that the country was inhabited by men but there is no evidence that they were ever domesticated and they became totally extinct some time before the arrival of the White Man in the New World—unless the Haitian Wild Horses constitute an isolated island relict of this breed. American wild horses did not, however, become extinct anything like so long ago as was once supposed, for there were numerous traditions about them among the Amerindians which were recorded by the Spaniards in the earliest days of their contact with these peoples. Most notable among these is a record from Panama of Amerinds who declared that horses imported from Europe could not be kept there because their ancestors had tried it and all the animals had been killed by some sickness contracted from the bites of Vampire Bats (see page 60). The so-called Wild Horses, or Fuzztails, of the North American West today are feral animals whose forebears escaped from domestication since the white man colonized the continent. There is no such thing as a pigmy breed of horses. The grotesque dwarfs of small size now fairly common in shows are individually raised on a ranch in Mexico and do not come from a "Lost Canyon" in Texas.

Przewalski's Horse is a small heavy-set animal with an upright crest stretching from the top of its head to just before the shoulders. It is of a reddish brown color, somewhat lighter below in summer, with a dark median back line and a shoulder line, dark crest, tail and lower limbs. In winter its fur is longer and lighter and the back and shoulder stripes are absent. It was once common all over the plains of Eurasia but today is confined to a limited area around the Altai Mountains in Siberia and in western Mongolia where it unfortunately is interbreeding with feral domestic horses and will thus probably soon become "extinct" as a pure wild species.

Asses (*Equus hemionus and asinus*)

The name donkey is best reserved for the domesticated breeds of the smaller, long-eared Equines, all of which are thought to have been derived from the Nubian Wild Ass. There are three recognizable kinds of wild asses in Asia and two in Africa. The best known is the Onager, a very horselike animal, of a rufous color and lighter below, that is found in sundry forms throughout Persia, Afghanistan, and the drier parts of northern India. It has a triple light dorsal stripe and usually a shoulder stripe. It travels about in small herds and is a semidesert animal. The presently most numerous wild ass is the Kiang of the Tibetan uplands. This is similar to the Onager but much larger, has very large hoofs, and is contrastingly red and white in summer but a dull brown in winter, at which time it is clothed in long almost shaggy hair. The third of the Asiatic types of Asses was once very numerous all over central Asia from Manchuria to Turkestan and from Siberia to Sinkiang but is now found numerously only in parts of Mongolia.

It is a smaller animal and displays typical sandy desert coloration and has a narrow black dorsal stripe.

The Asses of Africa are considerably different, have large upright crests and are generally greyer in tone. The Nubian type has a pronounced white muzzle, white rings round the eyes, and white undersides; the Abyssinian has a grey nose. Both have well-marked dorsal stripes and the Nubian a very distinct black shoulder stripe. They are desert and mountain animals.

Zebras (*E. burchelli, etc.*)

The Zebras are one of the most difficult groups of animals to arrange in any understandable order. No two individuals are alike and there is always a very wide range of variation in a herd, in any one locality, and in both sub-species and species as a whole. They will also interbreed with horses, producing zebrules (Plate 137) or zebrinnies. Actually there are four distinct types of zebra-striped Equines, one of which is now extinct, and one of which is the quite distinct *Dolichohippus*.

The Mountain Zebras form a distinctive small group having black noses and black and red stripes on the side of the muzzle. The ears have a white horizontal stripe and the rest of the body is vividly striped black and white, the stripes on the haunches being very wide. There are two kinds, one in South Africa which is now almost extinct and another in the coastal mountains of South West Africa and Angola.

The true, Common, or Burchell's Zebras (Plates 138 and 140) are numerous throughout a very wide range in Africa south of the Sahara. Dozens of different kinds have been described but it appears that all fall into one of four recognizable sub-species. The original type of the species was found in South Africa and is now extinct. It was pure white below and had white legs, and shadow stripes between the black stripes. As a whole, the Common Zebras are vertically striped in front, horizontally on the back legs, and diagonally on the rump and hind flanks. This causes a V-shaped junction-pattern about the middle of the sides. The second type—the Damaraland Zebra—has striped legs and profuse shadow stripes; it comes from a belt of territory south of the Congo and north of South Africa. The third type—Selou's Zebra—is very perfectly and closely striped black and white down to the feet, and lacks shadow stripes. It comes from central and southeast Africa. The fourth kind—the East African Zebra—is boldly striped and is found from Rhodesia to Abyssinia but inland from the coastal area and thence ranges north and west to the Sudan.

The extinct animal is the Quagga, a race of E. Burchelli which was fairly common throughout South Africa until exterminated by the white men mostly for food for their slaves and native labor. It was of a brown color with black and cream stripes on the head, neck, and shoulders only. The tail was rather long and black, the nose was black, and the undersides were white. The last known living specimen died in the Berlin Zoo in 1875 but wild examples are still constantly being reported from South West Africa.

All these animals are typically horselike grazers and

Onager

go about in large herds mingling with other game and affording the lion population their main article of diet. Zebras are shy and nervous but they are also rather pugnacious and have been known to defend themselves with considerable vigor, kicking both backwards with the hind feet and forwards with the front, and biting savagely. Their color pattern renders them extremely conspicuous against green backgrounds, but almost invisible in tall grass, and very hard to make out on open dry plains because of the breakup of their outline by the complex striping. They are invariably heavily infested with intestinal parasites of many kinds and it would seem that without these their digestion fails to function properly. It is interesting to note that the African insists a Zebra is a black animal with white stripes, whereas people of European origin automatically assume that it is a white animal with black stripes.

Grevy's Zebra (*Dolichohippus*)

This is the most horselike of the striped Equines (Plate 139), and has a very large head and huge ears. It is vividly striped black and white all over except along the mid-belly but the stripes are individually very narrow. There is a very tall, erect crest increasing in size from the shoulders to the top of the head, giving the animal the silhouette of horses on Ancient Greek pottery. This is a very shy animal that inhabits the desert-scrub belt that rings the southern and eastern flanks of the Abyssinian Massif. It is a browser rather than a grazer and can do with very little water.

TAPIRINES

As was said above, the Equines stand considerably apart from the other Odd-toed Hoofed Mammals. There is no doubt that the Tapirs and Rhinoceroses had a common origin and there is some evidence that the latter sprang from the former. The tapirs are very basic creatures having remained almost entirely unchanged for an incredible length of time and being, among Ungulates, just about as primitive as any hoofed animal could be. They have been a fairly numerous group over the ages but today there are only four species left, one in Malaya, and three in Central and South America, one of which is sufficiently different to be placed in a distinct genus.

Malayan Tapir (*Tapirus indicus*)

This astonishing looking beast lives in the deep forests of southern Burma, Siam, the Malay Peninsula, and the island of Sumatra. When young it is brown with yellow streaks and spots but when adult it is black except for a silvery white area over the back and flanks which looks just like the "coats" worn by small dogs in overcivilized cities. This color pattern which appears so idiotic and useless when the animals are seen in a zoo is enormously useful to the animal in its natural habitat. Being night feeders and semi-aquatic, herbivorous, and of large size,

these animals are abroad every night and in all weather in order to obtain a sufficiency of food. This means that they have to move about even in bright moonlight, at which time a wholly dark or wholly light animal is inevitably most conspicuous. This remarkable color pattern, however, breaks them up into half a dozen small portions in moonlight, so that they can hardly be detected at all when among the grey boulders of a dry stream-bed. During the day, moreover, they are little less obvious when asleep in such a place as is their wont. Tapirs are extremely compact, solid, and heavy beasts and although a prey to leopards and tigers, they can give a fairly good account of themselves when cornered. They have pronounced tusks and give a terrible bite like that of the larger pigs. They are also very thick-skinned and can careen through the densest undergrowth at the most remarkable speed, smashing through tangles that no other animal could face. One young is normal but twins are not uncommon. They browse on water plants and other succulent herbage and are themselves rather good eating.

The Central American Tapir (*T. bairdii*)

This is the largest of the Tapirs, equaling a small donkey in bulk and sometimes almost so in size. It is dark brown in color with a white fringe around its large ears, white lips, and sometimes a white patch on the throat and chest. The "fur" is very short and hard, and, in fact, almost bristly and piglike, and the animal is immensely muscled. This species ranges from central Mexico to Panama and possibly south into the northern mountains of Colombia. It is a montane form that makes narrow winding paths along the most precipitous slopes usually leading to water. These paths have been used by men, first as foot-tracks, and finally as paved roads, for they follow the contours and select the easiest gradients and are in many respects feats of pure genius from a surveyor's point of view. Although a retiring beast, which avoids Man and falls prey to Jaguars, this animal can be a most unpleasant antagonist since it has tremendous strength, can bite terribly, and has a habit of rasping and gripping with the corrugated underside of its short trunk. It will fight only when cornered or with its young and normally careens hysterically off into the undergrowth if alarmed. However, it appears to have rather bad senses of sight, smell and hearing and may be approached within half a dozen feet before it realizes that danger is at hand.

South American Tapirs (*T. terrestris*)

The most erudite studies notwithstanding, there appear to be more than one kind—or species—of small tapirs in South America (Plate 141). They have noticeable, erect crests of stiff hairs rising from the crowns of their heads and extending down the neck to the shoulders. The commonest type is dark brown all over and clothed in very short, hard, and rather sparse hair but there are pale-cream to fawn-colored individuals with very small ears now being brought out of some part of the upper Amazon or the uplands between that province and Colombia that seem to be of an entirely different

[*continued on page 241*

SAN DIEGO ZOO

Malayan Tapir

Mountain Tapir

SEVERIN

126. *European Polecat*

YLLA FROM RAPHO-GUILLUMETTE

127. *European Weasel*

MARKHAM

128. *Black-footed Polecat*

129. *Mink*

130. *American Badger*

PARK

131. *European Common Otter*

LA TOUR

132. *Hooded Skunk*

L. W. WALKER FROM DESERT MUSEUM

133. *Spotted Skunk*

HARRISON FROM NATIONAL AUDUBON

134. *Common Dolphin*

WILLINGER FROM SHOSTAL

135. *Common Dolphins under water*

WILLINGER FROM SHOSTAL

136. *Przewalski's Horse*

MARKHAM

137. *Zebrule (Hybrid)*

YLLA FROM RAPHO-GUILLUMETTE

139. *Grevy's Zebra*

ADPRINT LTD.

→

138. *Common Zebra* (*Bohm's subspecies*)
VAN NOSTRAND FROM NATIONAL AUDUBON

140. *Common Zebra* (*Grant's subspecies*)
MARKHAM

141. *South American Tapir*

142. *Great Indian Rhinoceros*

143. *Ceratothere or White Rhinoceros*

144. *African Black Rhinoceros*

145. *Collared Peccary*

146. *Wart Hog*

147. *Hippopotamus*
VAN NOSTRAND FROM NATIONAL AUDUBON

148. *Pigmy Hippopotamus*
LA TOUR

149. *Bactrian Camel*

150. *Alpaca*

151. *Domestic Llamas*

LA TOUR

152. *Muntjac*

PINNKY

153. *Wapiti or American Elk*

MUENCH

154. *Chital*

MARKHAM

155. *European Red Deer*

PINNEY

156. *Fallow Deer*

MARKHAM

[*continued from page 224*

species. Small tapirs of this type are rather common throughout the tall forests, secondary growth, and even the periphery of savannahs from Colombia and Venezuela in the north, to the Argentine in the south. They are semi-aquatic animals but may live high upon forested mountains where they also make "contour" paths of enormous extent and great precision. Although the smallest of the Tapirs they are still the bulkiest of all animals in South America and are somewhat persecuted by the larger cats and men: nevertheless, they are still very numerous, the author having counted over sixty at one time in a dried-out swamp in the hot season in Guiana. They make most excellent, docile, and rather intelligent pets if raised from extreme youth, but adult males may become aggressive with age.

The Mountain Tapir (*T. roulini*)

Along a strip of the upper Andes Mountains, from central Colombia to northern Peru, but restricted to a rather special kind of forest, there may be found the most exceptional of all Tapirs. This is a small species, slender in build, with an elongated body, long legs, and a very long, pointed snout that almost amounts to a trunk. The body is clothed in a fairly soft and thick coat of twisted black hair that may in old specimens almost resemble that of an adult Persian ram. There is no crest and the tail is rather pronounced. These animals live at incredible altitudes, having been recorded from over 15,000 feet, where they also make contour paths leading to any available water and sometimes over open areas above the tree line. Their ears are fringed with white hairs and their lips are clothed in short white hairs.

RHINOCEROTINES

The Rhinoceroses are somewhat more varied than the Tapirs but are also mere remnants of a past familial glory. Not so long ago, geologically speaking, the earth positively swarmed with rhinocerotine creatures of all sizes; today there are but seven recognizable forms left, as far as we know. These may be clearly separated into four genera—two in Asia and two in Africa. One genus is distinguished by having only one horn on the snout, the others have two, but one of these is really a very distinct animal—the Wide-lipped Rhinoceros of Africa. One form of the latter is also the second largest living land mammal.

One-horned Rhinoceroses (*Rhinoceros*)

There are two clearly defined species of this genus. The first is the so-called Great Indian Rhinoceros (Plate 142), which is now to be found only in certain limited areas of tall reeds in Bengal, Assam, and the river bottoms of eastern Nepal. It is an enormous bumbling beast, clothed in a pachydermatous hide that is covered with bosses and appears to be armor-plated, being thrown into folds with deep pleats running over the shoulders and ringing the limbs. It is a grass- and reed-eater and stays near water in which it bathes daily. It was once distributed much more widely over the moist parts of India but it is obviously marked for extinction, being far too specialized to thrive in a world containing large numbers of men as well as other petty annoyances. The "horn" is in fact neither bony nor horny but composed of congealed hair, and has been held in the highest esteem as medicine for centuries, particularly in China, and this has, more than anything else, contributed to the extermination of the animals. This species is notable for the fringe of stiff hairs around the ears and adorning the end of the tail.

There is another smaller species of this genus, known rather misleadingly as the Javan Rhinoceros (*R. sondaicus*) which is found in isolated patches throughout Burma, the Malay Peninsula, Sumatra, and Java. This animal does not have the bosses in the hide, its pleats are differently arranged, and the surface of the skin usually forms a sort of crocodile pattern resulting from small cracks. There was once a prolonged discussion in scientific literature as to whether there was both a one-horned and a two-horned rhinoceros in Sumatra, the presence of a one-horned species being doubted. However, it is still fairly numerous in the reed-filled bottoms, swamps and estuaries of that island. On the mainland it is becoming extremely rare.

Asiatic Two-horned Rhinoceroses (*Didimoceros*)

In this genus also there are two distinct species, the Hairy-eared, which used to be found throughout Assam, Bengal and upper Burma, and the Sumatran, which once ranged south from central Burma through Thailand and the Malay Peninsula and inhabited both Borneo and Sumatra. The first may now actually be extinct, and the second is greatly restricted in range and very rare on the mainland. The Hairy-eared Rhinoceros had a long and sturdy front horn and a small knoblike back one. The ears were fringed with long black hairs, the tail carried a long brush of black bristles, the whole body was covered with hard brown bristles, and the skin was thrown into small folds. The typical Sumatran form is the smallest of the rhinoceroses and inhabits mountainous areas, making paths like the tapirs, and mud wallows in which to rest. It has a small edging of short bristles around the ears, and the body is sparsely sprinkled with short black hairs all over.

The African Black Rhinoceros (*Diceros*)

The Two-horned Rhinoceros of Africa (Plate 144) is a much larger animal, reaching almost the bulk of the Great Indian one-horned animal. Its skin is naked and dark grey, and is not thrown into folds. The head is small, the front horn long, slim, sharply pointed and slightly recurved, the back one is conical and wide-based. The upper lip of this animal comes to a sharp point and hangs down over the lower and is highly mobile. These animals once ranged all over Africa south of the Sahara but outside of the closed-canopy forests. Today they have been exterminated in the south and much of the southeast but are still to be found from the southern limits of the Congo forest area on the west to the central east coast and thence north to Abyssinia and then west again north of the Congo to the western border of the Cameroons about Lake Chad.

Too much has been written about these animals and there is hardly a book about Africa that does not mention them, yet there is still a great deal to be learned about their ways. On the whole, though, the tradition that has grown up about them is fairly authentic. They are placid herbivores with rather poor senses and are notably indifferent to all other living things; but if approached too noisily they may decide that their territory is being trespassed and their privacy infringed, whereupon they endeavor to focus upon the menace. Usually failing to do so, they either return to their own business, or move off, snorting, only to stop and stare some more. If persistently pressed or if they actually locate the source of annoyance they may charge at a lumbering gait that is grossly deceptive. Nothing short of a stout tree stops this rush—a locomotive was once derailed by one—and if they hit and impale some animal on their horn they may carry it a long distance before making every effort to wipe it off. Occasionally they may get really peeved and wheel about as they lose momentum and then charge again. When going flat-out they actually canter, so that all four feet may be off the ground at the same time. Their young are sometimes killed by lions but the adults are bothered only by men and mosquitoes.

Ceratotheres (*Ceratotherium*)

The other so-called rhinoceros of Africa (Plate 143) is really a very different beast though it also has two horns. Both of these are conical but both point forward at an acute angle and both come to a slightly recurved sharp point. The animal is commonly called in English the "White Rhinoceros." It is not white in color, but pale grey when the skin is dry and seen under a bright sun. The name derives from the Dutch word *weit,* meaning "wide," and originally referred to the wide, almost square upper lip and muzzle of the animal. There are two populations of Ceratotheres in Africa, one in the southeast where the animals are now confined to two reservations in Natal, the other spread over a considerable area in a triangle between Uganda, French Equatorial Africa, and the Sudan. This is the second largest known terrestrial mammal. The body is elongated and the long head normally hangs down almost to the ground. There is a pronounced hump over the shoulders and the ears are rather large. Ceratotheres are grass-eaters and keep to the open plains, where they go about in small parties. They are even more placid and indifferent to others than the rhinoceroses.

Even-toed Hoofed Mammals

(*Artiodactyla*)

IT may be said with surety that this is the best known of all the orders of mammals and the one without which Man would certainly never have attained his present numbers or status upon this earth and the one without which he probably could not now continue to maintain his position. Man is an omnivorous mammal and must, apparently, eat a certain quota of meat in order to keep properly fed. Beans and other leguminous plants may, to a certain extent, supplement this need but in the absence of fish he must turn to the flesh of birds and mammals. Originally this food supply was obtained by hunting, but game, except on the vast savannahs of Africa, the prairies of America, and a few other areas was always limited and could support only a limited population of hunters. It was the domestication of several hoofed mammals of this order that made it possible for Man to save his energy wasted in hunting and devote more time to establishing civilization.

The number of Artiodactyle Ungulates that have been domesticated is greater than the average person realizes. To enumerate, there is the pig, a forest animal and probably first truly domesticated in the Orient; then the Bactrian and the One-Humped Camels; to a limited extent the Fallow Deer kept in parks and used as food in Europe; the all-purpose Reindeer; and one of the American Deer kept in corrals by the Toltecs, Aztecs, Mayas, and Quiches in Central America in pre-Columbian times. Then, among the Bovines, we have the original oxen or Aurochs first confined in Europe and from which many of our domestic breeds of cattle have been developed, the Yak in Tibet, and the Water Buffalo the general work-horse of all tropical Asia. Finally, we have the sundry breeds of goats and sheep which are apparently derived from several separate sources that were originally wild species inhabiting various parts of the Old World.

In the aggregate, the numbers of various domestic animals living today under man's supervision add up to billions and provide our needs with over a million tons of meat a year besides leather, wool, milk, and fertilizer. In 1949 there were seventy-eight million cattle, sixty million sheep, four and a half million goats, and sixty-two million truly domesticated pigs in the United States alone. Adding to these several million feral pigs and quite a number of cattle, it will be appreciated that this vast host of large plant-eating animals must produce a profound effect upon the whole land surface of the country. Artiodactyle Ungulates by their sheer numbers and ever-munching mouths played a major part in the distribution and regulation of vegetation before the advent of Man and since their multiplication by Man they have completely changed the complexion of half the world. Sheep and cattle tend to improve grasslands and pigs alter woodland composition by their rooting, but goats can completely denude a countryside of all vegetative cover and initiate desert conditions. Without the Reindeer, man could not have survived in the Arctic, and without the Yak the Tibetans would probably have long ago abandoned their country.

The Even-toed Hoofed Mammals are today a very numerous order—in fact the fourth largest, containing no less than a tenth of all genera of living mammals. If the extinct forms known from fossils be included they are second to the Rodents alone, and contain over a seventh of all described kinds of animals. Although they originated in very early times along with the first Carnivores, they are comparatively modern and the most highly evolved of mammals. They are also extraordinarily varied in size and form, ranging from the tiny rabbit-sized Chevrotains and Pigmy Antelopes to the huge Alaskan Moose and the positively monumental Bactrian Camel. What, moreover, could be more exotic than a Giraffe or more nondescript than a Suni?

That which is known or conjectured about the origins of this Order have been alluded to in previous chapters. Even sixty million years ago they not only existed as a separate group but were already divisible into a number of distinct categories, several of which still have living representatives. The classification of these animals is thus both fairly simple or extraordinarily difficult, depending upon the desired approach. For the expert making a proper study of the interrelationships of all known living and extinct forms there are almost insurmountable difficulties; for the non-experts who are simply interested in identifying the living forms and learning which belong together there is, for once, little difficulty. Nor, as with the Rodents, is a classic patience required, for almost all these animals are known by widely recognized popular names, they are almost all large and rather obvious beasts, and they fall into quite clear-cut categories. Patience is required only if one sets out to review the whole order at one sitting, for there are manifestly an immense number of different mammals that have cloven hoofs. There are actually nine major divisions of these animals, which are in turn grouped in various ways by zoologists. These niceties need not, however, concern us, except to note that the members of the first two groups do not chew their cud, but that the remainder do.

SUINES

The first and possibly the oldest and most primitive group of this order is of moderate numbers in point of types but of wide distribution. Their habits are surprisingly uniform also, all being primarily animals of the forests and woodlands, all being more or less omnivorous though subsisting basically on vegetable matter, and all digging with their muzzles. They are mostly gregarious and they have more than one and often very many more than one young at a time. The group is divided clearly and cleanly into two parts—the Pigs and the Peccaries—which are today confined to the Old and the New Worlds respectively.

PIGS (Suidae)

The habits of the Domestic Pig are sadly misunderstood and grossly misrepresented in our vernacular, folklore, history, and even in many treatises on husbandry. Its true habits epitomize those of the whole family. But first let it be mentioned that the Pig has proved to have an intelligence of the quality of our own —rated by the IQ—and second only to that of the Great Apes. The admonition that one "lives like a pig" ought to be taken as a high form of compliment since

these animals, unless prevented from doing so by Man, lead the most orderly and cleanly of lives. The fact that they customarily and very sensibly make mud wallows in which to cool off in summer, denude themselves of external parasites, and encourage the growth of favored food plants, does not mean that they like to be "dirty." Nothing is cleaner than the good earth and especially mud, as expensive beauticians and the discoverers of antibiotics will affirm. Just because a pigsty becomes a wallow does not mean that the pig desires it to be composed of excrement, rotting garbage, and decomposed bedding. In the wild, pigs do not excrete in their wallows.

The normal habit of all these animals is to roam the woodlands rooting into the surface soil for all manner of food from snails and earthworms to roots and fungi. At times they resort to their wallows and take a mudbath. For the rest, they are exceptionally clean and surprisingly devoid of external blemishes. They do, however, customarily harbor sundry internal parasites which readily pass to those animals that eat their flesh. In past centuries they were therefore rightfully excluded from the diet of various peoples by wise legal and religious decree.

Eurasian Pigs (Sus)

There are four very obviously different kinds of piglike animals distributed about an enormous belt of the great combined continent of Eurasia. All of these are found in the Oriental Region and all but one exclusively therein.

The exception is the Wild Boar (*Sus scrofa*) which, in one of many races, is found from western Europe and North Africa to the Andaman Islands in the Bay of Bengal and throughout Burma, to China where it still ranges north to Manchuria. It is not found in Russia or in central Asia north of the Caspian Sea or in the great Pamir-Himalaya-Chinese mountain complex. It is very common in India and is found in Ceylon. There are recognizable races in Europe, North Africa, India, China, and a small type on the Andaman Islands. These large pigs are all much alike, have long pointed muzzles and, in the male, pronounced, upturning tushes in both jaws. They are clothed in reddish, brown, chocolate, or black hairs that are stiff and bristly and usually form a crest on the top of the head and neck. In cooler climates they

European Wild Boar

have a dense underfur and in the tropics are half naked, having only scattered bristles. Several races have a grizzling of white around the mouth and on the chest. They are most competent and single-minded beasts and are intolerant of any interference. Even the babies will put up a determined defense and the males will attack with calculated strategy. Their bite is worse than that of any mammal with the exception of the Killer Whale, and actually much worse than that of the Great Cats though being a ripping rather than a slicing action.

The remaining pigs of the Old World are all concentrated in the Oriental Region. Perhaps the most outstanding is the smallest in stature and the most obscure. This is known as the Pigmy Pig (*Sus salvanius*) and is found in the hill forests of the central and eastern Himalayas. Adult boars stand only about a foot high at the shoulder and the piglets, which are striped a vivid yellow on dark brown, are about the size of a pack of cigarettes when born and look like tiny clockwork toys.

At the other extreme is the Giant Bornean Pig (*S. barbatus*), of which one race found in the lowlands of that island is the largest Suine known. These animals may measure over six feet in length and have enormous heads adorned with a crest on the crown and a kind of upturned moustache of bristles that hide the tushes.

Finally, there are a group of half a dozen small to medium-sized wild pigs (*S. verrucosus, etc.*) with pronounced warts on the side of their muzzles, and with light markings about their eyes and under their necks, that are found in the Philippines, Borneo, the Celebes, Java, and Sumatra. They are semi-naked, have small crests, and long, very pointed muzzles. All are forest animals and have the typical habits of pigs as a whole, and all appear to interbreed with feral domestic pigs.

African Bush-Pigs (*Potamochoerus*)

Throughout the densely forested parts of Africa from Senegambia to Uganda and south to the borders of Angola, and then again in the isolated patch of tropical forest in southeast Africa, there are countless droves or sounders of tall, short-bodied, narrow-flanked pigs with extremely large heads, long pointed ears often bearing plumes, long tails with a terminal tuft like a fly-whisk, and thin slender limbs. There is a small distinct species on the island of Madagascar, the only ungulate indigenous to that country and the only "modern" form of mammal thereon. These animals are found in almost all types of forest but always near water and mostly on the lowlands. Some forms are black with dirty cream ear-plumes, others are a brilliant iridescent coppery-golden hue with black eye-rings surrounded with white "spectacles," white moustaches and profuse pure white manes. They are the profoundest of all rooters in the proper sense of that word and can literally turn the forest floor upside down, even moving the fallen trunks of immense trees by working together in teams of up to fifty, all pushing with their long snouts on the same side. Under such logs all kinds of small life lurk and these—from poisonous snakes and rats to fungi and snails—are forthwith devoured with much crunching and lip-smacking. In fact, the very common giant forest snails form one of the principal articles of their diet.

The Forest-Hog (*Hylochoerus*)

Rivaling the Giant Bornean Pig in dimensions is an astonishingly large piglike mammal, now known as Meinertzhagen's Giant Forest-Hog (Plate 146), that somehow avoided discovery until this century. It first came to light in the presumably virgin equatorial forests of inner East Africa but has now been found to exist within this vegetational belt right across Africa as far west as the French Cameroon. It is a gigantic, rather elongated, low-chassied creature with large head and tushes and widespread ears around which sprout a profuse growth of stiff hairs. It has large warts below the eyes and is covered in sparse but long, curved, stiff bristles that become almost spines along the back. Nothing of a specific nature is known about its habits even today but it is said by the natives who hunt it to be almost predaceous.

Wart-Hogs (*Phacochoerus*)

This, the most exaggerated of pigs, is probably the commonest species alive today and is known to anybody who has ever sat through a travelogue or almost any other film made about or in Africa. Of the most repulsive mein, this squat-bodied, half-naked creature with fragile-looking legs and enormous flattened head, from the sides of which grotesque warts protrude, is found all over Africa south of the Sahara and outside the closed-forests, except in the mountains of Abyssinia and in the Kalahari Desert. It is now also eliminated from the enclosed areas of the Union of South Africa and is rare in West Africa except along a narrow belt between the northern edge of the forests proper and the southern edge of the deserts. It is not only a ridiculous-looking animal with a scraggly topknot of coarse hair and an absurd skinny tail with a small terminal brush that it holds straight up in the air when trotting over the open ground but it is also of somewhat crazy habits—at

Red River Hog

least to our way of thinking. It lives in large holes which it excavates itself but usually starting with an abandoned Aard-Vark's or other animal's retreat, and into this it backs when retiring to rest or avoid the heat. Its immense tusks grow outwards, then upwards, and finally inwards over the muzzle and are used much as we use a hoe. Wart-Hogs put on a great show of courage and may carry through their threats but usually retreat, tails on high, with an air of insolence that is hard to tolerate. When they lose their nerve, which is very often, they also lose their wits and become quite hysterical and often run over each other in their eagerness to decamp. They eat almost anything that comes to hand and have been known to take carrion.

The Babirusa (Babirusa)

On the islands of Buru and the Celebes in the East Indies dwell two closely related piglike animals of an aspect almost as extraordinary as the Wart-Hogs. The appearance of a Babirusa is perfectly displayed in the accompanying photograph. It is a purely nocturnal animal and travels about the lusher parts of the tall forest in small family parties, rooting in the soft earth by rivers and in swamps. The exaggerated tushes are not known to serve any particular purpose but, being present only in the males, are probably a sexual adornment though they certainly protect the animal's face when rooting. It is interesting to note also that the males do most of the digging, the females and young following along behind to chew on things that he unearths. Babirusas are extremely timid. They are also very good to eat.

PECCARIES (Tayassuidae)

Prior to the arrival of Europeans and even more so of the Chinese in the New World there were no indigenous Pigs in that hemisphere. Their place was taken and still to a large extent is occupied by somewhat similar-looking but totally different animals known as Peccaries. We stress the importance of the Chinese in the Americas because it was these people who brought the breed of domesticated true pig that so readily takes to the wilds and has supplied the main stock of the feral so-called "razorbacks" which infest the entire woodland area of southeastern North America, many parts of Central, and even some areas of South America today. The Peccaries, although now exclusively of the New World, were not apparently evolved there, for the fossilized skeletons of creatures that appear to be of this family rather than of true pigs have been found in both Europe and Asia and in most ancient strata. Somehow, however, the remnants of this family became isolated in the warmer belts of the Americas.

There are two known species of Peccaries; one, the White-lipped, which is the larger and which goes about in vast droves, is found—but rather rarely—all the way from the true forests of southern Mexico to the northern borders of the Argentine. It is of a reddish-brown color, piglike in body but with a very large and sharply pointed snout and has a pronounced white moustache, lips, and cheeks. Old males turn almost black and then become grizzled, and old females go dull brown. These

Wart-Hog

animals are highly dangerous since they not only travel in vast herds and spread out widely over the forest floor but are just as carnivorous as they are herbivorous and become truly predaceous when any living thing wanders into their midst. At these times, they have an uncanny method of communication and co-operation, silently surrounding the victim, keeping as far as possible out of view, and then closing in upon him. The author speaks from first-hand experience, having twice been treed for several hours by these animals.

The second species—known as the Collared Peccary (Plate 145)—is much more common and more widely spread, being found in almost all types of country—from

Babirusa

the deepest tropical swamps to barren uplands and the desert scrub all the way from the southwestern United States to the Argentine and occurring also on the island of Trinidad. Their habits are similar to those of the White-lipped Peccary, but they are not so aggressive, seldom if ever actually predaceous, but just as wily. When these animals enter a tract of land, be it forest or desert, they literally plough up its entire surface and eradicate all animal life and most else that is edible. Peccaries have a habit of clapping their jaws together when alarmed or angry and this is a severe warning to stay at a safe distance since they give a fearful bite. They have scimitar-formed tusks in both jaws as well as sharply pointed cheek teeth. Mothers defending their young may also slash with their pointed hoofs. Like the true pigs they appear to be immune to snake-bite and will eat snakes.

HIPPOPOTAMINES

The exact position in the scheme of life of these remarkable beasts was for long a puzzle to naturalists and zoologists; though the Romans seem to have correctly placed them alongside the pigs and the perspicacious Greeks called them Libyan (*i.e.,* African) monsters and left their genealogical status in limbo. There are two kinds of hippopotamines alive today, both found only in Africa, though there was until very recently a third species living in Madagascar and the islands lying between that sub-continent and the African coast. They must be of considerably venerable ancestry *per se* but, although sundry creatures are known from fossil remains that look like missing links between them and the original Suines, there is really no information on their earlier history.

PIGMY HIPPOPOTAMUS (*Choeropsis*)

The more primitive of the two extant species (Plate 148) is the Pigmy Hippopotamus (*Choeropsis liberiensis*) of West Africa, which is the size of a large pig, extremely rotund, and almost wholly aquatic. The head is comparatively small but the wide-gaping mouth is filled—rather erratically, it seems—with fang-like teeth. The animal is of mild disposition and so retiring it was only discovered at the beginning of this century. It inhabits a few lakes and tracts of deep forest where there are small rivers and stagnant pools and ponds, and never ventures far from water. It is a nocturnal feeder and cannot be out of water for any length of time because its moist and thin skin, which is perforated all over with large pores, rapidly dries up and cracks unless kept wet. The author watched a baby of this species weighing sixty-eight pounds actually absorb three and three-quarter gallons of water through its skin in an hour in a carefully controlled experiment conducted in a container of known capacity. Moreover, the animal was not unduly dessicated before being immersed and did not appear to drink any of the water.

COMMON HIPPOPOTAMUS (*Hippopotamus*)

The greater Hippopotamus (Plate 147) is still a rather common animal in most of the great rivers of tropical Africa and may occur in alarming numbers in some limited stretches of quite small streams as well as in isolated lakes and even ponds. It used to inhabit the Nile to its estuary in the Mediterranean, where it was known to the ancients as the "behemoth" or transliterations thereof—a word that appears to be of Assyrian origin and to mean monster. It is now unknown north of the cataract at Khartoum. Likewise, it was common in South Africa when the Hollanders first colonized that country but is now unknown south of the Orange River. There are, however, places in central Africa where as many as a hundred may on occasion still be seen at one time.

The bulk of a full-grown male hippopotamus can hardly be believed even when viewed at close range in a zoo. One wonders, when so observing the brute, how on earth its small stumpy legs, for all their girth, can support the body, yet the same beast in its native waters can overtake almost any man-powered small boat and even motorboats by paddling like a dog with these same ridiculous little limbs. The famous big-game photographer, Martin Johnson, recorded one of them rushing at the boat in which his camera was mounted at such a speed that it rose half out of the water, its navel plainly visible. Though of a mild disposition, the Hippopotamus is a rather fussy animal and adopts a highly proprietary attitude to its own chosen stretch of river. Normally, it moves out of the way of boats and then floats, just below the surface, with its periscope eyes protruding from the surface, observing the intruder, its small ears flickering constantly and vigorously. However, it may for reasons known only to hippopotamuses rush upon a luckless passing craft and either stamp it underwater or chew it up. And the bite of a hippo is a ghastly thing: the author has inspected the remains of a man bitten clean through the torso by one. On land, the animal is equally unpredictable and on occasion highly dangerous. Encountered at night when on its way back to water, it may not only charge but become a positive nuisance and most persistent in gaining the right of way; when on trek from one water-system to another, it often enters plantations and gardens and demolishes anything less than a fair-sized tree. What is more, it sometimes resents the protests of the human owners and goes mildly berserk and has been known to return hours later and demolish the out-houses of protesting citizenry. It can move much faster than its lumbering frame might suggest. There is usually one but sometimes two young at a birth and the babies are extremely small, being just the right size and always willing to be given a bath in a hand-basin.

These animals have always been almost a staple food of riverine and lacustrine tribesmen in Africa, being harpooned and speared to death from small canoes. Inconceivable as it may seem to us, one of the preferred ways for certain tribesmen to prepare this delicacy is to drag it ashore and leave it to ripen under the tropical sun for a few days and then open it up, plunge headlong into its interior, tear the softened carcase apart with their hands, and gorge themselves on the rotten flesh until so satiated that they cannot even walk home.

CAMELINES

The Camels and their very close relatives, the Llamas of South America, form another distinctive little group of herbivorous mammals that have extremities vaguely resembling hoofs. They are of most venerable origin. There is really nothing like them living today but they appear to merge with the tiny Chevrotains or creatures of that ilk near the dawn of the age of mammals. Although having complex stomachic apparatus and equally complex digestive procedures, they are not true ruminants in the understood meaning of that term. Whether this bizarre method of pre- and re-digesting food is of any real significance in the classification of animals is in itself open to considerable doubt and may indeed be nothing more than a form of mechanism developed by sundry kinds of animals to cope with sub-nutritious food in bulk. The Camelines, in fact, stand considerably apart from all the other hoofed mammals and, were it not for sundry forms known from fossilized remains which appear to link them in common ancestry with a vast group of which the only surviving remnants are the little Chevrotains, they might well be evicted from the Order of Artiodactyls.

At one time it was rather confidently believed that these animals evolved exclusively in the continent of North America and that the stages of this evolution were as fully extant as those pertaining to the Horses. Now, however, a host of other related creatures have come to light and the whole subject has become blurred. It still cannot be denied that there were in the past numerous cameline creatures in North America and that both the true Camels of the Old World and the Llamas of the New are descended from these. The end-products of this evolution are odd, to say the least, and happen to have proved of great value to man's economy on both sides of the world.

Camels (*Camelus*)

It is not certain whether there are any truly wild Camels left anywhere, though there are still large herds of the two-humped, long-furred species (Plate 149) to be found wandering about on the arid plains of central Asia. These are known to interbreed with animals of the same kind raised under pure domestication and they look in no wise different. Although there are also a considerable number of One-humped Camels wandering about unbranded and unclaimed in the deserts of northern Africa, Arabia, and the Middle East, nobody, and least of all the Arab peoples of those countries, believes that they are truly wild animals or descendants of such. In fact, the camels, though used by men for many centuries and perhaps millennia, have maintained a singularly stubborn attitude to true domestication and, despite their wide range of build and coloration, have managed to preserve their apparently inborn loathing for all other living things and particularly men and for all forms of labor. Given the sudden and complete

One-humped Camel

elimination of humans from the surface of the earth, camels would not pause for a moment in their measured tread and would doubtless continue to eat any dry scrub available on their own account instead of ours.

There are only two species of true Camels, one indigenous to and very numerous throughout Asia north of the Himalayan massif, with long hair and two humps; the other spread all over Africa today and being truly indigenous to the Near and Middle East, and having been introduced to India and Australia. It should be explained that the name Dromedary applies only to a particular type of the latter used for and specially trained to run, carrying a human rider. It is believed that the widely-splayed, soft-padded, cloven feet of both Camels were originally evolved to aid the animals in mushing through snow, and not for treading the loose sands of the desert. It is certain that, despite the relatively short thin coat of the One-humped species, these animals are as tolerant of cold as they are heat, while the Bactrian or Two-humped variety may be encountered plodding stolidly through north-Asiatic winter blizzards that would send an Eskimo scurrying to the shelter of his igloo.

Camels, of course, do not store water in their humps. Quite the contrary, these organs are fat-storage apparatus and were probably developed originally as body-heaters. The size, plumpness, and erectness of the hump or humps are an indication of the animal's state of health, recent food-supply, and general stamina. Those of the One-humped species should be long, full, tall and rigid; of the Bactrian bulbous and squashy to the point of mechanical collapse resulting from mere size. The water-storing of these animals is accomplished in their stomachs, which have numerous sac-shaped extensions in which liquids may be retained for a very long time. Nonetheless the classic cartoon showing one camel saying to another "I don't care what they say but I'm dying of thirst" is not too far from the truth—for these animals need water in large quantities and at regular intervals. Given liquid and a sufficiency of miserable,

prickly, dry scrub they can travel enormous distances, carrying huge loads, and sometimes at a tremendous pace. The famous Camel Corps established by the British in Egypt and marauding Arab tribesmen mounted on these animals have both descended on luckless communities with a speed that can hardly be believed, and the camel races staged in Morocco set records that make equestrian experts gasp.

Llamas (*Lama*)

There are also two camel-like animals indigenous to South America, one of which, the Guanaco, has given rise to two very distinct domestic or semi-artificial breeds. These animals are much smaller than true camels, lack humps, and appear to have originally been animals of the uplands. They have been of just as much use to men, and probably for just as long in South America, as the camels have in Asia. They are not only beasts of burden, carrying loads of over a hundred pounds at altitudes which no other animal can tolerate, but they provide wool for clothing, a dreary kind of milk, and good red meat as food. Just when they were first domesticated is not known but it must have been in the very earliest time of man's colonization of South America, for all manner of domestic breeds were on hand when the Spaniards reached the high Andes. These animals were then in current use by the hundreds of thousands, toting produce of all kinds back and forth all over the country, including ore from silver and gold mines. The Incas had neither money, wheels, nor an alphabet; yet they carried on a colossally wealthy economy by the use of these animals, had a road system for them that we have not yet surpassed, and sent messages over thousands of miles by a system called the *quipus* which originally consisted of tying complicated series of knots in the long fur depending from the undersides of these animals.

The original, and still wild Llama, is known as the Guanaco. It lives in considerable herds in all types of open country from the Alto Plano at fifteen thousand feet to the wind-blown prairies of Patagonia. It is a dull browser and rather shy. Apparently this is one animal that did resort to particular locations to die and the most extraordinary descriptions of this procedure have been given by none other than Charles Darwin and W. H. Hudson. From this animal the two domesticated creatures known as Llamas (Plate 151) and Alpacas (Plate 150) have been derived. The former comes in all manner of colors and color-patterns, and is a tall-necked, shaggy beast that spits at those who molest it, often using gravel as ammunition. From it wool is taken and its flesh is eaten. It is also the classic beast of burden. The Alpaca is a most appalling-looking entity, somewhat resembling a partly animated couch covered by a thick woolly blanket from which a cameline neck and head rises. It comes in a wide range of colors from bright red to black, white, and various browns. The wool is very long and matted.

The Vicuña (*Vicugna*)

Whereas the two camels are sufficiently alike to be placed in the same genus, there is in the chill uplands of

Vicuña

the central Andes at altitudes of from twelve to eighteen thousand feet, a small kind of llama so different in structure and habits as to warrant its separation from the Guanacos. This is called the Vicuña and was the royal animal of the Incas. It wanders about in small family parties on the alto plano and has never been successfully domesticated though always hunted for the exceptionally fine fleece that falls in a pendent bundle from the base of the neck of the males at one season of the year. The whole pelt is also exceptionally soft and silky and both fleece and hide were originally reserved for the Incan rulers and their families and entourage. After the coming of the Europeans the animal was nearly exterminated but is now protected by the Peruvian government.

TRAGULINES

The most obscure hoofed animals, known as Chevrotains or most misleadingly as Mouse-Deer, are but the remnants of a once very numerous, diverse, and ancient group of ruminating animals that go back to the dawn of the age of mammals and almost to the basic stock from which the majority of the Even-toed Hoofed Mammals sprang. Although somewhat deerlike in external form, the living members of the group are anatomically more like pigs on the one hand and camels on the other; nevertheless their ultimate ancestors seem also to have been the ancestors of the cervines, giraffines, bovines, antelopines, and caprines. Today, there are two kinds, one found in India, Ceylon, Malaya, and the Indonesian islands, the other in West and central Africa.

Oriental Chevrotains (Tragulus)

There are quite a number of species of these animals distributed from India to the Philippines, but most are no more than island races of two main super species. The most notable is the Indian (*T. meminna*), which is of an olive-agouti color, with lines of white spots along the flanks, three white stripes on its throat, and plain white undersides. It is about the size of a hare and minces about on the tips of its toes. The males have small tusks in the upper jaw but no upper front teeth. They inhabit tangled undergrowth and are nocturnal. The members of the other species, found farther to the east, are of various shades of brown without spots but with sundry arrangements of white bands on their throats sometimes augmented by large white chin and sub-ear spots. They have a median brown band down the light underside. In Malaya there are two distinct forms, one larger than the other, with longer and more pointed hoofs, that lives in damp lowlands. Two young are born at a time and follow the mother about for a considerable period, but for the most part they are solitary animals leading the lives of rodents.

Water-Chevrotains (Hyemoschus)

The African animal is known as the Water-Chevrotain and is represented by three distinct species, one in the western block of rain forest from Senegal to Dahomey, the second in the Cameroons forests, and the third in the Ituri forests of Central Africa. They are larger animals, dark brown above and white below but with

Water Chevrotain

confluent lines of vivid white or cream spots above and whole white lines below, along the flanks. They are semi-aquatic animals, never straying far from rivers, feeding on succulent aquatic plants and fallen fruits, and diving into the water to escape danger. They are extremely shy and very speedy animals that can outrun anything both in the open and in the densest undergrowth. They rest in holes by day if any are at hand.

CERVINES

The deer are in point of species and of kinds a very numerous group and they are still by far the commonest larger wild animals almost throughout the world. Despite constant slaughter by men for their meat and hides, they are still numerous enough to be hunted for sport and in some places for their antlers alone, which are believed by some people to have medicinal qualities. Modern deer, with short bony bosses on the foreheads and long deciduous antlers growing seasonally from these, are a comparatively recent development. They appear to have evolved from creatures much like the living Muntjacs and these in turn from animals like the little Musk Deer. Further back still, these merge with the primitive little Tragulines.

There are eight major groups of living Deer, two of which comprise well over ninety percent of all the known forms—these are the True Deer of Eurasia and North America, and the Hollow-Toothed Deer of North and South America. Deer occur indigenously in the Mediterranean zone of North Africa but are otherwise unknown in that continent. They have been introduced into New Zealand and Australia.

MUSK-DEER (Moschinae)

These are small deerlike animals with a short compact body, thick sturdy legs and very large hoofs. The males have long sharp tusks outhanging the lower jaw and directed backwards. The fur is long and thick, and the hairs are thick and spongy. They are a dull grey-brown above with a mixture of diffuse red and creamy spots. There are no antlers and no tail. They are mountain animals and were once found all over eastern Asia from the

Musk Deer

Sea of Okhotsk through Amuria, and Korea, Manchuria, inner China, and eastern Tibet to the Himalayas. However, they have a large gland in the skin of the belly called the *pod* which secretes a powerful aromatic substance that is the third-best fixative used for perfumes and on this account they have been and still are un-

mercifully hunted everywhere and have become extremely rare in many places. They have many goatlike qualities, depending on flight over the roughest ground as the best method of defense.

MUNTJACS (*Muntiacinae*)

There are a number of closely related, small, primitive deer (Plate 152) belonging to this sub-family that inhabit the Oriental Region, being found in southern India, Burma, southern China, on the island of Java, and possibly in isolated parts of the Indo-Chinese area. The body is of deerlike form but rises to a high-domed rump. The tail is short, the feet delicate, and the head very pointed. Males carry tusks that protrude below the chin. They are of various shades of brown, ranging from reddish to greyish or almost yellow. They are distinguished from all the modern deer by having long, bony pedicles arising from two bony ridges that extend up the forehead. These point upwards and backwards and the true antlers grow from the top of them. The antlers are short and straight and have only a short brow-tine. Muntjacs are forest animals and are very shy and retiring.

TRUE DEER (*Cervinae*)

We now come to an almost overwhelming aggregation of very similar animals spread all over Eurasia, North Africa, and the East Indian Islands. To these, well over a hundred different popular names have been given in various localities, yet the vast majority belong to only about a dozen valid species. Of these, three are sufficiently singular to be treated separately.

Père David's Deer (*Elaphurus*)

This is the most exceptional of True Deer, being unknown in the wild state and having been discovered by a French missionary in 1865 as a semi-domesticated herd kept in a hunting park by the Chinese Emperor outside Pekin. After much undercover connivance and some bribery, a few specimens were obtained and taken to Europe and began to breed in England. This was fortunate because all the remainder of the herd were slaughtered in the rebellions in China in 1900. The initial stock of twenty animals has now increased to some three hundred, and breeding pairs are located in several countries. They are medium-sized deer, with very odd, irregularly branching antlers, all the tines of which are more or less straight and go off at an angle of about forty-five degrees. They are reddish-grey above, white below and have light areas around the eyes, and light stockings.

Fallow Deer (*Dama*)

These are very beautiful small deer (Plate 156), the western form of which has for centuries been kept in parks all over Europe for purely ornamental purposes. This species appears originally to have inhabited the Mediterranean area. It is of a rich orange-reddish-brown color profusely spotted with white, and white below. The antlers are large and widely spread, with numerous tines, those at the top of the main tines developing into hand-shaped or palmate structures. The eastern species is much larger, has smaller and less

Père David's Deer

branched antlers that are not palmated, is very brightly colored and profusely spotted. It is found from Asia Minor through Iraq to Persia.

Axis Deer (Axis)

These are two closely related deer found in central and south India and Ceylon, and north India, Assam, Burma, Thailand, and Indo-China respectively. The first, known as the Chital (Plate 154), is a medium-sized deer, with large, widely-held but simply-branched antlers, and is of a rich reddish color with profuse white spots at all ages and in all seasons. The undersides are white. The eastern animal, called the Hog-Deer, is small, with antlers that sprout a brow tine and then fork but once. They are pale brown with white spots in summer but of a brindled dark brown in winter.

Red Deer (Cervus)

All the remaining True Deer belong to a single genus and the dozens of races to which all manner of names have been given compose only about a dozen species. It may seem odd to devote so little space to this mass of beautiful, outstanding, and well-known animals but despite their popularity, all deer are very much alike in appearance and habits. They are browsers and almost all inhabit woodlands. Some migrate seasonally, going up to mountain pastures in summer and descending to the valleys in winter. The males alone grow antlers, starting in the late spring and having them fully developed by fall, when the velvety skin covering is scraped off. The antlers then fall off in the early spring. The young, which are usually single but may be twins, are almost always spotted and can totter about as soon as they are dropped. Deer make pretty pets but are rather dangerous when the mating season comes around, and a lot of unsuspecting owners have been mauled and even killed by them. All deer are regarded as legitimate game by so-called sportsmen but, except in the wildest country, modern rifles have reduced this pastime to little more than an outlet for one of Man's drearier instincts. However, if not killed in considerable numbers and selectively every year, deer may nonetheless become so numerous in some areas that they eat out all the available food and starve out their own populations.

The first species, C. elaphus or the real Red Deer (Plate 155), are spread from the British Isles throughout Europe to eastern Siberia—where they are misleadingly called Wapiti—along North Africa, throughout the Near and Middle East to Afghanistan and then north of the Himalayas to Tien-Shan. These include the Maral of the Caspian Region, the Hangul of Kashmir, and the Shou of Tibet. The American species (C. canadensis) used to be found all over the northern continent but are now confined to the Rocky Mountain region. There are at least three species and these are called Elks (Plate 153), which is misleading, since that has always been the original popular name of the Moose in the Old World. The third large species (C. unicolor) is the Sambar of India, Malaya, Indo-China, Formosa and the Philippines. There are several varieties.

Thorold's Deer (C. albirostris) is a rare species from western China and eastern Tibet that has a white muzzle, lips, and chest. The antlers too are of a pale-horny to white color. The Barashinga, or Swamp Deer, of India and Assam (C. duvauceli) is sprinkled with white; the Thamin (C. eldi) is from the Indo-Chinese area. In eastern Siberia, Manchuria, Amuria, north China, Korea, the Japanese Islands, and Formosa there occur a number of small deer known as Sikas (C. nippon) that were regarded as sacred by the Japanese and are now very popular in European and American parks. In the islands of Sumatra, Java, the Celebes, the Moluccas and on other smaller islands are found the Rusas (C. timoriensis), which appear to be small forms of the Sambars, and are of a light brown color, usually yellowish below and with rather light horns. Finally, there is a spotted Red Deer in the Philippines (C. alfredi).

HOLLOW-TOOTHED DEER (Odocoileinae)

The other great group of Deer are exclusively American and were probably developed in the Western Hemisphere. There are six distinct genera and a large number of described species. One genus is found in North America, with some species extending through Central to northern South America. The other five genera are exclusively South American. They are distinguishable from the True Deer by a number of anatomical features and by their generally smaller size, lighter build and, on the whole, differently constructed antlers, which go first backwards and then turn forwards with all the tines pointing forwards. In habits they are also all very much alike but they are rather more selective in habitat than are the True Deer.

White-tailed Deer (Odocoileus)

This term may be used in either a general or a restricted sense. There are over a dozen species all having some white plumes on the underside of the tail—like almost all Cervines, for that matter—but one group of species has come to be called by this name popularly, and this happens to be the common deer of Canada and the greater part of the United States as well as Mexico. The appearance of this is too well known to Americans —especially in its immature form, to require description. The animal varies (Plate 160)—in size, color, and seasonal change, while the antlers of all are of course as variable as those of any deer. The largest races live in Canada, and the smallest, which only weigh about fifty pounds, on a few cuays off the coast of Florida. There are species scattered throughout Central America wherever there are "pine barrens," and some outliers of the genus are found as far south as the Guianas and inner Venezuela and Colombia. A distinct species called Mule Deer (Plates 157 and 161) inhabit the western part of the United States from Alaska to Mexico. A race of this deer is known misleadingly as the Black-tailed Deer. These are inhabitants of rocky areas, mountains, and deserts. All White-tailed Deer have very large ears and pronouncedly plumed tails which they use as signaling devices.

Marsh Deer (Blastocerus)

These, largest of the Hollow-toothed Deer, are found all over the Amazonian basin from the Orinoco and

Pampas Deer

Guiana watersheds to the borders of the Argentine. They are lustrous red in color but white below and have black stockings. Their antlers are large, many-tined, and widely set almost like the True Deer. They inhabit the denser wet lowlands and the clear floor of the closed-canopy virgin forests. Their ears are enormous.

The Pampas Deer (Ozotoceros)

On the southern savannahs and pampas of southern Brazil, Paraguay, Uruguay, and the Argentine, there is a small reddish deer with simple antlers composed of a main tine that forks at the tip and a brow tine. It has white spectacles, and the lips, chin, throat, chest, underside, and the insides of the legs are pure white. It has a signal tail, black above and white below. It is the most delicate and fragile looking of all deer and is exceptionally speedy and a terrific jumper, careening off when frightened by leaping clear over the eight- to ten-foot pampas grass-clumps.

Guemals (Hippocamelus)

The Latin name means literally "Horse-camels," which is of course quite absurd. These are merely somewhat specialized small deer that inhabit the high Andean ranges from Ecuador, through Peru and Bolivia to Chile and the Argentine. They are rather heavy-bodied, coarse-haired deer of a peculiar pepper-and-salt coloration and with virtually no tails at all. Their antlers are simple and each one forms a fore-and-aft "Y" with a long main and short brow tine.

Brockets (Mazama)

These small, delicate deer with simple, one-tined straight antlers take the place in South America of the Duikers of Africa (see page 260). There are many possible species but they are all much alike and all vary greatly in any one area. They are of a rich reddish-brown color above and lighter to white below. The short but fluffy tail is colored like the back above but is white below and is very conspicuous when held upright in the gloom of the forests. Their headquarters is the Amazon Basin but, surprisingly, one species is found also in the northwest Pacific forests of Colombia. While most South American deer feed at dawn and dusk, these little animals are wholly nocturnal and sleep in holes or arbors under fallen trees by day.

Pudus (Pudua)

These are the tiniest of all deer, adult "stags" of the largest species being about the size of a small terrier and the females being much smaller still. They come from the temperate forest zone of the Andes and range all the way from Colombia, where that zone is at a great height, to the southern tip of Chile, where it is at sea-level. They also occur on many of the islands off the coast of Chile. They are rather thick-furred and have sturdy little legs; their heads are small and the antlers

[252]

Pudu

mere spikes. In color they range from reddish-brown to pale grey. Several generations of these tiny mites were bred in a Paris apartment and were treated exactly like domestic dogs, which most people who saw them for the first time thought they were.

MOOSE (*Alcinae*)

We now turn from the smallest to the largest of the deer and to one of the largest of all Hoofed Mammals, the vast creatures of the northern hemisphere known to Americans as Moose but to everybody else as Elk. There are two kinds, or species, each divisible into a number of recognizable races that are separated geographically. One inhabits Eurasia, the other boreal North America. They are exceptionally long-legged animals with pronouncedly humped shoulders and a slight crest on the neck. The muzzle is long and the nose wide, fleshy and dependent, almost forming a trunk. Under the throat there is a fleshy dumbbell-shaped structure called the bell, covered in long fur. The antlers of the males are set at right angles to the forehead and are of enormous size. Those of the Siberian race are simply branched and curve backwards, but in the European form the main tine is palmated to a greater or lesser extent. In the American species, the antlers are always palmated and in the Alaskan form exaggeratedly so, forming two huge basin-like multi-fingered scoops. The record for all Moose antlers is over six feet seven inches across. Moose are found all across Eurasia from Scandinavia to Okhotsk; the American species range all over Canada with one race penetrating the U.S. via Nova Scotia, Maine, New Hampshire and Vermont, another via the Rockies to Montana and south to Idaho and California. The greatest specimens come from Alaska.

The Moose is a tremendously powerful beast and fighting is a recognized procedure in its life, with bulls looking for adversaries as soon as their horns are wiped clean of velvet in the fall. If they are molested by other animals at that time they may launch an attack, swiping with their horns, and pounding downward with their forefeet. They have been reported to have fought off Black and even Grizzly Bears by this means. They shed their antlers in early winter and start to grow them again in early spring. They are browsers, munch on leaves and bark, and are inordinately fond of water-plants—to obtain which they completely submerge. They are strong swimmers and have been encountered crossing wide arms of the sea. In winter they bog down in deep snow and sometimes resort to making compounds, many together, by trampling down the snow and piling it up around the edges into a bank, forming what is called a yard. This they inhabit while the food supply lasts; then they trek off to make another in a suitable location.

REINDEER (*Rangiferinae*)

There is always a great debate between Americans and Europeans as to the proper naming of these animals. The former resent the magnificent deer of their country being classed with, called, and made subsidiary to the rather paltry and very scraggly-looking little animal of Eurasia known for centuries by that name. None-theless, all the Caribou are nothing but large Reindeer, and the little moth-eaten creature from the northern parts of Europe and Asia is really a very remarkable beast—in fact, one of the most incredible of all mammals. All reindeer in the wider sense have horizontal backs, short tails, enormous hoofs that clack together when they run, and a pronounced dewlap depending from their throats. This is larger in the males than in the females. Their antlers—which are uniquely present in both sexes—are quite impossible to describe either in general or in the particular. They are wide-spread, slant back over the neck and then bend forwards, outwards and finally inwards again, but they meander and are bilaterally dissimilar, with numerous tines branching off at odd angles and often forming palmate structures. The brow tines are very pronounced and one—usually the left—is palmate and vertical. This grows down upon the muzzle. The complete rack is sometimes, and especially in the Siberian race, positively grotesque. The fur is thick and there is a dense undercoat.

Eurasian Reindeer (*Rangifer tarandus*)

Although there are herds of these animals (Plate 158) running wild on some of the least accessible peninsulas of northern Russia and Siberia, it is improbable that any of them are descendants of a pure line of entirely wild animals, for all the peoples of the Eurasian Arctic have used these animals in a state of semi-domestication since time immemorial. Most outstanding as reindeer people are the Lapps of northern Europe who until recently relied almost wholly on these animals for food, clothing and many other things. These small deer, only about six feet in total length and standing three and a half feet at the shoulder, can be ridden for days on end by a man of average weight and can pull twice their own weight on a sled over snow for forty-eight hours at a stretch. One pulled two men in a sled for sixteen hours at an average rate of 18 m.p.h. and another made an historic run of just eight hundred miles in forty-nine hours, pulling a Norwegian officer bearing a warning to his King of an uprising: it dropped dead on arrival. The animal has meat of very high quality, its skin makes fine leather and can be used with the fur on to make clothing or blankets for sleds, and the wool alone can be woven. The sinews make nylon-strong cordage and many small objects can be made from the antlers. Finally, its milk is so rich it has to be mixed with three times its volume of water before it can be drunk and from it cheese, whey, and an alcoholic beverage can be made. Yet, these animals feed mostly on mosses, lichens, and a certain amount of grass. The stunted vegetation of the arctic is, however, exceptionally rich in minerals and desirable biotic compounds.

Caribous (*Rangifer arcticus, etc.*)

Neither the Eskimos nor the Amerindians domesticated the reindeer of North America but both made great use of the animals and some tribes were almost wholly dependent upon them. There are three species-complexes of Caribou—the Barren Ground, the Woodland, and the Mountain. The first comes in a surprising

OTT FROM O.P.L.

Caribou

any kind but, in the males, having long tusklike upper canine teeth. They are marsh-dwellers and semi-aquatic. The fur is long, coarse, of a yellowish-brown color on the flanks but lighter on the back and white on the undersides. There is a distinct crest on the back of the head and the tail is very short but well furred. The most peculiar thing about them is that, like pigs, they have regular litters of young and as many as six or even seven young may be born at a time.

ROE DEER (Capreolinae)

The last sub-family of the deer, and one of the most distinctive, happens to be the best known of all deer to Europeans and to most peoples dwelling in Asia north of the Himalayan–Tibetan barrier, for it is the common small deer of the Eurasian temperate woodlands. It is known as the Roe Deer (Plate 159) and has been hunted since time immemorial, yet it still abounds in woods between the most highly industrialized areas of Europe. It is a small, rather compact deer with small upright antlers that lack a brow tine but start forking near the tip of the main stem. The antlers are usually knobbly about the base and this may extend almost to the tip of all the tines. Odd specimens are sometimes shot in which the horns form a partially combined mass, looking like brown coral or certain kinds of stalagmites. In summer, most of these animals are of a rich reddish, brown with an orange tone on the flanks and white below. In winter they go brown, with a yellowish grizzling and a large white patch developing on the rump. These small deer are found all across Eurasia from the British Isles to eastern Siberia and south to the Caucasus, thence north of the central Asiatic desert belt to inner China, Mongolia, Manchuria, Amuria and Korea.

SUSCHITZKY

number of fairly distinct shapes, sizes, and colors. The typical form is grey-brown above, creamy white below, with dark brown muzzle and legs. It is found north of the tree-line in Canada from Hudson's Bay west to the Mackenzie River. Distinct races are found on the great islands to the north, on the Labrador peninsula, on Greenland, in Alaska, and on Unimak Island. These animals migrate annually in a rather complex manner, the sexes splitting up at one period in the summer and all except the pregnant females making a subsidiary trek to the south in midsummer, before the whole group starts back to the forest's edge for the winter. The Woodland Caribou is a much larger, less shaggy deer, dark brown in color, with a light muzzle and heavily palmated antlers. It inhabits the evergreen coniferous forests all the way across Canada from Newfoundland, where there is a distinct sub-species, to the Rockies. The Mountain Caribou is the largest of all species, standing upright like a stag and being a very dark brown in color. It inhabits British Columbia and Alberta.

WATER-DEER (Hydropotinae)

In the lowlands of Korea and China there are to be found some small deerlike animals lacking antlers of

Chinese Water-Deer

GIRAFFINES

These, the most obvious of all mammals in almost all senses of that word, are one of the most difficult to treat from a descriptive point of view. There are two very different kinds of Giraffines remaining today, the Okapi of the forests and the Giraffes of the savannahs. In the past there were all manner of grotesquely formed animals of this group, some of which had quite preposterous horns. These animals are somehow linked to the deer and to the little Tragulines, far back in the mists of time but they have specialized in one particular way that has culminated in the towering Giraffes of today. As a result, their anatomy is extremely odd and it is still not entirely clear how their essential physiological processes are maintained, for the body goes almost straight up and down, so that blood circulation has literally to battle gravity and the nervous system has to contend with as much as a thirty-six foot circuit.

OKAPIS (*Palaeotraginae*)

Although a relative of the Giraffe, the strange animal known to science as *Okapia johnstoni* is so very different in habitat, habits, appearance, and immediate ancestry as to require its separation in at least a sub-family of its own. The animal is thought-provoking in many ways but most of all because, despite its size and rather startling appearance, it remained undiscovered until this century. That an animal the size of a large ox and purple in color could hide in very fair numbers even in what was then "darkest" Africa, not only points up the fact that the earth's surface has not really been known until very recently but also stands as a warning to any who would state categorically that such creatures as Snowmen, Marsupial Tigers, and sundry marine animals of monstrous size do not and cannot exist. The animal had been rumored for some years but when Sir Harry Johnston brought a part of a skin, showing the vivid back and leg stripes, to England in 1901 it was thought that it came from some form of forest zebra. It was some years before complete specimens were obtained. Now, these animals are displayed in several large zoos. The males have small, fur-covered horns and both sexes have enormous ears. They are browsers and keep to the recesses of the forests where they drift about in family parties. In color, they are maroon with a distinctly purplish tinge but are striped black and white on the limbs and over the rump, and have light facial markings.

GIRAFFES (*Giraffinae*)

All living giraffes are sufficiently alike to be grouped in one species; however, there are five very distinct kinds, each of which would undoubtedly be called a species if their individual members did not vary so much in color pattern and if some of the forms did not actually intergrade. Detailed description of the sundry forms is of particular interest only to zoologists and the chosen few who can take hunting trips to Africa. (Exact descriptions may be found in T. Donald Carter's *Hoofed Mammals of the World*.) All Giraffes have a pair of small fur-covered horns and in most there is a third

Okapi

bumplike protuberance placed medially in front of these on the forehead. In the South African Giraffe (Plate 164) it is absent. However, one form, the Baringo Giraffe from the borders of Uganda and Kenya has, in addition, another small pair placed between these three.

In color all giraffes are basically white upon which a complex pattern in various shades of brown is depicted much as a pattern is spread on a wallpaper. In those forms found north of the equator this does not extend to the legs; in the two forms found on the equator—the Congo and Reticulated—it does so sparingly down to the hocks; in the forms found south of the equator, the legs are fully marked to the hoofs. The pattern likewise varies from north to south and takes sundry strange forms. Fundamentally, it is composed of a reticulation of light lines forming a crocodile-hide arrangement, all the spaces between being filled with brown. The light reticulation, however, varies in width, and the brown may merge with it, so that in some types the animal is white with buff spots, each centered by a dark brown blob. Further, the brown blobs may take a leaflike or starlike form, as seen best in the Masai (Plates 162 and 163) and Rhodesian Giraffes.

Giraffes are spread all over tropical Africa from the east Sudan to South Africa and up the west coast to northern Angola. There is an outlier race in northern Nigeria. They strictly avoid the closed-canopy forest and stay on the drier savannahs and orchard bush where they browse on the uppermost leafage of the thorny, flat-topped acacia trees, some fifteen to twenty feet above ground. They live in loosely-knit communities, family parties keeping close together. They are mild-mannered giants, relying on their keen sight to keep out of danger and their speed to remove themselves from it. Although their gait appears awkward in the extreme, they can cover ground at a great rate in a long loping gallop with enormous strides. Males will do battle and use their heads and long necks as clubs, whacking at and sometimes stunning each other. Both mothers and fathers will defend their young against the

[255]

great cats with determination and there is an authentic record of one having kicked the head clean off a lioness' shoulders. They kick forwards with the front feet and with tremendous power, having a foot-wide hoof and an eight-foot shaft of solid bone and sinew behind this. Giraffes are not absolutely voiceless as popularly supposed: they have been heard to make a rather tragic gurgling whimper, and mothers make a whistling sound to call their young. Giraffes drink copiously but can go almost as long without water as can any camel. They cannot swim and have never been seen to wade even a small river.

ANTILOCAPRINES

Before the colonization of North America by Europeans, untold herds of hoofed mammals roamed the prairies of the central and western part of the continent from the southern boreal forest tree-line in what is now Saskatchewan and Alberta to the central plateau of Mexico. The most numerous of these were the mighty Bison; the next most numerous a rather delicate but somewhat ugly-shaped animal with laterally compressed upright horns bearing a small tine or point near the top at the front, known as the Pronghorn, or American Antelope (Plate 165). Nothing at all like these animals is known anywhere; they are a solitary leftover from pre-glacial times, when their tribe was much more varied. Nothing is known of the origin of these animals as a whole. They appear to be of exclusively North American origin, to have evolved there, and never to have spread to other continents. In a manner of speaking, they are a sort of minor experiment in "antelopes," initiated by Nature and then dropped. They undoubtedly have closer connections with the Bovine than the Traguline-Cervine-Giraffine group.

They are about the size of an average goat, have deerlike feet, rather sheep-shaped heads, erect ears placed on top of the head, and short tails. In color they are sandy brown, with light throat bands, and are white below. The rump has a large, circular, glistening white signaling patch composed of a huge rosette of long hairs that can be spread at will by means of special muscles immediately under the skin. All the pelage is rather hard and coarse and the individual hairs are pithy. The horns are the most exceptional feature of these animals. Although branched, they are hollow sheaths growing on a permanent bony core like those of oxen and antelopes. However, most surprisingly, they are shed annually like those of deer though by a different process. Instead of breaking off clean at the base, a skin covered with velvety hair grows out of the bony core, but inside the horny covering, and forms a second horny sheath underneath the original skin. As this thickens, the old outer horn splits and is burst off by the actively growing "velvet" inside. The whole process is complex and hardly seems worthwhile in view of the fact that the horns appear to be mere adornment, those of the females being only tiny unforked spikes.

Pronghorns are very fleet animals but bad jumpers and are perfectly adapted to life on level open prairies. They were almost exterminated at the turn of the century but have now made a strong comeback on reservations. They are, however, very intolerant of confinement.

BOVINES

All the remaining hoofed mammals with an even number of toes have until recently been lumped together in one vast family named the *Bovidae,* or Oxen. This procedure is manifestly unreasonable and one which not only defeats the ends of classification *per se,* but also ignores many obvious anatomical facts and disregards popular concepts which, although they so often prove to be untenable in face of proper scientific study, still warrant some consideration. The animals in question comprise at least two hundred distinct species among which are such well-known creatures as the cow, the sheep, the goat, and all those popularly known as "antelopes" and "gazelles." They fall into three great natural groups— the Bovine or Oxlike, the Antelopine, including the Gazelles, and the Caprine, or Goatlike. We take first the Bovine or oxlike beasts and include with them not only the Nilghai of India but also its close relative, the tiny Chousingha or Four-horned Antelope, the forest-dwelling Duikers and the oxen of Africa that are of Antelopine shape and have spirally twisted horns.

TRUE OXEN (*Bovinae*)

The origin of our domestic breeds of cattle is a subject that has just as much right to constitute a separate science as does the study of Man's origins. There is no doubt that a large species of ox, known until the end of the Dark Ages in Europe and the Near East as the Aurochs, played much if not the sole part in the ancestry of European cattle. However, there may have been other smaller species of wild oxen involved in this that became extinct even earlier. In tropical Africa and in Asia there were undoubtedly wild species that were early domesticated or at least herded and which contributed to the establishment of the huge-horned Ankole (Plate 168) and the humpbacked, dewlapped Zebu (Plate 166) respectively. In fact, these animals may be intrinsic species all members of which were long ago corralled by Man. There are, however, still several truly wild species of cattle distributed over tropical Asia and Africa.

Cattle (*Bos*)

There are four very different kinds of wild cattle known—the Gaur, Banteng, Couprey, and Yak. The largest and most oxlike is the Gaur (*Bos gaurus*), an enormous creature, the bulls standing six feet tall at the shoulder. These animals are dark brown in color and have white "stockings." There is an elevated ridge along the back from the shoulders to the rump and a small dewlap under the neck. The horns sweep backwards and ultimately turn inwards. They are found from central India, through Assam and Burma to Thailand, Indo-China and all down the Malay Peninsula. They are forest animals and prefer hilly country, where they roam about in small herds with one or two bulls in charge, but they will descend into the lowlands and invade cultivated land to obtain fresh grass. They are shy animals but lone bulls can be very objectionable and, like all cattle, they have to be treated with respect.

There is either another closely related animal or a semi-domesticated variety in eastern Assam and northern Burma known as the Gayal or Mithan (*B. frontalis*) (Plate 169), of slightly smaller size and with horizontal horns that do not recurve at the tip. Although they run quite free, all these creatures appear to be owned by tribesmen, who milk and tend them like cows.

The most astonishing species is the Coupray (*B. sauveli*), which is as big but not quite so heavy as the Gaur and which was not really discovered until 1936. This enormous creature turned up in central Cambodia and so stunned zoologists that they at first offered all kinds of useless explanations of its occurrence, including the notion that it might be a hybrid though no suggestion as to between what two animals was offered. It is greyish to black in color, with light grey stockings, has a very long tail with a pronounced terminal plume, and both sexes have an exaggerated dewlap depending from the neck. The horns of the females look like those of ordinary cows but those of the males set outwards from the crown, then bend straight forwards, and finally turn inwards, upwards, and backwards. Most bizzare of all, just about the final turn they are adorned with a cuff of hairlike, shredded horn-fibers pointing forwards. Nothing like this structure is known in any other mammal. Unfortunately little else is known of these animals.

The Yak (*B. grunniens*) (Plate 167) is an exaggerated kind of ox that is indigenous to the uplands of Tibet and used to roam in enormous herds all the way from Kashmir in the west to China in the east but which has now, in the wild form, been hunted back to a few isolated areas in the eastern part of its range. Nevertheless, this animal has been domesticated in Tibet for centuries and produced there a smaller, more docile breed which, in its extreme form, is excessively long-haired and lacks horns. Yaks are used as beasts of burden and for riding and are herded just like other domestic cattle, being milked regularly and slaughtered for their excellent meat. Yak butter is exceptionally rich and is used in tea by the Tibetans as we use milk. It is interesting to note that these people also developed a method of communication by tying knots in the long fur hanging from the bellies of these animals just as the Incas did with their Llamas. Domestic yaks come in all manner of colors from white to black, brown, reddish, and parti-colored. The wild animals are of a pale greyish brown and almost white in winter. There is a pronounced shoulder hump on all and the fur hangs down to the ankles.

The last true wild species of cattle is known as the Banteng, Banting, or Bantin (*B. banteng*), and is very like a buffalo. It originally inhabited southern Burma, Malaya, Borneo, Sumatra, Java, Bali, and some associated smaller islands but is today greatly restricted in range. It is about the size of the large domestic cattle but has buffalo-shaped horns joined together at the base over the front of the head by a horny casque. They are of a reddish brown color, have white stockings, a very slight shoulder ridge and only a suggestion of a

Gaur

dewlap. They are forest animals and are rather obstreperous though more inquisitive than aggressive.

Buffalo (*Bubalus, Syncerus,* and *Anoa*)

These three kinds of animals should probably be given separate status since they are markedly different and not nearly so closely related as are the True Cattle. Nevertheless, they are, with the possible exception of the little Anoas, clearly recognizable to everyone and are universally known as buffalo. The best known is the Water Buffalo (*Bubalus bubalis*) of India and the Oriental Region generally. It is certain that there still are truly wild stocks of this animal, notably in Borneo, and possibly in the Celebes, Sumatra and Java, but there is considerable doubt as to whether any in India, Burma, Thailand, Indo-China or Malaya are other than feral. The animal has been domesticated for so many centuries and has been so widely disseminated throughout the tropics—even to the United States—and it is kept in such loose confinement in many countries that it is hard to say which herds are truly wild. It is a very large, dark grey, semi-aquatic ox with a small shoulder hump and a dewlap between the front legs; it has huge back-sweeping horns with transverse ridges. The head is large and held low but the neck is relatively slender. They are patient beasts that can be kicked on the snout by small boys but are deadly to strangers and particularly "for-

Indian Water-Buffalo

eigners." Many a tourist has been killed by these animals when attempting to photograph them at their daily tasks and there are countless tigers that have succumbed, presumably in stunned amazement, to a swipe from their horns. Their strength is amazing and a pair can pull three combined teams of trained draft-oxen backwards in a tug-of-war.

The Buffaloes of Africa (*Syncerus*) comprise two species, one a forest animal (*S. nanus*) which rather distressingly has a form living outside the forest, and a larger animal (Plate 170) commonly known as the Cape Buffalo (*S. caffer*), found exclusively outside the closed-forest, from the southern Sudan, all down the eastern side of Africa to South Africa. This is a very large, powerful and aggressive ox with heavy-based horns meeting on the mid-brow and then sweeping outwards, backwards, and finally forwards at the tips. It prefers moist and swampy ground but may suddenly appear in dry areas when suitable grass is available. It travels in herds and should be treated with great respect as it will mount a vicious attack, is clever, and extremely agile. It gores with its horns and then kneels on its victims but it also has another unpleasant weapon in its tongue, which is like a very sharp wood-rasp and can tear skin and flesh.

The forest buffaloes are smaller and come in a variety of shapes and colors, the best known of which is the wicked little Bush-Cow of West and west central Africa. This is of a rich reddish color and has smaller, more upright horns and large, drooping, hairy-rimmed ears. These animals also move about in herds and are adepts at concealment and at keeping quiet. One may pass right through a herd without realizing until later that they were there; and then again they may resent the intrusion and launch a sudden attack. Moreover, if they are fired upon and one of their number is wounded, they have been known to move off, keep quiet, and then stalk the hunter for hours and attack again and again from ambush. They are, of course, browsers rather than grazers though they will eat grass if it is available.

Outside the forests, in northern Nigeria and to the east thereof as far as the Lake Chad region, is found a large, dark grey-brown buffalo which has more widespreading horns than the Bush-Cow but which is never black in color nor has horns that curve downwards like the Cape Buffalo. It is classed as a race of the forest animal and inhabits reedy bottoms and is rather shy.

On the Island of Mindoro in the Philippines and on the Celebes are found separate species of the smallest of the buffaloes, known respectively as the Tamarou (*Anoa mindorensis*) and the Anoa (*A. depressicornis*). They are small oxen, the one standing about four and the other about three and a half feet tall at the shoulders. The Tamarou is dark grey with a white throat-diamond; the Anoa is dark brown. The horns of the former are buffaline but grow backwards in the plane of the face, are ridged, and slightly incurved at the tips; those of the Anoa are triangular in section at the base and almost straight. Although both animals, and notably the Anoa, are really very common in zoos, practically nothing is known about their habits in the wild. They live in dense undergrowth in forested areas—the Tamarou on the lowlands, the Anoa in mountainous districts.

Bison (*Bison*)

Very closely related to the wild Cattle, and able to interbreed with them, are the Bisons of Eurasia and North America. These two are quite distinct species. The first, known as the Wisent, which in prehistoric times roamed all over Europe and apparently what is now Russia west of the Urals, is today probably and finally extinct in the truly wild condition, having already been reduced before World War II to a single small population in Lithuania. A race of this animal lived in the Caucasus until 1928. At the end of the last war there were only about a hundred pure-bred Wisent left, including those in zoos. It is a large oxlike animal, sloping rapidly from the shoulders to the rump, having a cape of longer woolly fur, and fairly pronounced oxlike horns. It has a slight beard. It is essentially a forest animal and preferably of the coniferous forests and it both browses and grazes.

The American Bison (Plate 171), miscalled "Buffalo," is a larger and heavier animal with a more extensive cape, shorter horns, a pronounced top-knot, and a long woolly beard. The front legs of the bulls are furred as though wearing what used to be called "plus-fours." There were once several forms of this animal but only two remain. The larger Wood Bison of northwestern Canada is still truly wild, having survived untouched since prehistoric times; the western Plains Bison, was rescued from total extermination at the turn of this century when there were less than a thousand left. The latter animals were once also found in a strip of territory from Lake Erie south to Georgia but this eastern form was extinguished by 1800. The western herds roamed the prairies from Canada to central Mexico in endless millions, drifting about in a form of seasonal migration; they were followed by the great grizzly-type bears and by some Amerindian tribesmen who, combined, were just sufficient to preserve the optimum balance in the bison's numbers. The arrival of the white man with his mania for wholesale slaughter of anything alive—and, on this occasion, often for no other reason but the tongue of the animal—quickly altered the situation by direct, senseless, and wanton extermination. The animal is now making a wonderful comeback on reservations and in semi-domestication.

Anoa

DEER-OXEN (*Boselaphinae*)

The English title is a literal translation of the Latin name and is used for convenience but does not indicate that these animals are in any way missing links between the Cervines and the Bovines. They are, however, of somewhat deerlike bodily form. There are two very different animals now classed in this sub-family. Both are from India.

The Nilghai (*Boselaphus*)

The form of this large beautiful animal is amply shown in the accompanying Plate 172. They are of a bluish shade though the pelt is really grey with a little brown, but the females and young are usually of a reddish-brown color. Only the male bears the small, smooth, forwardly-curved horns, but both sexes have a small neck mane and a patch of long hair on the rather humped shoulders. The males have a pendent tuft of fur on the throat. These animals live in glades in the forests, in open forest, and on park lands and they prefer low hilly country. They go about in small bands and are still fairly common throughout northwest, central, and northern south India. They form one of the principal articles of diet of the Tiger and have always been extensively hunted by men but seem to be able to hold their own. Despite their long legs and necks and sharply sloping bodies, they have a very oxlike appearance; nevertheless, they are customarily referred to as "antelopes" in India.

The Chousingha (*Tetraceros*)

The only close relative of the Nilghai is a tiny deer-like animal found in the same area but with a considerably wider range, and known as the Four-horned Antelope. In this respect—the presence of four horns in the male, the female being hornless—it is unique among Bovines. These delicate animals are true grazers and never stray far from water but meander in pairs or small family parties through the forest glades and the open-canopy woodlands wherever there is suitable food. They are very apprehensive, shy, and swift, and dash for the undergrowth at the least alarm. In color, they are a rich reddish-brown above and white below. The development of four horns in any animal, and particularly such a small and obscure one as this, is an odd quirk of Nature for which we really have no "explanation," though the fact that the males alone carry them shows that they are probably for adornment.

TWIST-HORNED OXEN (*Strepsicerosinae*)

Closely related to the Nilghai are a fairly large group of African cloven-hoofed animals, distinguished by having large horns with a pronounced spiral twist. They are also distinguished by having a curious arrangement of widely separated, light, vertical stripes on their sides, sometimes augmented by a complex arrangement of dark median and bilateral, light, horizontal stripes along the flanks. For the most part they also have light and dark patterns on their faces and necks so that they are referred to collectively as "Harnessed Antelopes." They range in size from the mighty Giant Eland to the delicate little West African Bushbuck. There are three

ZOOLOGICAL SOCIETY OF LONDON

Chousingha

groups—a plains and savannah type, the Elands; a mountain type, the Kudus; and a forest type, the Bushbucks. The forest Bongo holds an intermediate position between the three.

Elands (*Taurotragus*)

There are two species of Eland, one found along the southern edge of the Sahara from Senegal right across Africa to the eastern Sudan and known as the Giant Eland; the other (Plate 173), from South Africa, north to the Congo border on the west and to northern Kenya on the east. They are huge oxen, standing almost six feet at the shoulder in the case of the bulls, with heavy twisted horns, a large dewlap, a tuft of long hair on the forehead, and pronouncedly humped shoulders. They have long oxlike tails with a terminal tuft, and they usually display the harness-markings of their group but some are plain-colored, and old individuals often lose a lot of their hair and look grey in color. The bull Giant Elands also have a small, black mane. This species is a browser; the southern form is a grazer. Elands go about in large herds and prefer the savannahs and open-canopy forests where there is plenty of shade; they are rather indifferent to other animals, including man, unless they have been constantly hunted. They will defend their kind from marauding lions and even a half-grown Eland can pound a lion to death with its forefeet if the cat happens to be insufficiently agile. Both sexes carry horns.

The Bongo (*Böocercus*)

Standing somewhere intermediate between the Elands, the Kudus, and the Bushbucks is a rather large, heavy-bodied animal with a fairly long, tasseled tail, and stout, flanged horns that make a single great twist. It is known as the Bongo. The horns have pale yellow tips and are carried by both sexes. In coat color, Bongos are bright orange-red with vivid white stripes, but are almost black below. Old bulls go dark chestnut brown and their stripes get yellow. They are rather rare and are very retiring high-forest animals and, although seldom seen or collected, they range all across Africa, living in both the western block of equatorial forest and from the Cameroons divide, east to the borders of Kenya. They stay in the deepest forest near water and usually in bamboo and canebrakes growing along creeks under the tree canopy. They are exceedingly hard to see in such locations.

Kudus (*Strepsiceros*)

There are also two species of Kudus, the greater (Plate 174) being much the commonest and still occurring all over south Africa, west to northern Angola, and east to Abyssinia. The Lesser Kudu is identical in general shape but the color and markings are much more vivid, and it has no crest on the shoulders and no throat mane. It is much smaller and there are white areas on the throat and chest. The Greater Kudu is a mountain animal and sometimes lives at very great altitudes among the giant heathers above the tree-line. The Lesser Kudu lives in the desert scrublands and the thorn forests of the semi-deserts and ranges from Abyssinia to southern Tanganyika. The horns of the greater species are enormous and have a very wide spread; they may measure five feet in length and make three complete twists. Females of both species are hornless.

Bushbucks (*Tragelaphus*)

There are five kinds of Bushbucks; to these the names Mountain Nyala, Nyala, Sitatunga, Mountain Bushbuck, and the Common Bushbuck have been given. The first is very like a small Kudu but its horns make only a little more than one twist and there is a crest of long hairs all along the back. This last character is a feature of all the Bushbucks. The Mountain Nyala inhabits the uplands of southern Abyssinia. The other Nyala is really a very different animal and is found in southeast Africa. It is grey-brown in color and the full harness pattern, though present, is very dim and blurred. The males carry a tall crest from the top of the head all along the back and on to the top of the tail, which is itself of considerable length and bushy. Under the throat there is a small beard that extends back as a pendent fringe, forming a false dewlap in front of the legs, and thence continues back along the mid-line of the belly where it is very long. The horns are widespreading and make a single twist. Females lack horns and all other adornments except for a very short dorsal ridge of longer hairs. They are bright chestnut in color.

The Sitatunga (sometimes separated as *Limnotragus*) is really no more than a forest and swamp form of the Nyala that has longer horns, a shorter tail, and lacks the crests and fringes. It also has extremely long, slender, sharply-pointed hoofs. It is found all round the Congo Basin from Rhodesia in the south, to the eastern Sudan and thence west to Senegambia.

The true Bushbucks of the closed-forest are the closely related Mountain and Common Bushbucks which are found throughout the territory within the ring of mixed forest inhabited by the Sitatunga. Thus, there are small bushbucks in West, Central and East Africa, the last being the Mountain form which ranges from Uganda to Tanganyika. They are delicate little

W. T. MILLER

Bushbuck

animals of a rich reddish-brown color with pronounced white harnesses that include a horizontal white line along the flanks and dark and light spots on the face and neck. The fur is rather long and forms a back crest. The undersides are white. The crest can be raised at will. The females are hornless. They are very "narrow" animals, their bodies being compressed like those of collie dogs. They are also terrific jumpers and have a habit of kicking their back legs like deer.

DUIKERS (*Cephalophinae*)

These very numerous, considerably varied, but generally rather nondescript little ungulates of Africa for long puzzled zoologists because they are so obviously not antelopes in the proper sense of that word and do not look like any other animals, except perhaps the Anoa. That they are Bovines is certain from their general anatomy but they are rather primitive and may represent the descendants of the animals from which all others of that group evolved. The Duikers can be divided and grouped in several different ways and they probably represent several genera but are customarily arranged in only three. Both sexes bear small, simple, straight, spiked horns. The back is arched and the hair of the rump is usually very long and can be somewhat elevated like that of the rodent Agoutis. The tail is short but oxlike with a terminal tuft. The ears are large.

Common Duikers (*Sylvicapra*)

Sundry species are recognized but neither any one

nor all of them collectively is, despite their popular name, any more common than other duikers. However, they inhabit an enormous area outside the forests and they were the first duikers that the white man came to know. They are a brindled grey in color with dark legs and white undersides but one is a pale yellowish brown and has a big tuft of reddish brown hair on the forehead. These animals are not entirely herbivorous in that they deliberately eat insects, snails, and other small animal food and have been seen licking and worrying bits of carrion, possibly for their salt content.

Forest Duikers (*Cephalophus*)

All across Africa from Gambia in the west to Kenya in the east and south to Angola and Rhodesia, throughout the true forests and to a certain extent in all woodlands, and open forests, there are almost countless numbers of duikers. In the aggregate, there are probably many more of these animals than there are big-game on the open plains but, being small, shy, and nocturnal, they are not so obvious and are, in fact, seldom seen at all, so that Europeans have sometimes lived in areas where they are very common for years and yet never seen one alive. The forest duikers are various colors of brown from a dull olive to a rich chestnut red or bright orange. They vary in size from about that of a terrier to a small donkey. The largest is the Yellow-backed Duiker. This is of a very dark brown, somewhat lighter on the face and chest, and has a vivid yellow, wedge-shaped area from the mid-back to the base of the tail composed of coarser hair that can be opened up somewhat like the rump-patch of the Pronghorn. The strangest of all, in coloration, is a small species (*C. zebra*) from Liberia and adjacent parts of West Africa that has a bright orange coat marked with wide blue-black vertical stripes. Some of the smaller species of Forest Duikers are also partly omnivorous and in captivity some will readily select chopped meat in preference to fruit, vegetables, or horse-feed. Another curious characteristic is their habit of climbing tangles of low bushes and vine-covered trees to get at special leaves.

Blue Duikers (*Philantomba*)

In the West African forest block, and again from about the Niger River east to the Ituri and south to the Congo River there are to be found rather numerously some small, pale grey duikers with a distinct bluish tinge to their coats. They have light grey throats and are pure white below. When seen alive or freshly killed alongside any of the Forest Duikers their very different appearance becomes obvious. This is nevertheless very hard to describe and even harder to define. In habits, moreover, they must be very dissimilar, for they inhabit the same range as the forest forms but never mingle with them and appear to keep to the deepest vegetative recesses near water. They are the tiniest of the Bovines and in some areas they live in holes, though presumably ones dug by other animals.

ANTELOPINES

The Antelopes proper are an extremely numerous group and, with about half a dozen exceptions which

Duiker

will be noted in due course, are exclusively African. Even the keenest sportsman is apt to retreat in mystification when confronted with anything like a complete list of them, and zoologists have been hard put to it to classify them. However, they fall into five major groups, each containing several outstanding types. Countless sub-species and local races have been described but with these we will not concern ourselves since they are based only on slight differences in color and horn structure. The Antelopes are savannah, grassland, scrub, and desert animals and none is found in any kind of closed-canopy forest. Like the Bovines, they have solid, permanent bony cores to their horns and these are sheathed with horny coverings.

HORSE-ANTELOPES (*Hippotraginae*)

This sub-family comprises three types of large hoofed mammals with bodies shaped very much like those of thoroughbred horses. They are inhabitants of the open acacia parklands, go about in herds led by one or more large bulls, and are very fleet.

Sabre-horned Antelopes (*Hippotragus*)

Of these there are four known species, the most magnificent being the particularly rare Giant Sable Antelope (*H. variani*) of Angola, where it is only known from one locality. It is a purplish blue-black animal with heavy forequarters, a long thick neck bearing a crest from head to shoulders, a deep chest and enormous horns rising straight from the top of the head

Sable

and sweeping backwards in a half-circle. The Common Sable (Plate 176) is smaller, dark brown to black in the males and brown in the females, and contrastingly white below. There is a slight throat mane and the horns are much less developed. It ranges from the coastal plains of Kenya to the Transvaal. The third species is the red-brown-coated Roan Antelope, which is as big as the Giant Sable but has very much shorter horns that are only half the length of the Common Sable and long narrow ears bearing a terminal plumelike tuft of long hairs. It is found all around the great Congo Basin outside the forests from Northern Nigeria to the eastern Sudan and Abyssinia and thence south to the plains of East Africa, the Orange River, and round the northern edge of the Kalahari into Angola. The last species is the Bluebuck, which looked like a small Roan but was of a bluish-grey-brown color and inhabited a small area in what was Cape Colony, in extreme South Africa. The last example of this species was killed in 1799 and only half a dozen specimens are preserved in museums.

Rapier-horned Antelopes (*Aegoryx* and *Oryx*)

Of considerably smaller stature but even more horsy form are a group of scrub and desert antelopes distinguished by having long, either straight or slightly back-curved, rapierlike horns that, like those of the Sabre-horns, are heavily ringed from the brow to between half and three-quarters of their length. Although they hold their heads high, these animals do not run with them arched as do the Sabre-horns. They gallop like horses. Their tails are oxlike with a terminal brush.

With horns intermediate between the Sabre and Rapier types is the White Oryx (*Aegoryx algazel*), a true desert animal with huge hoofs that ranges all across the lower Sahara from the Atlantic coast south of Rio de Oro to the borders of Ethiopia. It is almost white in color, with chestnut-brown facial and other markings similar in design to those seen in the Oryxes. It travels about in small bands and appears to be able to exist without water, at least for very long periods.

The true Oryxes are three in number and are known as the Gemsbok, the Beisa, and the Beatrix Oryx. The first has the longest horns and inhabits the Kalahari Desert and adjacent dry areas in Southwest Africa and Angola. It is of a sandy-grey color with vivid black facial and foreleg markings. The second is very similar but lacks some of the black chest-marks of the Gemsbok. It is a very common animal on the plains of East Africa from the eastern Sudan south to Tanganyika. The smallest species is the Beatrix which is found only in two separate isolated areas in Arabia. It is a true desert animal, white in color with dark brown stockings and oryx-like facial markings.

Screw-horned Antelopes (*Addax*)

Closely related to the above three genera is the Addax, which is also a native of the Sahara but used to be found on the north as well as the south side of that desert and still appears from time to time in southern Algeria and Libya. It is a dusty grey in color with black facial marks. Its horns spread wide and are twisted like a screw, sometimes making more than three turns. It can exist absolutely without water apparently indefinitely but will drink gallons of it both in the wild and in captivity if it is available.

DEER-ANTELOPES (*Alcelaphinae*)

This sub-family may be rather sharply divided into three types with one additional, somewhat intermediate form—namely into the Hartebeests, the Damalisks, and the Gnus, with a rare animal called Hunter's Hartebeest (*Beatragus*) linking the first two. They are all animals of the savannahs and grassfields, with horse-shaped but extremely narrow, slender bodies, long legs and long necks. Their heads are grotesquely long and

ZOOLOGICAL SOCIETY OF LONDON

Arabian Oryx

[262]

Addax
ZOOLOGICAL SOCIETY OF LONDON

narrow in the Hartebeests, typically antelopine in the Damalisks, and rather like buffaloes in the Gnus. Their tails are oxlike and their horns often of crazy design.

Hartebeests (*Alcelaphus*)

The general form of these animals may be seen in the photograph of Coke's Hartebeest (Plate 177). There are half a dozen clearly recognizable forms, two of which constitute distinct species. They are all of a somewhat reddish-brown color and most of them have darker to black legs and faces, with the tail plume usually black. They all have elongated faces topped by upright horns which grow together at their bases from a tall bony pedicle clothed in skin and fur. There is a slight shoulder hump and the body then slopes away rapidly to the tail base. The horns of all of them are lyrate when seen from the front but, looked at from the side, they are shaped like an "S," going first backwards and upwards, then forwards and upwards, and finally backwards and upwards again. The most exaggerated are those of the Lelwel which is found from the Sudan to Kenya and west into French Equatorial Africa. The first hartebeest to be known was commented on by Roman governors of Barbaria, having in their day been very common all across North Africa from Morocco to Egypt. It was called the Bubal and it became extinct only in this century. It was once considered to be a horned horse. Other species are distributed all over Africa in the zones lying between the true deserts and scrublands on the one hand, and the true forests on the other, but mostly on the grasslands, where they are one of the commonest forms of game. One species is known as the Kongoni.

Damalisks (*Beatragus and Damaliscus*)

Hunter's Hartebeest (*Beatragus hunteri*) is a rare antelope now found only in a small area from the Tana River to southern Somaliland. It is pale brown with a white facial mark and white tail. Its horns grow separately at the base and are long and intermediate in curvature between those of the Hartebeests and the Damalisks, going first forwards, then backwards, and finally straight upwards. Its face, however, is that of a hartebeest. The Damalisks have even more slender and sloping bodies than the Hartebeests but their faces are like those of the more normal antelopes, and their horns, lyrate when seen from the front, make a graceful upward and forward sweep, then curve backwards and turn slightly upwards at the tips. They are not jointed at the base. There are five recognized species known by the names Korrigum, Topi, Sassaby, Blesbok, and Bontebok. The first comes from West Africa, the second from East Africa, the third from southeast Africa, and the two remainder from South Africa, where they remain only in limited numbers on reservations. There are less than a hundred Bonteboks left, and the Blesbok is only numbered in a few hundreds. Damalisks are all of a reddish-brown color with sundry lighter and darker markings.

Gnus (*Connochaetes and Gorgon*)

These animals, so popular in crossword puzzles, are among the most remarkable of all ungulates and behave

in a manner than can truthfully be described as hilarious in the old and purist sense of that term. They are of sturdy, horselike build, with heavy forequarters and slender limbs but they have very long, plumed tails with terminal hairs that almost reach the ground. Their foreheads and necks, both above and below, and their shoul-

COMMANDER GATTI AFRICAN EXPEDITIONS

Tanganyika Oryx

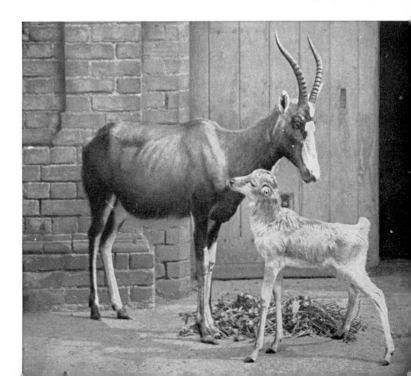

Blesbok

Sassaby

ders are adorned with long hair forming a sort of exaggerated crew-cut and a scraggly mane. Their horns, seen from the front, look like those of a Cape Buffalo but are more slender; from the side they are seen to rise a little off the crown, then grow straight out laterally, and then curve forward and at the same time make a sweep upwards. There are three forms—the White-tailed of South Africa, now existing only in semi-domestication on farms and heavily protected, the Brindled, and the White-Bearded (Plate 178). They wander about the open grass plains in large, loosely-associated herds, each component of which is usually led by an old female. They are great bluffers and extremely intelligent, waiting till man or other offensive beast comes within a certain distance, then getting up and standing to face the intruder. If he persists in approaching they start a series of imitation charges, kicking up dust, dashing about with heads down and tails lashing but finally making off in orderly retreat. After going some distance they then suddenly about-face, form ranks, and stand to stare again. Sometimes they will suddenly all go galloping off at a great rate and at a tangent either to left or right, and end up on the intruder's flank at the same distance, and then stop to stare some more.

The South African species is dark brown with a white tail and black frontal fuzz and mane. The Brindled is blue-grey in color with dark brown vertical stripes on the side of the neck and flanks, black mane and tail. Its range is north of that of the White-tailed Gnu, from western Angola to Mozambique and north to Nyasaland. The White-bearded species is similar to the last but has a white throat and dependent neck fringe and is found all over the grass plains of East Africa. For the benefit of crossword puzzlers, these animals are also known as Wildebeests.

MARSH-ANTELOPES (Reduncinae)

The members of this group of antelopes are actually more deerlike in form than the last. What is more, they behave more like deer. They vary in size from that of a small horse to a sheep and are found all over Africa but only in a very special natural niche. This is hard to define properly, as it does not constitute one of the major vegetational zones but rather a particular environment within two or three zones. Thus, these animals are not found in the forests or on the savannahs or among the scrub but in and around marshes, swamps, reed-beds and canebrakes wherever they occur within those zones. Some species will wander far from water or even marshy ground, provided there is tall grass, and one species customarily lives on open mountain grassfields. They fall into four distinct groups.

Waterbucks (Kobus)

There are two closely related Waterbucks, one (Plate 175) found all down the eastern third of Africa from the borders of Abyssinia to the Transvaal, the other, known as the Defassa, cropping up in suitable localities all over the rest of Africa south of the Sahara from Gambia to the eastern Sudan and south around the eastern edge of the Congo Basin to Angola. They are large, heavy-bodied antelopes with widespreading horns well ridged at the base. They are clothed in rather long coarse hair that forms a thick ruff on the throat, chest, and along the underside. Water is not essential to their well-being but they never stray far from reed beds or tall grasses and they do not enter the closed forest. If caught young they make very delightful pets and will wander freely around the house.

Lechwes (Onotragus)

The Lechwes are much lighter-bodied animals with long, rather thin horns that sweep gracefully outwards and backwards, then upwards and finally either inwards or outwards. These horns are deeply ridged almost to their tips. Their hoofs are extremely large, long and slender, a most admirable adaptation for animals that are semi-aquatic. They assemble in large herds and stay near water, often spending all day submerged up to their necks among the reeds. They are also clothed in long coarse hair which forms a distinct fringe down the throat. There are two principal species-groups, one in South Africa, the members of which are a rich reddish brown in color though the males of a given race may be almost black, those of another yellowish-red with black limbs. The other, which occurs only in the great

Bahr-el-Ghazal swamps of the eastern Sudan, is dark chocolate with white facial markings and cream-colored shoulders. This animal is often most misleadingly called Mrs. Gray's Kob.

Kobs (Adenota)

There are two species of these antelopes known commonly as the Kob and the Puku respectively. A number of distinct forms of the first have been described but they are all very similar. They are orange-red in color except for the males of one race which are black and have white ears. Kobs are on the whole rather smaller than the Lechwes and have smaller horns that grow in a lyrate form. Their tails are very much shorter and the coat, instead of being hard and coarse, is short, sleek, and soft. They occupy a slightly different niche from that of the Waterbucks or the Lechwes since they move about in small herds on the open meadows which usually spread over the low ground outside and above the swamps proper. The Kob is found all across Africa north of the central forests, from Senegambia to the upper Nile and south to Uganda. The Puku is confined to an area between Nyasaland and the Zambesi River.

Reedbucks (Redunca)

These, the smallest of the Marsh Antelopes, are likewise spread all around Africa from the far south to Abyssinia and thence north to the Sudan and west all the way to the Atlantic. They have rather long hoofs, small, forwardly curving horns, and short but very bushy tails which are white below and which the animals use as signaling devices, holding them upright

Blackbuck

Reedbuck

when in flight. They are of a dull greyish to light brown color and the coat is rather stiff. A considerable number of forms have been described but only one is really different. This, called the Mountain Reedbuck, lives among tall grass high upon the mountains and especially in the armchair-shaped cols and saddles that are so common on Africa's softly rolling ranges. In these, considerable parties of these animals may live in an extraordinarily small area but never be seen except at sundown, when they venture forth to graze on the surrounding shorter grass.

The Rhebok (Pelea)

In South Africa proper, there is a close relative of the Reedbucks that lives perpetually among the rocks and low, tangled, dry growth of the mountains. This is known as the Rhebok and although of much the same small deerlike form as the Reedbucks, has small, upright, straight horns and very large pricked ears. It is clothed in a pale grey, woolly fur. Rheboks move about in small herds and feed blatantly out on the open grass.

BLACKBUCK (Antilopinae)

This is the only antelopine ungulate that does not fit into any one of the great sub-families though it appears to stand near the Gazelles. It is also in other respects most distinctive. Curiously, the name of the whole group has devolved upon this animal through a process of slow elimination, all the others having been assigned to other groups. This does not mean that it is typical of the Antelopines as a whole. It is popularly and very widely known under the title of the Blackbuck and has been known to Europeans since early classical times. It is an inhabitant of India, living on the open plains all over the country from the foothills of the Himalayas to Cape Comorin, and from Western Pakistan to lower

Assam. The form of the animal may be seen in the accompanying photograph.

The corkscrew-formed horns are unique among antelopes since they are round in section, make many complete turns and may grow to almost three feet in length. The males are black-and-white in color, the young and the females are pale brown. Only the males carry horns. They are diurnal feeders, unlike most other antelopes, which, contrary to popular belief, do more grazing and browsing at night than during the day. When other animals are sleeping or resting in the shade, the Blackbuck are cropping steadily away at the grass in the hottest sun, the young playing about energetically and the adults often making sudden runs for no apparent reason but probably to get away from flies when they become intolerable.

Blackbuck have been the target of sportsmen, real and otherwise, since time immemorial and have been hunted with everything from Cheetahs to machine-guns —until this last revolting practice was summarily stopped. Yet they have survived in great numbers, and this mostly because of their incredible speed, which, on occasion, even defeats the Cheetah.

PIGMY ANTELOPES (Neotraginae)

Least known and in most of their forms the most obscure of all Antelopines are the tiny creatures known variously as Klipspringers, Oribis, Sunis, Beiras, Dik-diks, Grysbok, Steinbok, and the Royal Antelope. All these beautiful little terrier-sized ungulates are closely related. The males always bear tiny spikelike horns. The smallest—and probably the smallest of all ungulates—is a form of the Royal Antelope that comes from West Africa and is only about eight inches high. There are eight distinct genera of Pigmy Antelopes.

Klipspringers (Oreotragus)

The most delightful and perhaps the best known of all these tiny hoofed creatures are the Klipspringers, which are found in rocky areas all over tropical Africa. In some species the familes have tiny horns. They are rather thick-legged creatures, with dense coats composed of thick stiff hairs, have tiny round hoofs, mere stubs of tails and large rounded ears. They are terrific jumpers and more sure-footed than any goat although they trip about on the very tips of their toes. Most of them are a yellowish grey in color.

Oribis (Ourebia, and Raphicerus)

These are somewhat larger and more typically antelopine animals though still of very small size. The males alone have spikelike horns that rise straight from the dome of the head. They have very large ears and extremely slender legs, and they too mince along on the very tips of their small hoofs. The animals of the genus Raphicerus are known as Steinboks and Grysboks. The former is highly specialized, having no false-hoofs. The Oribis are sandy-colored and have little fringes of long hairs round their knees. The Steinboks are red in color and the Grysboks are of a curious reddish-agouti pattern, the hairs being a mixture of red-brown and white. The Oribis are spread from South Africa to Abyssinia and then west across the continent to Senegal: the Steinbok is found from Tanganyika to Kenya; the Grysboks are South African.

Sunis (Nesotragus)

Along a strip of coastal territory in Kenya and thence south and inland to northern Tanganyika there is found another kind of tiny antelope known as Sunis. There are apparently several species, all of a brown color above and white below. They are virtually tailless and stand about a foot high at the shoulders. They have rather long necks and small ears, and live in tangled vegetation, feeding in open places at night.

ZOOLOGICAL SOCIETY OF LONDON

Royal Antelope

Klipspringer
SAN DIEGO ZOO

The Beira (*Dorcatragus*)

This tiny, delicate antelope with huge ears, little upright, straight, spiked horns spaced widely on the top of the head, and with a small bushy tail is the first of what may be called the "bunch-bodied" Pigmy Antelopes. It is of a stippled steel-grey color and pure white below and it has white "spectacles" around the eyes. It is a rock-dweller and is found in mountainous districts of Somaliland and eastern Abyssinia. The legs are long and the body very short.

Dik-Diks (*Madoqua and Rhynchotragus*)

The last two Pigmy Antelopes are extremely short-bodied and both have exaggerated muzzles, the second having a regular trunk almost of the proportions of the tapirs. These are some of the most prodigious jumpers of all mammals, almost equaling the longer-legged rodents in this respect and appearing, when alarmed suddenly at night in a bright light, like some kind of toy made for practical jokers, flying into the air at all angles and to extraordinary heights without any apparent preliminary contraction of the leg muscles. When going flat out they completely ignore anything but rocks or trees and just flip through the scrub as if airborne. They are of a strange color arrangement almost exactly equivalent to that of the rodents of tropical America known as agoutis; grizzled above turning greyish on the flanks, and rich red-brown with an orange wash below and on the limbs.

Royal Antelopes (*Neotragus*)

Scattered about Africa there are also a number of minute antelopine creatures to which the high-sounding name of Royal Antelopes has been given. They are distinguished, among other details, by a small tuft of long hairs on the top of the head between the small spike horns. Their coloration is generally reddish-brown above and lighter, to white, below and they are of extremely delicate build. They make a screaming noise when alarmed. These smallest of antelopes come from West Africa; other species are found on the Red Sea coast and in northeast Africa.

GAZELLES (*Gazellinae*)

We here come to the largest group of Antelopines and the one which has the greatest number of forms. Most of these belong to the genus *Gazella* and all are, except to zoologists and keen sportsmen, very much alike. There are, however, some types so odd as to warrant special note. Most of the Gazelles are desert animals but they also inhabit the grassy plains, the scrub zone, and parklands. They are small, delicate antelopes with spreading horns that are heavily ringed at the base.

Impalla (*Aepyceros*)

The largest of the group are very widely known as Impalla and are unlike most gazelles in that they inhabit many types of country, associate with all manner of other game, and have large, wide, spreading horns in the form of a lyre when seen from the front. They are found all over South Africa, west to Angola, and north to Uganda. They are the all-time champion high-jumpers and are the light-bodied, pale brown antelopes that are

Gerenuk (female)

almost always seen in movies of Africa, flying off in all directions through the air at the most unexpected moments. The females lack horns.

The Gerenuk (*Litocranius*)

By far the most astonishing of the Gazelles and one of the oddest-looking antelopes obtains its scientific name from the fact that its skull (*cranium*) is indeed almost as dense as a stone (*lithos*). It is a very long-legged animal with an almost giraffine neck, very small head, large ears and, in the male, small lyrate horns. It browses on thorn bushes and stands on its hind legs to obtain leaves that are out of reach of other more normally shaped animals. This is most advantageous to the Gerenuks because all the foliage at lower levels is sometimes completely eaten out by these numerous rivals of smaller stature. Like almost all gazelles these animals are a pinkish brown in color, darker on the mid-back and pure white below. They are found from Abyssinia to Tanganyika in dry scrublands.

The Dibatag (*Ammodorcas*)

Almost as exaggerated as the Gerenuk but of much smaller stature and having simple, forwardly-curving horns. The tail is exceptionally long and ends in a club-shaped tassel. Dibatags are colored like the true gazelles. As in the Gerenuk, the female lacks horns. They are found only in Somaliland and parts of Abyssinia.

The Springbuck (*Antidorcas*)

This little antelope differs considerably from the rest

Springbuck

of the Gazelles in a number of respects. First, both sexes bear horns; secondly, there is a pure white crest from the mid-back to the base of the tail and this can be raised and partially opened at will. Thirdly, the sandy-colored back is separated from the white undersides by a striking, dark brown band along the flanks and there is a pronounced, white rump-patch. These little antelopes once drifted across the uplands of South Africa in countless millions and when they were forced to emigrate from one area to another because of their excessive increase and the consequent dearth of food, armies of them, pressed shoulder to shoulder, would take days to pass a given point although they moved at a constant pace. Some early descriptions of these emigrations are at the same time both terrifying and tragic, for the animals overwhelmed the whole earth but were slaughtered and died by the thousands. In one case they even pressed on into the sea so that a long stretch of coast in Southwest Africa was piled with their bodies. Today, they are greatly reduced in numbers but are still found in small herds in various parts of South Africa.

The Addra (*Addra*)

The last exceptional form of gazelle is called the Addra and is found all over the Sahara north of the savannahs. With very long legs and neck it has almost the proportions of the Gerenuk. The tail is short and the horns lie straight back from the plane of the face immediately after rising from the crown; they then turn outwards, and finally curl up at the tips. In color, these desert animals are almost entirely white with a shading of pinkish buff over the back.

True Gazelles (*Gazella*)

There are over fifty described forms of this genus, all looking very much alike. Their general form and coloration may be seen in Plates 179, 180, 181, and 182. Collectively, they are found all over Africa outside the forest zones, and one, the Dorcas, occurs in Palestine and Syria, another in eastern Arabia, a third in western Arabia, a fourth on the plains of India, and still another, all across central Asia from Turkey in the west to the Gobi Desert of Mongolia in the east. There is a small species in the Atlas mountains of Morocco, Algiers and Tunisia known as the Edmi. All have back-curving lyriform horns that are strongly ringed, except the Edmi, which has small, upright spikes, and Loder's Gazelle of the same areas and east to Egypt and the Sudan, which has very tall horns that rise straight up and then curve gently outwards. Grant's and Robert's Gazelles of the East African plateaus have the longest horns, those of the first being lyrate, of the latter curving backwards, then outwards and finally downwards.

Goat-Gazelles (*Procapra*)

The True Gazelle of central Asia (*Gazella subgutturosa*), or the Goitered Gazelle, so-called because of its long-haired dewlap, is a distinctive species and points the way to the first of the Goat-Gazelles (*Procapra gutturosa*), which is also known as the Zeren. The next stage in this series is the Goa (*P. picticaudata*), a long-haired form from Tibet. Next comes a race of this species, known as Przewalski's Gazelle, found in Mongolia, which leads us on to certain animals of the next group, that may be called Gazelle-Goats—namely the Chiru and the Saiga. Through these animals there is thus a very close affinity between the Antelopines and the Caprines. The Zeren is a pale, sandy-colored animal, without the facial markings of the True Gazelles, that inhabits the upland grassfields of Mongolia and a belt of territory westwards to the Altai Mountains. In winter it grows a long coat of a curious pinkish fawn color. The Goa is somewhat similar but has a large white rump-patch, almost no tail, and a heavy-furred ruff down the throat onto the chest. It inhabits the high plateaus of Tibet. Przewalski's race of this animal is larger, with a dark brown coat in summer, and comes from the uplands of inner China and south Mongolia. All these animals go about in small herds and were once much more numerous for, although they are fleet and can keep out of range of arrows, they seem never to have understood the range of firearms.

CAPRINES

This is the last of the great subdivisions of the Even-toed Hoofed Mammals and, although by no means so numerous in specific forms as the Bovines or the Antelopines, is just as diverse. As the general title implies, these are the goatlike animals, but it is almost as difficult to separate goats from antelopes as it is traditionally supposed to be to separate the goats from the sheep. Almost all the Caprines are animals of the mountains and they are distributed indigenously throughout Eurasia and North America but only two or three species are found in Africa, and none is found in South America or, of course, Australia. They are distinguished by having heavy and sometimes massive horns composed of an outer, horny sheath on a solid and permanent bony core. These horns are almost invariably strongly cross-ridged, and they usually curve and twist. Except in a very few cases, horns are present in both sexes but are much smaller in the females. There are five major groups of Caprines.

GAZELLE-GOATS (Saiginae)

The two Asiatic animals comprising this family were formerly placed with the Gazelles, as Antelopes, but recent anatomical research has shown that they are not only much closer to the Caprines but are in many respects very typical members of that group.

The Chiru (Panthalops)

This is a sheep-sized and sheep-shaped animal clothed in thick woolly fur and having a short tail and slender legs. The male alone carries horns that rise straight from the head at a slight outward angle to each other and then curve inwards so that their points end pointing horizontally inwards. These horns may grow to more than two feet in length. In color the animals are reddish brown, with black faces and black fronts to their legs. They have goats' eyes. They live on the high plateau of Tibet between the twelve- and eighteen-thousand-foot levels and graze right up to the snow. The muzzle of this animal is curiously enlarged on either side, giving it a ridiculous appearance when seen head-on. The purpose of this structure is not determined but may be connected with the animal's acute sense of smell in the highly rarefied air in which it dwells.

The Saiga (Saiga)

The only known relative of the Chiru is an even more preposterous-looking beast that has a muzzle that swells upwards to such an extent that the nostrils point straight down. It is much the shape of a small sheep but has shorter and more slender limbs. The horns are much smaller than those of the Chiru and grow erectly, and have a very slight forward and then backward bow; they are yellow in color and look just like amber. Saigas are clothed in dense fur in summer but this grows long and woolly in winter when the animals are white in color. The summer coat is yellowish-brown above and white below. Today, Saigas are found throughout a belt of desolate prairie and steppe country from the Caspian Sea east to the borders of the Gobi Desert but they once swarmed all over the steppes from Eastern Europe to Mongolia and south to the foothills of the central Asiatic Massif.

ZOOLOGICAL SOCIETY OF LONDON

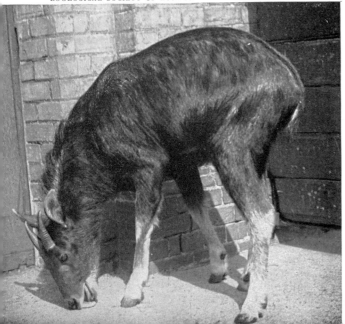

NEW YORK ZOOLOGICAL SOCIETY

Saiga

ROCK-GOATS (Rupicaprinae)

There are four kinds of Rock-Goats very widely distributed over the Northern Hemisphere. They are true rock-climbers and rock-dwellers though this is not to say that they are actually any more sure-footed than a number of the true goats. Nevertheless, their performances are sometimes quite unbelievable even when recorded on film, slowed down, stopped, and carefully analyzed. They appear to be able to defy gravity by walking along absolutely perpendicular rock-faces, using only half-inch-wide toe holds and with their bodies leaning out over space. They can gallop up and down impossible precipices and yet none of them could be called graceful in form and none gives the impression of being particularly sure-footed on ordinary terrain.

The Goral (Naemorhedus)

The smallest of the Rock-Goats, the Goral, is found in various racial forms all the way from the Himalayas to Amuria and Korea via the mountain ranges of inner China. It is a dull olive brown animal with dark brown legs and a yellow blaze on its throat. It has small, upright but backwardly curving horns. It prefers mountainsides where some vegetation manages to retain a foothold and in these it lurks. It careers along precipitous faces without apparent regard for itself or anything else. Gorals have been filmed skipping from one loose stone to another on a scree but so swift were they that they were gone before the dislodged stone began to roll down the slope.

Serows (Capricornis)

There are several species of Serows distributed over a very wide area of the Oriental and Far Eastern Re-

Serow

gions wherever there are rocky mountains, and on the islands off the Asiatic coast. Species are found in Sumatra, Malaya, Burma, Assam, and west along the Himalayas almost to Kashmir. They also occur in Thailand, Indo-China, Hainan, Formosa, and Japan. There are other species in the mountains of China. They are larger than the Gorals, dark grizzled grey to black in color, and clothed in long shaggy hair. The northern forms and those in Japan have a thick woolly undercoat. Their horns are somewhat compressed from side to side, grow backwards in the plane of the face, and curve downwards almost to the neck. They are agile climbers but no match for the Gorals and they prefer thicker cover where they are exceedingly hard to find or see, although they may be heard. In Sumatra, they feed on the montane grassfields and sometimes descend to the edge of the forests. Two young are the order, and they trail about after their mothers, who remain separated from the males while raising their young.

Chamois (*Rupicapra*)

These are European animals, of which there are sundry local races isolated on the higher and barer mountain ranges of Spain, Italy, Jugoslavia, and in the Alps, Carpathians, and the Caucasus Mountain ranges. These were the first Rock-Goats known to Europeans and they early impressed them with their marvellous mountaineering abilities, but these are actually the least audacious of the group. They prefer the upper pine forests and, although feeding above the tree line and making frantic dashes across open rock faces when pressed, they do not normally frisk about thereupon but prefer the protection of the thick cover. They are brown in color with a yellowish throat blaze and darker below and on the limbs. Their horns grow straight upwards almost parallel and then form a sudden backward hook. The tail is short and rather fluffy. It was from these animals that the famous "shammy-leather" was originally derived but most of the commercial product of this name today is actually either kid, sheep or buckskin. The supple leather of the chamois still can not, however, be surpassed either for its softness or for its amazing water-absorbing qualities which still makes it ideal for drying off varnish, notably on yachts and other decorated seagoing vessels.

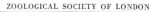

Takin

The Rocky Mountains Goat (*Oreamnos*)

These are the largest of the Rock-Goats and are distributed all down the Rocky Mountains from Alaska to Montana in western North America. They have deep, but short and very narrow bodies with heavy shoulders, long necks, and large heads. They are clothed in long, pure white hair with a thick undercoat; there is a ruff of long fur on the shoulders. The males have a full beard and all but the young appear to be wearing "plus-fours." The small horns rise abruptly and then curve backwards. Despite their much greater size, these animals are almost as intrepid in their behavior as the little Gorals and will perform prodigious feats on apparently perpendicular rock faces. They never descend to the trees like the Chamois but stay up on the barest and bleakest mountain tops in all seasons and in all weather. They feed on mosses, lichens, and whatever other stunted growth is available. The females retire in parties together to bear their young and they then stay away from the males for some weeks. During the rest of the year they go about in small herds made up of all ages.

OX-GOATS (*Ovibovinae*)

Two very odd ungulates that puzzled zoologists for many years have been shown to have much in common from a structural point of view and to be aberrant forms of Caprines. The appearance as well as the habitat of the Muskox was from the first against its ready recognition as a member of this group, and the Takins are so altogether singular that nobody knew where to put them before their anatomy was studied in detail. Both are leftovers from apparently more numerous and varied families of animals, and both once had a much more extensive range. *Ovibovinae* actually means "ox-sheep."

Takins (*Budorcas*)

It is fortunate that we are able to display a photograph of one of these beasts since they are almost impossible to describe. Both sexes bear identical horns which, as may be seen, are almost like those of some buffalo. There are several races of Takins that inhabit a considerable area of mountainous territory from the eastern Himalayas through Assam and northern Burma to east Tibet, and Szechwan, Kansu and Shensi Provinces of China. Nevertheless, they are not truly montane animals in that they prefer thick woods and the giant bamboo forests, and descend regularly to the valleys—albeit those at considerable heights—to feed on swamp vegetation, lush grasses, and to browse on bushes. They are ungainly in movement and amble along with their bloated noses close to the ground like hounds tracking, but they can navigate the roughest terrain at a steady and surprising pace and they are adept climbers. The common western form is a dismal, yellowish grey-brown in color but the races from China are bright yellow with dark muzzles and limbs, and the Golden Takin is one of the most brightly colored of all mammals—shining metallic golden all over the head, neck, shoulders, and upper forelegs, with the hind parts passing into pale gold, then grey and finally black.

The Muskox (*Ovibos*)

This strange, cow-sized animal is an inhabitant of the bleakest, most windswept areas of the treeless Arctic tundras and snowfields. It once occurred all across northern Canada from Labrador to Alaska but is today reduced to a few herds between Hudson's Bay and the Mackenzie River, and to rather more numerous populations on the great islands to the north, from Banks Island in the west to Greenland in the east. There are three races—one, the White-fronted, on the islands, second, the Coastal, and third, the Barren-Ground which lives farther south and away from the coast. Muskox are clothed in long shaggy overcoats and a thick underwool. Their feet are huge and their hoofs widely splayed to assist in moving over snow. Their horns curve back along the sides of their heads and form a bony casque over the forehead. These animals have a habit of forming a hollow circle for protection, all pointing outwards with lowered heads, while the young and females remain inside the ring. They thus present a solid wall of sharp horns that afford ample protection from marauding wolves but not from men armed with modern rifles. They stay all year on their bleak plains and somehow manage to find an ample diet of lichens, mosses and other stunted growth, scratching for it through the snow.

TRUE GOATS (*Caprinae*)

We here meet a situation comparable to that which we encountered in dealing with the dogs. Although we manifestly have millions of animals that we call "goats," we know of no such creature in the wild state today, nor have we the fossilized remains of any animal from which these animals are surely descended. The wild goats living today are all somewhat exaggerated forms that can clearly be classified as Tahrs, Markhors, Ibexes, or Turs, and it is hard to conceive of any of these giving rise to our everyday domestic goats. As the latter are of considerable range of shape and have a wide variety of horn structure it is very probable that they have a multiple origin and the species from which they were originally created were probably captured and taken over *in toto* by early man, all wild examples being hunted out. The domestic goat appears to have originated in the Near or Middle East and is obviously an animal of the dry scrublands. The remainder are all mountain animals.

Tahrs (*Hemitragus*)

There are three species of Tahrs found in the western Himalayas (Plate 183), the mountains of Southern India, and certain areas of southeastern Arabia respectively. The first is the largest, the last the smallest. They are typical goats, with small, backwardly arching horns and would be the most like candidates for the ancestors of our domestic breeds were it not for their surprisingly isolationist attitude. The Himalayan animal is of a reddish color, the Indian, brown, and the Arabian, greyish. The first is thickly clothed and has a shaggy mane around the neck, over the shoulders and hanging down to its knees. The Indian has rather long fur and a grizzled white saddle across the back; the Arabian is sleek-

ROUSE FROM FISH AND WILDLIFE SERVICE

Muskox

haired. Their tails are short and can be elevated. The horns grow very close together and are almost semi-circular in section, their flat faces facing each other inwards. In habits, they are also typical goats, being very agile climbers and eating almost everything vegetative.

Markhors (*Capra falconeri*)

Although placed in the same genus as the Ibexes and the strange Tur, the Markhors warrant separate description. They are magnificent, upstanding animals with a heavy mane of long shaggy fur clothing the neck and shoulders and hanging down over the chest. Their horns are greatly twisted but the twists may be open or tightly closed. The structure of the horns is hard to describe and even harder to visualize: the best way to understand it is to take a long tapering strip of paper, hold the ends and give it several twists: before it makes a complete cone it will be seen that a slit winds up the structure. If you could then fill this slit with a gutter you

ZOOLOGICAL SOCIETY OF LONDON

Kabul Markhor

[271]

will have the solid form of a Markhor's horn. In one race the horns are perfectly straight, though individually twisted, and form an upright "V." Markhors inhabit the mountains in and around Afghanistan.

The Tur (Capra caucasica)

This unusual, sheeplike goat is found in the Caucasus Mountains. It is short-haired and of a rich brown color with dark stockings. It has a small, forwardly brushed beard and huge horns with a round front edge and flat back face. These grow out of the forehead, wide apart; go upwards a little and then sweep backwards and outwards and finally downwards and inwards so that their tips, which curve upwards again, almost touch behind the head.

Ibexes (Capra ibex, etc.)

The remaining wild goats may all be called Ibexes and although they are widely distributed in isolated populations they are remarkably similar apart from the size of their horns. These, moreover, rise high from the forehead, diverge at an angle of about thirty degrees, and then describe a quarter-, semi-, or in some species almost three-quarter-circle. They are very heavily ridged all

SAN DIEGO ZOO

Arabian Ibex

along the front edge. Ibexes have sturdy, compact bodies with heavy forequarters, short, tufted tails and ridiculous little beards hanging from the middle of their lower jaws. They are a greyish-brown color with darker undersides, except in the southern races, which are lighter below and in some cases almost white. Some have a dark stripe along the back and others white

bands on the forelegs. Ibexes are mountain animals and are found on the Italian Alps, in Spain and Portugal, on some of the Aegean Islands and on Crete, in Turkey and the Caucasus, and thence east through Persia to Pakistan and Afghanistan, and all along the Himalayas to Nepal. To the south, they range through Palestine to the Sinai Pensinsula and Arabia, all along north Africa, and south again down the mountains bordering the western coast of the Red Sea to Nubia and Abyssinia.

SHEEP (Ovinae)

The old saw about separating the sheep from the goats is not as funny as it seems to Europeans and Americans where the domestic breeds of the two animals are quite distinct. The author narrowly escaped serious trouble in one part of Africa after giving an animal to a native chief under the impression that it was a sheep—a most acceptable gift—only to learn that the majority of that dignitary's retainers declared it to be a goat, which happened to constitute a dire insult in that country. Nor can science be called upon for any clear definition, except to point out that the goats have a transverse slot across the pads under their hoofs; but, then, there are breeds of sheep in Africa that also have these. Once again, there is no wild "sheep" known today though there are numerous feral populations in various parts of the world. However, the truly wild sheep are not so unlike some domestic breeds as to preclude the one having been derived from the other. Two kinds of wild sheep are given separate generic status, the rest form the genus Ovis but have various popular names.

The Aoudad (Ammotragus)

This is otherwise known as the Maned or Barbary Sheep (Plate 185) and is found all around the Sahara, being the only indigenous sheep of Africa. It is a very large species, the rams having most powerful forequarters and a long, plumelike fringe of rather soft hair hanging from the chin to between the front legs, and also wearing a pair of hairy "shorts" on the front legs. The horns are large and very thick, with a distinct keel at the base and small cross-ridging. They sweep outwards, backwards, and then inwards. The females bear small slender horns. Aoudads are rock-climbers and are almost as nimble as the Rock-Goats, the young especially being able to walk along perpendicular faces, using the tiniest ridges as footholds. But even the great rams go into the most impossible places, holding their heads sidewise to balance their great horns.

The Bharal (Pseudois)

This animal (Plate 184) is also known as the Blue Sheep and is an inhabitant of the Himalayas from central India to central China. It is a beautiful blue-grey color above and white below, including the inner sides of the limbs. A black band separates these two colors all the way from the head to the front edge of the forelegs, then along the flanks and down the front of the back legs. The horns of the ram go outwards and backwards, then inwards and upwards like those of the Tur —which the Bharal greatly resembles. In fact, it stands about halfway between the sheep and the goats.

[272]

[continued on page 289

157. *Black-tailed Mule Deer, in velvet*

MUENCH

158. *Domestic Reindeer*
MARKHAM

159. *Roe Deer*
MARKHAM

160. *White-tailed Deer Fawn*

LA TOUR

161. *Mule Deer*

L. W. WALKER FROM DESERT MUSEUM

162. *Head of Masai Giraffe*

COMMANDER GATTI AFRICAN EXPEDITIONS

163. *Masai Giraffe*

COMMANDER GATTI AFRICAN EXPEDITIONS

164. *Cape Giraffe*

MARKHAM

165. *Pronghorn*

166. *Zebu Bull*

PAUL MILLER

167. *Yak*

MARKHAM

168. *Ankole Cattle*

YLLA FROM RAPHO-GUILLUMETTE

169. *Asiatic Water-Buffalo*
LA TOUR

170. *East African Buffalo*
COMMANDER GATTI
AFRICAN EXPEDITIONS

171. *American Bison*

172. *Nilghai*

MARKHAM

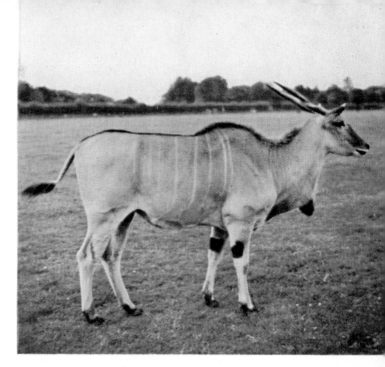

173. *Common Eland*

MARKHAM

174. *Greater Kudu*

VAN NOSTRAND

FROM NATIONAL AUDUBON

175. *Waterbuck*

176. *Sable Antelope*

177. *Coke's Hartebeest*
MARKHAM

178. *White-bearded Gnu*

179. *Arabian Gazelle*

PINNEY

181. *Impalla*

MERLYN SEVERN FROM LANE

180. *Grant's Gazelle*

MARKHAM

182. *Red-fronted Gazelle*

MARKHAM

183. *Tahr*

LA TOUR

184. *Bharal*

MARKHAM

186. Mouflon

187. Rocky Mountain Sheep

185. *Aoudad or Barbary Sheep*

MCKINNEY FROM SHOSTAL

←

188. *Tree-Hyrax*

MARKHAM

189. *African Elephant*

SIMON

190. *Indian Elephant*

LA TOUR

[*continued from page 272*

True Sheep (*Ovis*)

The horns of the True Sheep, and those of the domestic breeds also for that matter, although rising from the front of the head, actually grow outwards with a back-handed twist; that is to say, they sweep slightly forwards, upwards and outwards, then turn under and backwards like a right-handed screw on the right and a left-handed one on the left. The amount of development of the horns varies according to the species and, of course, also the individual. Their greatest development is seen in the Argalis (*O. ammon*) from one of which a horn measuring over seventy-five inches around the outside curve was taken. Argalis are spread throughout east Central Asia from the Pamirs to eastern Siberia. They have light rump patches and slight chest ruffs of lighter tint than the body, which is some shade of brown. Their faces are usually greyish. On the southern slopes of the Himalayas in Afghanistan and thence south to Baluchistan the wild sheep are known as Urials. They are short-haired and mostly of a reddish hue with a very dark chest ruff but light limbs and undersides. They vary greatly, however, and there are types with white ruffs and dark legs. Their horns are tightly curled so that the tips come opposite their eyes.

The third group of wild sheep inhabit the Mediterranean area—one, the Mouflon (Plate 186), comes from the islands of Corsica and Sardinia; the other is spread from Turkey and the island of Cyprus to the Caucasus and the north Persian Mountains. They are red in color with a light greyish saddle, and are white below. The females are hornless but the males bear large pairs that curve back almost on to their shoulders. They are rock-dwelling animals like the Aoudad but they graze on upland pastures and take shelter in any available woods.

The last group of wild sheep are North American. There are two distinct species but races are found in Kamchatka and the Stanovoi Mountains of extreme eastern Siberia. The first is known as the Canadian, Rocky Mountain, or Bighorn Sheep (Plate 187) and ranges from British Columbia to Lower California and the western Sierras of Mexico. There are numerous races but all are some shade of grey-brown with darker forequarters and lighter undersides, nose and rump patch. The horns are massive and tightly coiled. The second species is, in its extreme form, very distinct, being of lighter build and pure white all over. However, there are grey forms and these appear to intergrade with certain races of the Bighorn that happen to have rather small horns, so that the white animals, known as Dall's Sheep, may themselves be no more than a race of the Bighorn that has only recently become separated from the main stock.

Sea-Cows   (*Sirenia*)

SINCE time immemorial these curious aquatic beasts have been a puzzle to observant and thinking people. Very early Indian writings mention them as milk-giving "fish" and the Greeks considered them to be mermaids, a notion that persisted almost to the present century. And, it is true that Seacows have a habit of standing straight up in the water with head and shoulders above the surface, and that the females when doing this may hold nursing infants with one crooked foreflipper to their rather prominent breasts, which are placed on the chest. Here, however, any possible likeness to the traditional mermaid most certainly ends, for their bald heads, fat wrinkled necks, and profuse moustaches more resemble the features of the famous cartoon character "Old Bill" than the long-haired pinup-girl devised to lure frustrated mariners from their plotted course. Seacows are nonetheless very singular.

Until the present century the science of zoology remained almost as much in a quandary over these animals as did mariners. Although the original modern classifier of animals—Carl von Linné, or Linnaeus—as long ago as the middle of the eighteenth century made the astonishing suggestion that they might have some connection with the elephants, they were for long tacked on to the end of the Whales. Nobody was happy about this arrangement because it is obvious that they are constructed upon quite a different plan from those mammals. Sirenians or Seacow-like creatures left their bones to be fossilized in rocks of what is called the Eocene period, estimated to be about fifty million years ago. Nothing of them is known from earlier times and no fossil yet found gives any real indication of their relationships. We have to fall back upon an assessment of their over-all characteristics and then search around for something that looks like them. The result of such investigations is that the consensus of experts now approves Linnaeus' bizarre suggestion. They are distantly related to the Hoofed-Mammals but probably sprang from a side branch of the general Carnivore-Ungulate group (see page 220) that also leads to the Hyraxes and the Elephants.

There are, today, two living kinds of Seacows or Sirenians—the Manatees, and the Dugong. Until the end of the eighteenth century there was a third, very large, purely marine form, living about certain islands off the coast of Kamchatka in the North Pacific. This creature was known as Steller's Seacow and it was exterminated by the sealers and whalers who slaughtered it for its flesh. This was an enormous creature, weighing upward of three and half tons and measuring as much as thirty feet in length. It had a flat, horizontally arranged paddle-like tail and wide foreflippers without nails but fringed with stiff bristles. Its head was very small and in place of teeth its jaws were furnished with hard horny pads. With these it cropped seaweed in estuaries around the coasts of Copper and Bering Islands. Its body was covered in an immensely thick, dark brown, much-wrinkled hide that looked like the bark of a tree. It was a very bad swimmer though it could not come out on to the land, and it was often washed ashore when a high sea was running on to the coast. There were estimated to be only a few thousand left when the poor creature was discovered in 1741 and they were all gone in a little over twenty years.

Manatees (*Trichechus*)

There are two species of living Manatees distinguished, among other minor details, by having seven and six neck vertebrae respectively. The former inhabits the Atlantic coasts of the Americas from Florida around the Caribbean to the Orinoco, the Guianese, and the Amazonian river mouths, and is also found among the cuays and islands, in salt, brackish, or freshwater lagoons, and smaller river mouths almost everywhere about those regions. They do not venture upon the high sea but may occur in atoll lagoons as much as eighty miles from the coast. The other species inhabits West Africa and, although it is found all along the coast wherever its rather special food grows, it is more essentially a freshwater animal, ascending the large rivers for hundreds of miles.

The general appearance of these animals may be seen from the accompanying photograph but in the water they look altogether different, being strangely supple though they could never be called graceful. They are lazy, slow-moving beasts of a distinctly amiable nature so that they may be readily approached in an engineless boat, and some, kept in the Botanical Gardens in Georgetown, British Guiana, readily come half out of the water on to the bank and take handfuls of grass from tourists. They are indeed very bovine and can make a pathetic groaning noise at night not unlike a cow pleading to be milked. Manatees are dark grey in color when wet but their naked skins dry to almost white. The tail is a circular paddle and the rounded fore-flippers are highly movable, having a noticeable "elbow" joint. They pluck their food, which consists of a form of greenstuff known as sea-grass that grows along the coasts, and of various aquatic plants in the rivers, with their upper lips, which are split down the middle and which work against each other sidewise. As the lips are provided with strong muscles internally and a profuse crop of stiff recurved bristles externally the whole device becomes a very powerful grasping organ. They use their flippers to stuff the greenstuff into their mouths and then chew it with great deliberation.

One young is born at a time but mothers may sometimes be seen holding two infants, and the native fishermen assert that this is a form of "baby-sitting," mothers exchanging young for care while they dive to feed. They are fairly inquisitive beasts and will often surface to inspect small boats anchored in shallow water for fish-

SEVERIN

Manatee

ing. Being usually unknown to the average northerner visiting the warmer temperate and tropical coasts where they are found, they often cause great consternation by this habit. Their appearance is indeed ridiculously human on occasions.

The Dugong (*Dugong*)

This silly-sounding title is the Malayan name for these animals. They are found all around the Indian Ocean from central East Africa, and India and Ceylon and Burma, to Indonesia and the northern coasts of Australia. Their numbers have, however, been reduced almost to the point of extermination in many areas by ruthless hunting for their oil and for their excellent-tasting flesh. Dugongs are more purely marine than Manatees and do not ascend rivers to any great extent. They also meander along coasts in shallow water and browse on sea-grass and seaweeds. To crop this they extend the whole upper lip and curve it over the food, pressing it against the lower lip and then pulling it off: they do not use their flippers to stuff it into their mouths.

The Dugongs have a few peglike crushing teeth and the males have two large upper tusks. In general form they are not unlike manatees but the head is proportionately larger, the flippers shorter, and the tail is sickle-shaped when viewed from above and is not unlike that of some whales. They are dark grey above and white below. Although no Manatee normally exceeds eight feet, large adult Dugongs may measure nearly ten feet. They are also rather easygoing, sluggish beasts that drift back and forth with the tides to feed; they used to be found in enormous herds and were so trusting in places where they had not previously been hunted that it is recorded in the accounts of early voyages that they could be patted on the head from small boats. Nonetheless, they have been seen to gang up on sharks in shallow water and drive them off by butting them with their heads and the males will apparently endeavor to bite if molested under water. They have been mercilessly persecuted, being harpooned or caught in mile-long nets and then drowned, for they are air-breathers and have to surface to fill their lungs at rather short intervals.

Dugongs have also been reported to be found in certain of the great, landlocked, freshwater lakes of the East African Rift Valley but, if this is so, we would like to know how they got there, as these animals are today purely marine although those bodies of water have never been openly connected with any sea and it is most improbable that the Dugong ever attained them via the Nile. They may, however, have originally been freshwater animals and have evolved in Africa, only later descending now extinct rivers to the sea.

Hyraxes

(*Hyracoidea*)

ANOTHER small group of mammals that has caused zoologists considerable headaches, and generations of ordinary people much confusion, is that of the little rabbit-shaped animals now usually called Hyraxes. At other times they have been called by various other popular names, the best known of which was Coneys, a name that has also itself been applied to various other mammals and notably to rabbits. Hyraxes are apparently the animals referred to in the Bible as Coneys. They look rather like short-eared rabbits and the best known of them, found in the Mediterranean region lives in holes among rocks and leads much the same life as some species of those animals. However, Hyraxes have teeth like rhinoceroses, and small hoofs on their toes.

When first investigated scientifically, they were thought to be a kind of Rodent and were placed near the South American Guinea-pigs. Later, they were grouped with the Elephants, Rhinoceroses, and Hippopotamus and were called collectively with those animals Pachyderms, which was ridiculous because that word means "thick-skinned." When it was realized that the other three types were not related but that each be-

longed to a separate order, the little hyraxes went along with the rhinos on account of their teeth and thus became associated with the tapirs and horses. Finally, it was realized that, although they are hoofed animals and definitely belong to the great Carnivore-Ungulate division of mammalian life, they really stand completely alone but, like the Sirenians, somewhere near the Elephants.

Although retiring in habits and mostly nocturnal, Hyraxes are the most atrociously aggressive of beasts and will stand up to and bite anything that molests them. They are also one of the rowdiest of mammals, giving out with whistles, screams, and chattering burbles that have to be heard to be believed and then actually seen coming from the animal to allay the last suspicion that they are not made by disembodied human souls in anguish. There are two principal kinds of Hyraxes that live among rocks and in tall forest trees, respectively. They have strange corollas of contrastingly colored fur on the mid-back that can be opened at will to disclose a gland. They are virtually tailless and are clothed in thick rather coarse fur. They are very quick in their movements and are rather good jumpers. The feet have pads below but do not look like those of climbing animals. Their remarkable abilities in this respect are possible only because the animals can draw the central portion of their thickly padded feet up into a dome, thus making a vacuum cup and have pads that are always moist. With these devices they can cling to almost perpendicular rock faces. There are four toes on the front and three on the back feet. In addition to their rhinoceros-like cheek teeth, they have a pair of upper front teeth developed into daggerlike slashing weapons.

Dassies (*Procavia*)

A rather large number of species have been described from a wide area, covering most of Africa south of the Sahara, and even a belt of territory north of it on the southern face of the Atlas and associated mountain ranges in Algeria. One, the first and best known, and referred to above as the Coney of the Bible, is isolated from the rest in the Sinai Peninsula, Palestine, and Syria. This is a pale fawn-colored, rock-dwelling form,

Dassie (Procavia)

that has a large yellow dorsal glandular patch. It lives in crannies among rocks and appears to feed only when it considers no possible danger is abroad, and usually about dawn and dusk. Although it continues to move about by night, it may be seen sunning itself at midday. There are other species distributed down the east side of Africa from Abyssinia to the Cape. There are two distinct kinds in Abyssinia, one small, with coarse hair, and another very large, with long silky fur and a dark glandular patch. The largest species is grey-brown in color and lives in South Africa. There are also hyraxes of this type scattered about Africa wherever there are mountainous or elevated open rocky areas. They are all singularly alike in habits, having about four young at a time in fur-lined nests, living in small communities, being excessively wary, and most active at dawn and dusk.

Tree-Hyraxes (Dendrohyrax)

In forested areas, the place of the Dassies is taken by a group of closely related animals (Plate 188) that dwell almost exclusively in trees and that seldom come to the ground. Several distinct species may be recognized, each more or less isolated in one major forest tract. Thus, there is a large dark grey-brown species in the southeastern forests south of the Zambesi, and another with a bright yellow underside, which lives at elevations of 7000 to 10,000 feet in the montane forest about Kilimanjaro. A third type lives in West Africa and has shaggy black fur, each hair on the back being tipped with white. One species of Tree-Hyrax, however, does not normally live in trees but inhabits the open grasslands north of the forests in west-central and west Africa. Tree-Hyraxes sometimes live in holes in tree trunks but more usually inhabit the dense parasitic growth that festoons the larger branches of the forest giants in closed-canopy growth. The limbs of some of these trees may be several feet in diameter and carry a tremendous load of creepers, orchids, and other epiphytic plants and may even have fair-sized epiphytic trees growing upon them. In this natural herbaceous border live all manner of small animals from earthworms to tree-frogs, crabs, and rats, and upon these the larger animals prey despite the fact that this whole little world is elevated as much as a hundred feet into the air. The Tree-Hyraxes are omnivorous, taking insects as well as vegetable matter and they grub industriously in these tangled masses.

Hyraxes have given rise to a large body of folklore and some strange beliefs. One African tribe insists that they embody the departed spirits of white men; Hebrew law prohibited their being eaten on the ground that they chew the cud but do not have cloven hoofs—actually they do not chew a cud and have doubly cloven hoofs; and the author was once considerably startled by being told by an African Chief through a series of interpreters that these animals were the "Little brothers of Elephants." How the African reached this conclusion could not be ascertained but the people of that country are often consummate naturalists and have an uncanny grasp of the true relationships between different kinds of animals which may look totally dissimilar.

Elephants
(Proboscidea)

ALL the ten to fifteen thousand different kinds of mammals now inhabiting this earth probably display some feature or have some habit that is unique if only we knew more about them, and many times in this book we have remarked that some particular kind is one of the oddest of all mammals. About no animals can this be more truly said than about elephants. In almost every way they are most singular and in many respects they are unique. In fact, they have much more right to be taken out of the animal kingdom than we ever had. They are probably the most highly evolved form of life on this planet, meaning that they are furthest removed from the primitive basic mammals without having become grossly specialized. That they walk on the tips of their fingers and toes, have "knees" that, like ours, bend forwards, and have developed their noses into grasping organs does not render them either grotesque or dangerously overadapted to their environment and way of life. Like men, they are really fairly basic but for a few highly useful adjuncts. Like us also they have enormous brains—in point of fact, but not of course in proportion, bulkier than ours. It is in their mentality, however, that Elephants are so outstanding. These animals have lived with men for millennia but have steadfastly refused to become domesticated in that they will not breed in true captivity—though of course individuals have been born in such circumstances.

All sorts of exaggerated and bizarre qualities have been attributed to them, most of which have never been confirmed by modern scientific observation, yet even the slightest intelligent association with these huge creatures will startle one into believing almost anything about them. Elephants vary in personality like any other animals: there are among them intelligent ones and blithering idiots. As a whole, they have good memories but they forget just like us. Much more strange is the undoubted fact that some at least actually learn to understand human speech. By this is not meant the kind of automatic response of a dog or chimpanzee to certain sounds or the timbre of a particular voice. A good elephant man will literally introduce you to his charge and once you have been touched lightly by the animal's trunk you may almost invariably—if confident and acquainted with the language to which the animal is accustomed—stand perfectly still and talk the great beast into performing a wide variety of actions up to the limit of its hearing, and this even if you use different terminology from that to which it has become accustomed. Even naturalists often fail to appreciate the further fact that elephants have very pronounced personalities of their own and that, like us, they are prey to complex emotions. The Apes may be said to be temperamental and thus unpredictable, but elephants are even more like human beings, and particularly highly civilized human beings, in that they may resolve petty frustrations or plain liver-disorders by the transfer of their pent-up feelings into quite unrelated channels. Thus, they are really extremely dangerous creatures because they not only reason but may, like us, act illogically. Many if not all elephant-trainers are highly aware of this and usually take all possible precautions, especially at the time of *musth* or periodic sexual excitement, which comes upon elephants at regular intervals.

An enormous amount has been written about both the Oriental (Plate 190) and the African (Plate 189) animal but if anyone is interested in reading more extensively about these animals we would suggest the writings of the Englishman J. H. Williams, or the Indian, Dan Gopal Mukerjee. The former lived with working elephants in Burma; the latter—although he wrote and died in America—was brought up among wild ones in India and was steeped in the uncanny lore of his ancient caste, the members of which understand aspects of animal behavior that are completely beyond the realm of our current, purely materialistic understanding. From the writings of these two, it will be seen that no exact but compact description of the habits of elephants can be attempted.

To generalize, however, it may be said that both kinds of elephant-like animals, are gregarious, tribal, and social creatures, having a wide complex of community laws, rules, and regulations, a marked discipline and many well-established customs. They are nomadic, either traveling back and forth between two extreme points every year in accord with the seasons, or moving around a set course over longer periods that may take them as much as ten years to complete before arriving back at any one point. Females assist each other at the birth of the young, many duties being performed by these nurses rather than the mother. Young males may only mate after a specified period and, although sexual dueling is in order, again in this certain set rules of procedure are observed. Old or imperfect males, or gross individualists may leave the tribe and live alone and then appear to be permanent exiles. If only temporarily ill or out of touch with their own tribe, they may, however, be permitted to join others temporarily and then rejoin their own group.

The gestation period, contrary to most popular belief, is 18 to 22 months in the Indian elephant and about 21 in the African; the young are weaned in about 5 years, and their rate of growth, the age at which they reach puberty, their life-span, and their gerontic progression is exactly similar and equal to that of man. The average life-span of a captive elephant, and as far as we know of wild ones, is sixty years and none over seventy-two years are authentically known although the life records of thousands have been kept by the great teak companies of Burma and by numerous organizations in India. The famous "Jumbo" was killed by a train in Canada at the age of forty, and "Napoleon's Elephant" turns out to have been a series of three individuals none of which survived more than forty-five years! Indian elephants may, like men, live to greater ages and possibly even

to be centenarians but so far we have no record of such. The African animal may be different. This species has only recently been tamed and domesticated, and apart from the Union of South Africa, white men have not been anywhere in its domain long enough to dispute or confirm the assertions of experienced native hunters who affirm that certain clearly recognizable lone bulls have been known to frequent a particular area for centuries. The African's idea of time is not ours, however, and until comparatively recently a fifty-year-old man was very ancient indeed in that continent.

As has already been said, the two animals called elephants are really quite different beasts, though they are but two offshoots of a very much more numerous group constituting the order Proboscidea. This has a most venerable ancestry, going back over fifty million years to small tapir-like animals. The intervening gap is rather better filled than in the case of other mammals except for the horse and the camel, and from almost countless fossil remains we can piece together a huge family tree with some extraordinary side branches. Not only were there vast hairy beasts like the Mammoths, and sort of depressed and elongated parodies of present-day elephants—the Mastodons—but all manner of large creatures with tusks in both jaws, in the lower jaws only, or without tusks. The living animals are two of three genera in a single sub-family of this great assemblage. The third member was the Mammoth.

The African Loxodon is the largest living land mammal; the Indian Elephant, however, is not the second largest. This place is taken by a race of the African Ceratothere or White Rhinoceros (see page 242). The largest Loxodon is recorded as standing twelve feet six inches at the shoulder (though the measurements of this specimen, in a museum in England, have been disputed); the record seems to be 12 feet 8 inches. For the Indian it is 10 feet 8 inches. The size to which the tusks of either species may grow is open to considerable doubt, all manner of extravagant claims having been filed but, from the records of the long-established ivory merchants of London and Hamburg, through whose hands millions of tusks passed over the centuries, it would seem that the largest Indian elephant is 8 feet 9 inches, measured along the curve, and weighs 161 pounds. The record African tusk is 11 feet 5 inches and weighs 293 pounds. Of course, a great deal of Mammoth ivory dug from the frozen muck of the Siberian tundra also came on the market and this is of far greater dimensions.

The fable of elephant graveyards at a few localities in both Africa and the Orient may be totally rejected. No such place has ever been found, and several cases of elephants dying or recently dead from natural causes are on record. In fact, the bones of elephants are just about the only remains of defunct animals that are ever found in the tropical wilds and this because they are just so big that even fungi and bacteria take some time to demolish them.

Another myth that must be dispelled, though the general public will probably never accede, is that of the "Pigmy Elephant." The most elaborate searches and the most profound researches have failed to bring to light any trace of such an animal. It is true that there are in parts of the Congo rather hairy races of Loxodons that tend never to reach the average dimensions of the open-country populations, but neither these nor the smallish races from West Africa are in any manner a separate species, and none, when adult, is notably smaller than any others. There are, of course, elephants that grow to monumental proportions as in men and other animals, and there may presumably be runts that never reach even the average of their sex; but of a Pigmy species or race there is so far no trace.

It is also untrue that bull-elephants are untamable and more especially that the Loxodon cannot be handled at all. The famous Jumbo was a male African Loxodon, and even a tourist trip to the Congo today will make it possible for anyone to see considerable numbers of the most enormous bull Loxodons fully broken to farm and lumber work despite carrying sets of tusks so huge that they actually touch the ground when the animals are lined up for inspection. Hannibal's elephants, which he used as tanks and bulldozers on his march through Spain and over the Alps upon Rome, were African Elephants which were apparently then still common in the Atlas mountains.

Several kinds of elephants and related Mastodons inhabited North America coincident with man and perhaps even with the first Amerindians, for stone weapons have been found in association with their bones at more than one spot. Mammoths, of course, were contemporary with Palaeolithic man in Europe and were depicted by him on the walls of caves.

Index

NOTE: Numerals in boldface type refer to pages showing black and white illustrations. Boldface numerals enclosed in parentheses refer to color plates which are numbered separately.

Photographic Acknowledgment

The publisher wishes to acknowledge the co-operation of the Chicago Zoological Park, Brookfield, Illinois, and the World Jungle Compound, Thousand Oaks, California, in allowing Mr. Cy La Tour to make the following color photographs: Chicago Zoological Park—Plates 19, 47, 91, 95, 101, 144, 148, 183, 189; World Jungle Compound—Plates 51, 52, 98, 105, 151, 169, 190.

Other photographic sources are acknowledged in the credits accompanying each photograph.

THIS BOOK has been printed and bound by Kingsport Press, Inc., Kingsport, Tennessee. *Color engraving by Chanticleer Company, New York. Designed by James Hendrickson.*